Classification
and Biology

Classification
and Biology

R. A. Crowson D.Sc. (London)

Lecturer in the Department of Zoology,
Glasgow University

Heinemann Educational Books Ltd
London

Heinemann Educational Books Ltd
LONDON EDINBURGH MELBOURNE
TORONTO AUCKLAND JOHANNESBURG
SINGAPORE HONG KONG IBADAN NAIROBI

SBN 435 62984 0

Published by Heinemann Educational Books Ltd
48 Charles Street, London W1X 8AH

Printed in Great Britain by
Butler & Tanner Ltd, Frome and London

Deduct from a rose its redness, from a lilly its whiteness, from a diamond its hardness, from a spunge its softness, from an oak its heighth, from a daisy its lowness, and rectify everything as the philosophers do, and then we shall return to Chaos, and God will be compelled to be Eccentric if he creates.

WILLIAM BLAKE, marginalia to Lavater's *Aphorisms*

stablishment of Truth depends on the destruction of falsehood continually.

WILLIAM BLAKE, 'Jerusalem'

l'effort classificateur des Naturalistes...

1. *non content d'être devenu un travail aussi relevé que n'importe laquelle des analyses scientifiques du Réel;*

2. *est en voie de découvrir pour lui-même, et d'ouvrir aux autres Sciences de Nature, un domaine nouveau de recherche;*

3. *cependant que son object propre (la distribution naturelle des êtres) se découvre peu à peu comme le terme commun et suprême ou converge, par sa partie speculative, tout l'effort scientifique humain.*

TEILHARD DE CHARDIN 1925,
'L'Histoire Naturelle du Monde', *Scientia* **37**: 15-24.

Happy is he who lives to understand
Not human nature only, but explores
All nature—to the end that he may find
The law that governs each; and where begins
The union, the partition where, that makes
Kind and degree, among all visible Beings;
The constitutions, powers and faculties,
Which they inherit—cannot step beyond,—
And cannot fall beneath; that do assign
To every class its station and its office,

Through all the mighty commonwealth of things;
Up from the creeping plant to sovereign Man.
Such converse, if directed by a meek
Sincere, and humble spirit, teaches love:
For knowledge is delight; and such delight
Breeds love: yet, suited as it rather is
To thought and to the climbing intellect,
It teaches less to love, than to adore;
If that be not indeed the highest love!

WORDSWORTH, 'The Excursion'

The molecular questions, what is it?, where does it occur? and what is it doing?,
are taxonomic questions. The higher questions, when?, where? and how did it come
about?, are phylogenetic questions. Taken together they constitute the spirit, if not
the essence, of chemosystematics; indeed, of all biology.

B. L. TURNER, in *Science Progress* **56** [214]

Preface

In recent years a number of more or less excellent books have appeared on the subject of botanical and/or zoological classification, so that some apology may be expected from the author of yet another. The present work was commenced in 1959, before most of the other comparable works had appeared; it differs from all the British and American ones in adopting a strictly phylogenetic approach to the subject, like that of Willi Hennig, though my own views on the subject were formed independently of Hennig's and before I had read his work. The problems of zoological, botanical and paleontological classification are considered *in extenso*, but microbial systematics is only briefly touched on, even though it is in this field that the greatest problems and the greatest opportunities for systematics remain. The excuse is, of course, the limitations of the knowledge and capacities of the author. A feature of the work which is rather unusual, at least in the Anglo-Saxon world, is the philosophical approach to the subject outlined in the first two chapters, which convinced empiricists will probably consider superfluous and wrong-headed. In defence, the author would plead that his views on these matters are deeply and sincerely held, that they are relevant to his approach to all the matters treated in later chapters, and that they are different from those of the majority of present-day biologists; intellectual honesty thus requires them to be explained at the outset.

November 1969   R. A. C.

Contents

Contents

1 The Classification of the Sciences

For our purposes, science may be defined as the rational and objective study of the external universe by human beings. Whether the study of man himself is included as part of science will depend on how we interpret 'human beings' in this definition. If we regard humanity as in essence an assemblage of isolated individuals, then any one of them may regard the rest as part of the external universe and thus as 'material' for scientific study; on the other hand, if, as seems preferable to me, humanity is regarded as essentially one body and science as a collective rather than an individual function, we can hardly avoid maintaining in some form or other the traditional distinction between the sciences and the humanities. The problems of classifying human beings will not be considered in detail in this book, though it will appear that if the criteria developed for other animals were applied to our own case, the Chimpanzee, the Gorilla and perhaps the Orang-utan would join us in the genus *Homo*.

Classifying things is perhaps the most fundamental and characteristic activity of the human mind, and underlies all forms of science. The capacity of our brains to function as computers seems to have been a rather late discovery in human cultural history, there being no evidence that anything which could be called mathematics was developed before late Neolithic stages, up to perhaps 7,000 years ago—though doubtless simple counting of flocks and persons came considerably earlier. Even now, the human mind solves mathematical problems in quite a different manner from that adopted in mechanical computers. Confronted with the symbols '9 x 8' the reader will doubtless recognise a portion of the multiplication table and instantly produce the answer '72', whereas the procedure of a mechanical (or electronic) computer would amount to adding together, at very high speed, 9 groups of 8. In the same way, a more advanced mathematician, confronted with a complex differential equation, will seek some way in which it can be transformed, according to established rules, into a form for which there is a recognised technique of integration—to put it in another way, he will seek to 'recognise' it as a more or less disguised member of a classificatory group. The solution of any type of equation by an electronic computer is normally by 'trying out' a vast number of different values for the unknown(s) until it hits ones which give the right result—the human analogue would be to plot a graph of the function and find the roots thence. The human mind solves mathematical problems essentially by feats of classification.

Reasons for the primacy of classification as a function of the human mind are not difficult to imagine. Our ancestors were social as soon as (or before) they became human; they probably lived in more or less numerous and ordered groups and utilised a great variety of food materials. Effective group organisation would depend on the ability to recognise, and react appropriately to, a large number of other individuals and on the system of communication by means of language. The essential elements in language are nouns and verbs—the use of which involves the classification of things and of actions or events. Widely polyphagous feeding habits would be safe only for organisms which could learn to distinguish many kinds of eatable and uneatable things.

There have been many authorities who have asserted that the basis of science lies in counting or measuring, i.e. in the use of mathematics. Neither counting nor measuring can however be the most fundamental processes in our study of the material universe—before you can do either to any purpose you must first select what you propose to count or measure, which presupposes a classification. And no amount of mathematical treatment of your results can confer on them any greater reliability than there was in the classification by which you selected the things to count or measure. As we shall see, there have been recent attempts to bring numbers and mathematics into the actual process of classifying organisms, usually by way of counting the number of characters in common or differing between them. However, even if it were possible, as it would need to be if 'numerical taxonomy' were ever to become a scientifically rigorous process, to formulate an objective definition of a 'unit character', such a definition could only be formulated in terms of comparisons between suitable similar organisms —no conclusions about unit characters could be derived from the comparison of a cat with a mushroom. The 'unit characters' which are the basis of material taxonomy must be abstracted from a prior non-mathematical classification. Mathematics would thus appear as a possible method of refining an existing classification, not of constructing one *de novo*.

The sciences themselves may be classified in various ways. In an interesting article the Harvard philosopher, Tejera [188] wrote:

> Classifications of the sciences in our time appear, on the whole, to be a matter of parcelling out different areas which it is assumed are to be investigated in the same way. We seem to have lost the sense that different sciences might not be the same in the amount of system or degree of explicitness of which they are capable, or that they might in fact differ in kind from other sciences. This is due to the speculative bias of Western civilization which, since Aristotle, has preferred and developed the sciences which he called theoretical and which today are called deductive.

This I think is largely true, but even today a general, if confused, awareness of differences is manifest in such current phrases as 'the physical sciences', 'the biological sciences', 'exact sciences', 'observational sciences', 'experimental sciences', etc. Underlying all these there is, I think, a real and deep distinction, which commonly escapes recognition in its true form today. The most fundamental division among scientists is, I believe, that into natural historians and natural philosophers [37].

A scientist who seeks to explain the actual universe, or any (non-human) part of it, in detail as it is, in terms of its concrete history, can properly be called a natural historian—he is literally concerned with the history of nature. The natural philosopher on the other hand is devoted to the search for fundamental laws of the highest possible generality—laws which he hopes will apply throughout space and time. For the natural historian, the laws of natural philosophy are not ends in themselves, but tools for the understanding of the actual universe. The scientific work of the natural philosopher has its beginning and its end in the laboratory; the classic pattern is for some initial experimental result to suggest a hypothesis, which is elaborated in the study, and results in predictions which are finally verified (or falsified) in further laboratory experiments. A hypothesis in natural history on the other hand is (or ought to be) based in the first instance on some observed facts in nature, and though laboratory experiments may play an important part in its intermediate development, its final testing ground is again in nature itself.

The characteristic mode of thought of the natural philosopher is deductive reasoning from given postulates; if the results of the reasoning process prove not to be in accord with experimental evidence, he will consider that he has disproved one of the postulates. The mode of thought typical of the natural historian on the other hand is much more akin to that of Sherlock Holmes than to those of Newton, Einstein, Bohr or Rutherford. The individual steps in his arguments can rarely claim the infallibility of mathematical deduction, and he is as a rule suspicious of long chains of reasoning. Dealing in probabilities rather than certainties, his main intellectual skill lies in perceiving how a number of only moderately strong individual probabilities can be combined to produce an overall near certainty.

The distinction between these aspects of science is essentially one of human predilections—if you are predisposed to believe in uniformity, you will seek to explain away apparent differences in terms of underlying similarities, and if you become a scientist you will find your natural place as some kind of natural philosopher; if, on the other hand, you delight in the differentness and uniqueness of things, you might become as artist of some sort, or a natural historian.

The remark of Professor Oakeshott [146b] that 'the Rationalist is always in the unfortunate position of not being able to touch anything, without transforming it into an abstraction; he can never get a square meal of experience' is, in the scientific world, particularly applicable to the natural philosopher.

Tejera's criticism of the one-sided natural-philosophical outlook occurs, significantly, in an article on 'The Nature of Aesthetics'. The natural philosophers as a class show the speculative and deductive bias mentioned by Tejera, they are addicted to those philosophies which preach fundamental unity and which exalt the observing mind as against the natural universe; Platonism, Idealism, Monism, Positivism, and even Spiritualism, have all numbered their adherents among the physicists. Philosophising of this sort is comparatively rare among natural historians, whose private inclinations are more likely to be towards Aristotelean dualism, old-fashioned materialism, or Wordsworthian pantheism.

The division of science into natural history and natural philosophy is not, of course, new; the very terms in which it is expressed derive from Bacon's 'De dignitate et augmentis scientiarum' [7]. Bacon's *historia naturalis* corresponds almost exactly to natural history as we have defined it; for him it represented that part of science in which the faculty of memory (*memoria*) was dominant, whereas he considered reason (*ratio*) to be the dominant faculty within the sphere of *philoso-phia naturalis*. Bacon recognised that the distinction was one of pre-dominance only, the part played by reason in natural history, like that of memory in natural philosophy, he recognised as vital even though it was subordinate. The third of Bacon's fundamental human faculties, *poesis* or imagination, was also recognised by him as playing an essen-tial part in both *historia naturalis* and *philosophia naturalis*, though its sphere of dominance (creative art) lay outside the bounds of science altogether.

Bacon's attribution of a dominant role to *memoria* in the domain of *historia naturalis* was no doubt just. The natural historian, like Sherlock Holmes, needs to have a memory richly stored with diverse and little-known information. Progress in natural history almost invariably results from the perception or intuition of a hitherto unsuspected con-nection between two pieces of information, no doubt the same basic process as Koestler's 'bisociation' [118]; this is only likely to happen when both pieces of information are stored up in one and the same mind. Thus modern trends in education, which increasingly emphasise 'principles' rather than facts, are detrimental to natural history; this is one more manifestation of the excessive predominance of natural philosophy in current scientific attitudes.

Herbert Spencer, in his pamphlet *The Classification of the Sciences* [176], makes the same distinction very explicitly, with a threefold division of science into Abstract Science (=Mathematics), Abstract-Concrete Science (=*philosophia naturalis*) and Concrete Science (=*historia naturalis*). Spencer, who had psychological theories of his own, did not follow Bacon's correlation of different kinds of science with different human faculties. He was at pains to make it clear that his division of the sciences was in conflict with the teachings of August Comte, whose doctrine of a natural, logical and historical sequence of the sciences in the order Mathematics—Astronomy—Physics—Chem-istry—Physiology—Botany—Zoology has had a very wide influence lasting up to the present time. Comte's seriation of the sciences, it will be noted, significantly ignores Geology.

More recently than Spencer and Comte, the German philosopher WilhelmWindelband, in his 'Geschichte und Naturwissenschaft' [208], distinguished between what he called Nomothetic and Idiographic sciences. A nomothetic science was, forWindelband, one whose ulti-mate aim was the formulation of laws of the utmost possible generality, and in the thinking of whose practitioners the tendency to abstraction (i.e. Bacon's faculty of *ratio*) predominated; idiographic sciences on the other hand aim at the fullest possible understanding of actual things and events, and their practitioners make much use of a faculty which Windelband called 'Anschaulichkeit' (usually translated as intuition). Idiographic science obviously corresponds to *historia naturalis*, though Anschaulichkeit can hardly be equated with *memoria*; in Baconian terms it might represent a synthesis of *poesis* and *memoria*.

A more recent author who has used the phrase 'natural history' in a rather similar sense is Teilhard de Chardin [186*b*], who wrote: 's'il fallait trouver un nom général à la Science speculative, telle qu'elle tend à se constituer par l'alliance des disciplines les plus abstruses et les plus raffinés de notre siècle, il conviendrait sans doute de l'apeller l'Histoire Naturelle du Monde'.

A similar distinction was made by Bergson, who distinguished philosophically between 'laws' and 'genera'. He pointed out that ancient science was more concerned with genera than with laws, leading to a confusion of the physical with the vital. He goes on to point out that 'There is the same confusion in the moderns, with this difference, that the relation between the two terms is inverted: laws are no longer reduced to genera, but genera to laws; and science, still supposed to be uniquely one, becomes altogether relative, instead of being, as the ancients wished, altogether at one with the absolute. A noteworthy fact is the eclipse of the problem of genera in modern philosophy. Our theory of knowledge turns almost entirely on the question of laws: genera are left to make shift with laws as best they can. The reason is, that modern philosophy has its point of departure in the great astronomical and physical discoveries of modern times. The laws of Kepler and of Galileo have remained for it the ideal and unique type of all knowledge.' If natural philosophy is the science of laws, natural history might be defined as the science of genera.

Natural philosophy, as we have defined it, will include physics and chemistry only. The twentieth century 'take-over' of chemistry by physics is popularly believed to have been final and total, but in this as in many other respects popular opinion has been misled by propaganda. It is, in fact, very far from being the case that all phenomena studied by chemists are quantitatively predictable in terms of the fundamental laws of physics; for example, no physicist could calculate, from fundamental physical laws alone, whether, when two given solutions of metallic salts are mixed together in a test-tube, a precipitate will form or not. There is as yet no rigorous theory of the liquid state, nor of solubility. A chemist might be able to make reasonably accurate forecasts of the solubilities of hitherto unprepared compounds, but this would be on a basis of empirical chemical generalisations, not laws of physics. In its continued reliance on empirical generalisations, chemistry shows some similarity to natural history, but the fact that the elements and compounds with which it deals either exist or could in principle be prepared anywhere and at any time in the universe, takes it out of the domain of history and places it in the same category as physics.

The fact that in one branch of astronomy—Celestial Mechanics—predictions of a high degree of accuracy can be made on the basis of physical laws has had the result that in the common mind astronomy is often separated from the rest of natural history and considered as an 'exact science' akin to physics. Yet the fact that the earth has its moon whereas the similar-sized planet Venus has none is no more to be deduced from the laws of physics than is the difference between a primrose (*Primula vulgaris*) and a cowslip (*Primula veris*). Generalisations about Cepheid variable stars have the same logical status as ones concerning Rodentia, reversed faults, cold fronts, or Compositae, and quite a different one from laws of physics.

The major divisions of natural history, as here defined, are Astronomy, Geology, Meteorology, Virology, Bacteriology, Botany and Zoology. Sociology and Psychology are excluded as being concerned with Humanity rather than Nature; they are more closely bound up with History, Politics and Art than with the natural sciences. Our divisions are, of course, concerned only with pure science; technology and applied science (Engineering, Electronics, Agriculture, Medicine, etc.) stand outside the scheme; their methods may be those of science, but their aims belong in the sphere of Humanity. It is currently fashionable to associate together, under the name Biology or 'Biological Sciences', Virology, Bacteriology, Botany and Zoology, on the grounds that certain phenomena—particularly the basic structure and mode of replication of nucleic acid molecules—are common to all of them. Though this is no doubt true, it is also true that the numbers, diversity and organisational complexity of plants and animals are sufficient to make both Botany and Zoology fully worthy to rank as independent sciences, and they will be so treated in this book.

The laws of natural philosophy claim to be both universal and exhaustive—anything that happens in the universe is supposed to be in principle explicable in terms of them. If this is so, it may be asked, what need is there for natural history at all? Should not the physicists be capable of explaining and predicting all natural phenomena? The best test case for these questions is perhaps to be found in meteorology. All the basic phenomena of weather—heat conduction and convection, evaporation and condensation of water, absorption of radiation, electrostatic charge and discharge, winds, etc.—are strictly controlled by known physical laws. Mankind has a great interest in the reliable forecasting of the weather—why do we not see meteorological prediction taken over by a group of physicists and converted into an 'exact science'? The recently publicised attempts to apply electronic computers to weather forecasting do not, it must be made clear, involve the taking over of meteorology by physics—the programming of such computers is based on empirical laws of weather forecasting, not on laws of physics.

It is, I believe, the intrinsic limitations of the human mind which render us incapable of making an exact science of meteorology. That such limitations exist is only grudgingly, if at all, admitted by the leading scientists of our day. The most eminent zoologists believe that the brain of man originated merely as an improvement on that of other anthropoids, and that it reached its present structure in serving the simple needs of a Stone Age society; it seems very strange to me that, accepting such postulates, they are so unwilling to admit any limitations of the power of this organ in themselves. In physics, these limitations are most clearly revealed in our inability in general to solve, and often even to formulate, mathematical equations describing the behaviour of systems of three or more mutually influencing components. Given all the parameters of such a system at a given instant, and the precise rules governing the interaction of its components, neither physicists nor any other human beings could in general calculate the precise condition of the system at a specific future instant. The classical manifestation of this is in astronomy, where the behaviour of systems of three or more massive bodies, each moving under the combined gravitational influence of the others, can be predicted only by methods of successive approximation which are laborious and never produce an exact result. If some problems of this type can be tackled by means of analogue computers, this does not really affect the argument. In meteorology, the variables involved in weather are mostly mutually-influencing, and certainly more than three.

The following two paragraphs, quoted from an article by Professor Hayek [87], make the same point very explicitly:

It is, indeed, surprising how simple, in these terms—i.e. in terms of the number of distinct variables—appear all the laws of physics, and particularly of mechanics, when we look through a collection of formulae expressing them. On the other hand, even such relatively simple constituents of biological phenomena as feed back or cybernetic systems, in which a certain combination of physical structures produces an overall structure possessing characteristic properties, requires for its description something much more elaborate than anything describing the general laws of mechanics.

Probably the best illustration of a theory of complex phenomena, of great value though it describes merely a general pattern, whose detail we can never fill in, is the Darwinian theory of evolution by natural selection. It is significant that this theory has always been something of a stumbling block for the dominant conception of scientific method. It certainly does not fit the orthodox criteria of 'prediction and control' as the hallmarks of scientific method. Yet it cannot be denied that it has become the successful foundation of the whole of modern biology.

Professor Hayek is, of course, using this argument in an attempt to justify the 'scientific' status of sociology—which, for reasons advanced earlier in this chapter, I am not disposed to accept.

The celebrated theorem of Gödel (see Ladrière, [122]) in effect denies the possibility of subsuming the whole of mathematics in a single coherent formal system, and also denies that any single coherent formal system could be complete and self-sufficient, in the sense of giving the capacity without reference to any principles outside itself, of deducing all possible relations lying within its own field. The scientific implications of this have not been fully realised as yet. Such a physical theory as wave mechanics might perhaps be considered as a formal system in the Gödelian sense; if so, the dream of many physicists, of explaining all the phenomena of the material universe in terms of wave mechanics will have to be dismissed as impossible. Ladrière goes further than this, in asserting that 'il y a toujours plus dans le domaine de la vérité intuitive que ce que l'on peut représenter dans le domaine des dérivations possibles' and that 'Il y a donc place pour des formes de raisonnement qui ne se laissent paz réduire à des procédures uniformes et toujours efficaces, du type de celles que l'on peut représenter au moyen des machines. Il y a plus dans la pensée que ce qui peut être enfermé dans les limitations exactes du calcul.'

In science, the sphere of natural history is precisely that of 'formes de raisonnement qui ne se laissent pas réduire a des procédures uniformes et toujours efficaces'. The procedures of systematics are of the essence of natural history, and it would seem to be a further corollary of Gödel's theorem that the attempt to transform the procedures of systematics into 'procédures uniformes et toujours efficaces, du type de celles que l'on peut représenter au moyen des machines' is misconceived and doomed to failure.

In general, the exact predictive power of natural philosophy can be demonstrated only in very simple systems, such as are rarely found in nature, at least above the atomic level, though they are common enough in the world of human artefacts. The power of the physicists is better shown in the laboratory and in the factory (the ultimate source of their strength) than in their dealings with nature. If the outlook of natural philosophy dominates science today, this is due less to the superior appeal of its characteristic types of philosophy than to its very un-Platonic links with technology. The power and prestige of modern biology are very similarly dependent on its links with medicine.

The historical and particular character of meteorology is apt to be overlooked by those who think of it merely as the body of general principles required for weather forecasting, just as the historical character of zoology is denied by those who see it as the general theory underlying medicine. The 'laws' of meteorology, however, are inductively derived from historical experience; this experience, as embodied in climatic records from a vast number of stations, is of greater scientific value than any 'laws' which have so far been derived from it. In any case, these laws relate only to the atmosphere of a single planet; if ever human beings land on Mars or Venus they are not likely to find such laws of much use in foretelling the weather they may experience on those planets. Similarly, if they find there anything analogous to living organisms, it is hardly to be expected that the understanding of these will be much helped by the most 'fundamental' recent discoveries in molecular biology; the terrestrial generalisations which are most likely to be applicable are those of Darwin.

There is no necessary antagonism between natural history and natural philosophy, the two could (indeed I think they should) be regarded as the opposite sides of the scientific medal; it ill becomes a natural historian to try to belittle the great achievements of his colleagues in natural philosophy. On the other side, Rutherford's celebrated apophthegm 'all science is either physics or stamp-collecting' may represent an extreme attitude, but unfortunately cannot be dismissed as altogether unrepresentative of opinion among physicists; in the current climate of opinion, the natural philosophical aspect of science is too often the only one which is communicated to the non-scientific public, e.g. in the recent work of Braithwaite [18a].

Even so penetrating a critic of current intellectual attitudes as Oakeshott [146d] has implicitly accepted the equation of science with natural philosophy, in his statement that 'if we speak still more strictly, there can in fact be no "scientific" attitude to the past, for the world as it appears in scientific theory is a timeless world, a world, not of actual events, but of hypothetical situations.'

Similarly, a recent writer on the philosophy of history [195d] states that 'we tend to employ the word "scientific" only where we have to do with a body of *general* proposition. A science, we should say, is a collection not of particular but of universal truths, expressible in sentences which begin with such words as "whenever", "if ever", "any" and "no". It is a commonplace to say that scientists are not interested in particulars for their own sake, only in particulars as being of a certain kind, as instances of general principles'. He goes on to say that 'To define history as *the* study of the past, and attempt to ground its autonomy as a form of knowledge on that point, can thus not be defended. But, of course, history is, in some sense, *a* study of the past. What past? The answer is the past of human beings.' Walsh does not, however, explain in what category of knowledge he would place knowledge of the non-human and pre-human past. History itself he finally concludes to be 'a humane discipline rather than a science'.

In those branches of science which are essentially natural-historical in character, and particularly in botany and zoology, a common reaction to the present public attitude is for the more able and ambitious workers to try to persuade others (and perhaps also themselves) that they too are really natural philosophers. 'If you can't beat them, join them' might be the motto. The following quotation from an eminent biologist exemplifies this attitude.

> The scientific method as I have outlined it is a tapering hierarchy of hypotheses, the more general counting the less general among their consequences, the least general—the ordinary colligative inductions—finally touching down in a multitude of particular statements about fact. The structure of science is therefore logico-deductive . . .

In the essay 'A Note on the Scientific Method' [142] from which this quotation is taken, Medawar does not explicitly claim himself to be a scientist practising the scientific method as he defines it, but the claim is implicit in his essay (in which incidentally he fails to explain for less learned readers the meaning of the phrase 'colligative inductions'). Medawar's definition is one which squarely equates science with natural philosophy, and we note that even the physicists, who have long tried to live up to this sort of definition, would have some difficulty in embodying the whole of their subject in such a 'tapering hierarchy of hypotheses', One might legitimately ask why Medawar, if he is really able to look on his subject (Biology, or Zoology, or Immunology, or whatever he prefers to call it) in this way, neglects to write a great book expounding it on these lines—for assuredly no one has hitherto even pretended to present any branch of natural history in the form of a tapering hierarchy of hypotheses.

An excellent critique of the attitude Medawar is expressing can be found in an article by T. A. Goudge [78] on 'Causal Explanations in Natural History', of which the closing paragraph is as follows:

> I submit, then, that natural history does employ causal explanations, and that the most typical, being the sort I have attempted to describe, do not bring an individual phenomenon under a general law. Furthermore, these typical explanations seem to be wholly appropriate to the subject matter with which natural history is concerned. The method used in constructing them is wholly sound. There is no superior method (e.g. that of physics) which ought to be substituted for it. Hence no warrant exists for saying that all explanations found in historical sciences are 'weak', or for suggesting that these sciences are still at a rudimentary stage of development. Physics is not necessarily the adult form of every science, and natural history is not adolescent physics.

For the natural philosophers, the ultimate test of a scientific law is its predictive power; I think this criterion could well be applied to the generalisations of natural history too. In general, the principles of natural history are of the nature of inductive generalisations—they are not to be established, as many laws of physics have been, by a single crucial experiment, but by the accumulation of a large number of supporting instances. A phenomenon which is no doubt connected with the current eclipse of natural history is the anti-inductive bias of nearly all recently influential philosophers. A favourite word among the English-speaking ones has been 'rigour', which can be translated as 'relying exclusively on strict deductive methods'. Rigour of this sort is not characteristic of natural history, which is characteristically ignored by these philosophers when looking for scientific illustrations of their theories. As we shall see, classification plays a vital part in the inductive processes of natural history.

In general, the predictions of natural philosophers can, in favourable circumstances, achieve a degree of accuracy and certainty beyond the reach of natural historians, but in the sphere of natural history, where natural historians can predict only with some degree of uncertainty, natural philosophers would as a rule be unable to predict at all in terms of their fundamental laws. It is rather notable that those modern biologists who are most concerned to represent themselves as 'really' natural philsophers are conspicuously reluctant to venture in print any predictions which might be verifiable (or falsifiable, according to the principles of Popper). Their research publications commonly describe some aspect of the working of a single organ or system in some kind of animal (much more rarely a plant)—as we shall see, there is no valid way in which a generalisation can be drawn from such evidence without making similar observations on other species, which these experimentalists very rarely do. As a rule, they claim or imply that their results are of wide general significance, but leave it to lesser mortals to discover to what actual organisms and to what extent they apply. Humbler scientists are thus denied the chance of disproving a prediction by the great man because, like the newspaper astrologers, he carefully refrains from making any which are precise enough to be clearly falsifiable.

As an example of prediction in the sphere of natural history, I can quote from my own work on the higher classification of the insect order Coleoptera. When defining a new series Cucujiformia [36] I characterised it in the adult stage by the absence of functional spiracles from the eighth abdominal segment. The Cucujiformia, as defined by me, will comprise more than 160,000 species—well over half the known Coleoptera. My generalisation about the spiracles was based on the examination of only a few hundred species, and might thus appear as a rather bold piece of induction. I had, in fact, predicated a character for something like 10,000 genera on the basis of its presence in about 200 of them—and so far no exceptions have been brought to light. One possible exception, in the small family, Jacobsoniidae, was noted by me in 1955, but subsequently I was able to find larvae of a species of this group in New Zealand and to establish that it did not belong in the Cucujiformia. The original few hundred species on which the generalisation was based were not of course randomly selected, but chosen carefully to represent as fully as possible the main adaptive types and phylogenetic lines of Cucujiformia. As we shall see, one of the main benefits to be gained from the understanding of systematics is a stronger basis for generalisations of this type. The quoted example was, of course, a particularly favourable one; in other instances generalisations of this sort, made by myself or other systematists, have often proved to be subject to some exceptions, however, in cases like these, a prediction which proves to be right in only 95 per cent of instances is still worth making.

One manifestation of the current fashion of regarding biology as a form of natural philosophy is the attempt to distinguish a field of 'fundamental biology' comparable to fundamental physics. As the fundamental theory of modern physics is largely concerned with the smallest and simplest components of matter—with sub-atomic particles and wave-equations—it is natural to look for biological equivalents of it in the study of the ultimate components of living things, the complex molecules of which cells are made. Thus, for example, recent advances in knowledge of the chemical structure and mode of operation of the chromosomes are presented as the biological equivalents of basic discoveries in atomic physics. On the basis of this analogy, enormous (future) practical importance is often claimed for these discoveries. A truer physical analogy for them might be found in meteorology—suppose that a team of physicists discovered the precise process by which gaseous H_2O molecules came together to form water or ice particles in the atmosphere, this might be represented as a tremendously important advance in 'fundamental meteorology' and as offering unlimited future possibilities of weather forecasting and weather control. In reality I believe that there is not, and cannot in the nature of things be, in meteorology any more than in botany or any other division of natural history (or for that matter human history or sociology), a body of fundamental theory comparable to that which is the proudest achievement of the natural philosophers.

The presentation of any particular advances in biological knowledge to the general public as comparable to major advances in basic physics is bound to arouse expectations of resultant major practical advances in medicine and agriculture. Such expectations may have the immediate effect of stimulating the flow of money to research institutions but if, as seems inevitable, they are disappointed, the ultimate results may be less happy for biology. A warning might be drawn from the fate of the Mendelian geneticists in Russia. In early post-revolutionary days, young and enthusiastic geneticists like Vavilov were preaching chromosomal genetics as a major 'revolution' in biological theory, from whose application immense practical benefits were to be expected. As a result, the young Soviet Government endowed genetical research institutions with unprecedented generosity—but the ultimate fate of Vavilov's school was less happy.

Biology, which now threatens to swallow up and replace botany and zoology in our society, had its effective birth as an anti-religious polemical weapon in the nineteenth century, and this aspect of it is by no means obsolete yet—as is manifest for example in P. B. Medawar and F. H. Crick. In pre-Darwinian days Life had generally been thought of as a transcendental, God-given property, not amenable to scientific analysis; Christian apologists like Paley had appealed to it as the permanent witness to superhuman creative powers. In deliberate opposition to this attitude, propagandists like T. H. Huxley and Ernst Haeckel alleged that life was just another property of organised matter, comparable to magnetism, electricity or heat—and as such should form the subject matter of an analytical science. Both these men were true prototypes of the leaders of modern biology, in that both were preeminently publicists and propagandists—and, in such original scientific work as they did, zoologists. For polemical purposes, it seems both easier and more effective to show, or appear to show, that an animal is a machine than to do the same for a plant. It may well be that some of the leaders of present-day biology will ultimately be adjudged to resemble Huxley and Haeckel in yet another respect—that the lasting value of their scientific work will prove to be less than commensurate with their contemporary fame and influence.

There are indeed those who claim that our contemporaries have discovered, or are about to discover, the Secret of Life, in the process of 'coding' genetical 'information' in the chromosomes. A generation or so ago, similar claims were being made on behalf of the discoveries of the 'Krebs cycle' in biochemistry; we may expect another generation to discover a new 'secret of life' for itself. In the days of Huxley and Haeckel, it was widely thought that a similar discovery was on the verge of being realised by those investigating the properties of a mysterious substance known as protoplasm. It is by now very clear that life, unlike magnetism, electricity, or heat, is not a property which inheres in molecules; it 'emerges' from complex systems. If inanimate analogies to it are required, they are best found in things like rivers or spiral nebulae. A river, or a spiral nebula, is a dynamic system with properties characteristic of its class, as is a living organism; geographers do not, however, seek to abstract from rivers a property known as 'fluviatility', nor do we hear of astronomers claiming to have discovered the secret of 'galacticity'. The concept 'life' is abstracted from the real and useful distinction which can be drawn between living and non-living things, and particularly between a 'live' and a 'dead' organism. It is not, however, a concept which can be given the sort of precision which natural philosophers demand of any basic principle, for example, is a crystallised virus, or a dehydrated but revivifiable animal placed in liquid air, alive or not?

Life is not an elementary scientific concept in itself, nor can it be analysed into elementary scientific concepts. Organisms, however, are objective data for science, and are capable of analysis to any desired extent, just as are rivers or spiral nebulae. Compared with these latter, organisms are far more highly integrated and organised systems; in the complexity of their subject matter, botany and zoology far exceed any other sciences. The traditional division of the study of living things into botany and zoology, unlike the fashionable 'disciplines' of biology, respects both the unity of the organism and the extreme diversity of organisms.

The popularisers of natural philosophy often dwell on the intellectual excitement and satisfaction which is liable to accompany the initial inductive conception and subsequent experimental proof of a new piece of fundamental theory. If, however, as we are often told by these same popularisers, physics is steadily approximating to the model of a tapering hierarchy of hypotheses, the more general including the less general among their consequences, then the amount of it which ranks as 'fundamental theory', small as it is, is liable to decrease still further in future. The glamour attached to fundamental theory casts a deep shadow in which the activities of the great majority of physicists (and even more, of chemists) are obscured. With physicists becoming ever more numerous, the proportion of them who can ever hope to taste the above-mentioned intellectual delights will become ever smaller.

In natural history on the other hand the structure of the subjects is radically different—in none of them is there anything really analogous to 'fundamental theory' in physics, any more than there is in sociology (even if Karl Marx really is analogous to Darwin). None has ever for-mulated, and none is ever likely to formulate, a set of principles from which the whole of botany or zoology could be deduced. If natural history lacks the glamour of fundamental theory, it is free also of the shadow. Practically any branch of natural history offers unlimited opportunities for inductive generalisation. Some such generalisations may have a very wide scope, like Darwin's Natural Selection, or the chromosomal basis of heredity, and naturally it is these which are emphasised by those who would present biology as a form of natural philosophy. It must be pointed out, however, that exceptions to them might conceivably occur in things which we would still call 'living organisms', yet such exceptions would not alter the status of the generalisations.

It seems fair to claim that natural history offers its devotees vastly greater opportunities to taste the delights of successful inductive generalisation that are now to be found in physics or chemistry. Generalisations about plants or animals may not have the glamour (or technological possibilities) of advances in basic physics, but they offer satisfaction of a not dissimilar type to the human mind. Anyone who has ever felt in himself the workings of Windelband's 'Anschaulichkeit' or Bacon's 'poesis' will be convinced that the rigorous deductive methods favoured by the positivists are not the only, nor even the highest, modes of scientific thought.

2 The Function of Classification in Natural History

Systematics, which might be defined as 'the ordering of genera' in Bergson's sense of the term is characteristically a pre-occupation of natural historians; it is not so much that the natural philosophers are able to dispense with it as that they find the task already done for them, and are apt to accept classification as a datum rather than as part of their science. In the mind of a modern experimental physicist, perhaps the largest and most frequently used body of classificatory knowledge is that concerning pieces of apparatus and experimental configurations—a great part of his professional skill consists in the ability to recognise a given assemblage of apparatus as, say, a 'maser' or a particular type of heat-engine. However, a biological systematist overhearing recent arguments between physicists concerning the status of certain newly discovered 'elementary particles' will recognise that even today natural philosophers may be confronted with problems not unlike his own.

In all branches of natural history—Astronomy, Geology, Meteorology, Botany and Zoology—classification, whether it is of stellar spectra, of igneous rocks, of pressure-distributions, of algae or of worms has always occupied a central position, comparable perhaps to that which natural philosophers accord to mathematics. As we have seen, in physics (and even more, in chemistry), classification plays a real if subordinate part, and natural history may similarly make considerable use of mathematics at times. The basic principles guiding the thought of a natural historian are not, however, normally expressible in algebraic formulae.

The words classification, systematics and taxonomy are now commonly treated as synonyms, an example of the confusion and carelessness in the use of words which is prevalent in so much modern scientific writing. Words, to quote a poet 'decay with imprecision, will not stay in place', and however true this may be of his own art, it is at least equally so in science. Originally and properly, classification would have denoted the activity of placing things in classificatory groups, whereas systematics would be the body of general theory underlying this activity. The word taxonomy, though proposed long ago, has come into general use comparatively recently, and has suffered the fate which commonly befalls new terms in science. Old established words, with a long tradition of use behind them, generally have more or less stable meanings, not greatly influenced by changing fashions. A new word has, however, no such inherent stability, it is liable to far more rapid and radical changes in its application than are old ones. The promulgator of a new scientific term may, and usually does, provide a carefully worded definition of it, but there is no sanction by which others can be compelled to use the word only in accordance with this definition. Words are public property, subject to democratic rule. In practice, taxonomy has become a vaguer synonym of classification and systematics (also of nomenclature); the word does not appear to me to perform any very necessary function in the language, and could well be dropped. The tendency to write 'systematics' when what is really meant is classification, as for example, when the researches of Dr X are described as being on 'the systematics of Compositae', exemplifies a very common vice of modern scientific writing, like the use of 'fluid' instead of liquid, of 'morphology' instead of structure, 'biology' instead of habits, etc. The general principle seems to be that you should always use the most abstract possible terms for things—systematics sounds more abstract, more natural-philosophical, and so more scientifically 'U' than classification.

It is important at the outset to distinguish between classifying things and naming them. The use of a name (other than the 'proper name' of a person, house, river, etc.) implies the recognition of a group, the class of things referred to by that name, and such groups are the elements of which classifications are composed. Classification, at least in the traditional sense in which the word is used here, involves the incorporation of such groups into a rational, hierarchical system in which each group has a unique place. Classificatory groups (taxa) are of various grades, or categories as they will be called here, and for all categories except the lowest, each taxon will include one or more taxa of the next lower category.

From the point of view of dialectical philosophy, Classification might be regarded as the third term of a Hegelian triad, in which are resolved the two antitheses of Similarity and Difference. For classification to be possible, the entities with which it is to deal must be at once similar and different. My dream of last night, a particular pebble, your house, F. H. Crick, last Christmas Day, Konrad Lorenz's left ear, the earth's magnetic field, Darwin's *Origin of Species* and Bruckner's 8th symphony compose a collection of entities which have insufficient mutual similarities to be usefully classifiable; all the neutrons in the universe, on the other hand, are, according to current physical theory, individually indistinguishable and therefore equally incapable of being classified.

The problem of denotation of classificatory groups is commonly called nomenclature, though we may note that such groups need not bear names in the strict grammatical sense of the word. For example, in the Dewey and similar systems used in the classification of library books, the various groups are denoted by numbers, or combinations of numbers and letters. Although, as we shall see, there are important differences between the problems of classifying books and those of systematics in botany or zoology, these differences are not such as to necessitate different systems of 'nomenclature' in the two cases. Methods of denoting classificatory groups by formulae rather than names have been seriously proposed for use in biology, though at present there seems little likelihood of any of them being adopted generally. In the present work, following the classical philosophical tradition that meanings are logically prior to words (almost the opposite doctrine to that of semanticist philosophy), we shall consider first the problems of classification, leaving the consideration of nomenclature to the last section of the book.

If the invention of nomenclature can be traced (figuratively, at least) in the second chapter of the book of Genesis, the invention of true systematics is commonly attributed to Aristotle, in whose philosophy it played a central part. The syllogism, another reputed Aristotelean invention, is based on evidently classificatory statements such as 'All (or some, or no) A is B', and in developing it the Stagyrite* was really demonstrating the use which could be made of a natural classificatory system. We could quote a zoological syllogism as an example: 'All mammals have mammary glands in the female sex (major premiss); the extinct Multituberculata were mammals (minor premiss); therefore the Multituberculata had mammary glands in the female sex.' The conclusion here, unlike those in the usual textbook example of syllogisms, tells us something we do not know, and probably could

* As pointed out by Toulmin [193a], 'Aristotle the zoologist certainly wanted to couch substantial arguments in syllogistic form.'

not ascertain, by direct observation. Although Aristotle nowhere clearly indicates the difference between 'natural' and 'artificial' classifications, his use of the word 'essence' implies that the sort of classification he was thinking of was natural.

The definition of a thing, said Aristotle, should consist in mentioning its essence—those qualities which it must have in order to be entitled to its name (i.e. classificatory group). He clearly believed that things which are alike in their essence will also be alike in many other, non-essential, characters concerning which useful syllogisms can be constructed. In our zoological syllogism, the skull characters of the Mammalia could be considered as the 'essence' of the group, the reason why Multituberculata are attributed to it. A natural system may be defined as one in which things may be placed in their appropriate groups by reference to a comparatively small number of characters, corresponding to Aristotle's essence, and in which the members of a given group will normally resemble each other in characters other than those needed in order to refer them to that group. *Classification, based on this principle will herafter be referred to as Aristotelean.* The second half of the definition will not in general apply to artificial classifications, in which category we can probably place the Dewey and similar systems used in libraries. I have not seen it suggested that books included in a particular group under the Dewey system are liable to resemble each other in characters other than those needed to refer them to that group (it is, for example, a continual trouble to librarians that the sizes of books bear no relationship to their classificatory positions), nor have I encountered any attempts to reason syllogistically on the basis of such systems.

Evidently, the distinction between Aristotelean and artificial classifications will not be an absolute one; a given set of objects could generally be classified with varying degrees of 'naturalness', though there will probably be a maximum possible degree for any given set. How 'natural' in the Aristotelean sense a classification can be will no doubt vary with the character of the things being classified. Most types of natural object are susceptible to more natural classification than are books or other human artefacts; the development of Aristotelean systems of celestial bodies, geological structures, minerals, types of crystal structure, etc., has played an important part in the development of astronomy, geology, physics and chemistry. In dealing with living organisms, it seems probable that classification can reach its highest degree of naturalness—presumably as a result of the highly integrated character of organisms—and systematics correspondingly occupies a more important place in botany and zoology than in other sciences. This, I believe, is 'in the nature of things'; it is not, as the followers of Comte would allege, a symptom of the relative 'immaturity' of the life sciences.

An Aristotelean essence, defined as a series of characters more or less strongly correlated with each other (and with other characters) in their incidence, may seem a satisfactory and unambiguous basis for the definition of a group. It must be made clear, however, that for plants and animals correlations of several characters may be used to associate together forms which may not be of common phylogenetic origin— for example, the old 'Amentiferae' and 'Ungulata'. Moreover, it is possible for one and the same group to have more than one 'Aristotelean essence', in the sense that one set of correlated characters might associate it with another group A while an independent set of equally correlated characters might provide a basis for uniting it with a third group B. This is in fact the classical situation for a controversy between systematists about the classificatory position of a given group. Aristotelean systematics offers no general objective principle for settling such disputes; as we shall see, the only objective criterion in deciding questions of this sort is phylogeny. Our present accepted systems are as a rule based on some sort of compromise between Aristotelean and phylogenetic principles. Major divisions continue to be based on Aristotelean essences except where there is good evidence that, as in the case of the old Ungulata, the correlations result from convergent development. The adherents of the *status quo* are thus able to stigmatise both phylogenetic systematists and the adherents of the Sneath–Sokal 'numerical taxonomy' school as dogmatic and unpractical extremists.

Those modern biologists who question or deny the fundamental importance of classification are mainly to be found among the biochemists and experimental physiologists or geneticists; such people characteristically speak and think of living organisms as machines—the use of the word 'mechanism' is one of the most tired and tiresome clichés in their writings. It is significant that the philosopher Descartes, who may be considered as the originator of this and other recently influential modes of thought, nowhere in his writings showed any understanding of the nature and possibilities of systematics.

We have indeed every reason to believe that in all their details living organisms function in accordance with the laws of physics and chemistry, but this is probably as far as their analogy with machines extends. A machine could be defined as something put together, out of separately fabricated parts, to serve certain pre-determined ends or functions. Living organisms, as far as we know, are never put together from parts (except in the very special sense exemplified by lichens and the 'graft hybrids' or 'chimeras' of the horticulturalists),the parts are differentiated from a pre-existing whole. The only function which could reasonably be assigned to organisms is ultimate survival, which is hardly comparable with any function of a human-fabricated machine.

The objections to the machine analogy could be put in another, seemingly quite different but perhaps ultimately related way. Even the simplest of Protista, considered as machines, would be incredibly complicated; their essential unit parts would be molecules, whereas our machines are built with pieces of macroscopic (in the physicists' sense) matter. These molecules, furthermore, interact in largely liquid systems. Physicists have no rigorous theory of the liquid state, as they have for solids and gases; the molecules of a liquid can neither be considered to move independently and at random as in gases, nor are they bound in rigid lattices as in solids. The laws of chemistry are essentially statistical, and will not enable the behaviour of individual molecules to be predicted; you may if you like describe living things as 'chemical machines'. but their behaviour could not be predicted in terms of the laws of chemistry. The parts in human-fabricated machines normally interact in such a way that the effective 'chains of causation' are linear or simply branched—each part, while it may act on one or more other parts, is itself usually acted on by only one other part. The causal interconnections in living organisms are liable to be much more complex, each part or factor being liable both to act on and be affected by several others. Thus we are continually confronted with analogues of the 'problem of three bodies' discussed in the previous chapter. It is possible, as was demonstrated by Ross Ashby's 'Homoeostat', to put together pieces of physical apparatus in such a way that three or more variables become mutually influencing. Such systems, as pointed our by Ross Ashby [6] show interesting analogies to living organisms, not least in the fact that physicists are unable to predict their future behaviour accurately.

There are, of course, parts of some living things in which something like a linear chain of causation exists, e.g. in the motor nerves and the effectors which they control in the higher animals—naturally it is to such systems that mechanistically minded biologists direct attention, while ignoring much more fundamental processes like growth. The construction of an 'electronic cabbage' which would grow would be

B

a far greater scientific achievement, and far more illuminating for biologists, than the much-publicised 'electronic brains'.

As noted in the last chapter, Professor Hayek [87] emphasises the importance of the complexity of systems in determining the ways in which they may usefully be studied, but does not, however, consider the 'three-body problem'; he defines complexity simply in terms of the necessary number of constituent elements for a system to exhibit the characteristic properties of its class. And he totally fails to note the prime importance of classification in dealing with complex systems of the types he is discussing. Admittedly, the value of classification in sociology is far less evident than it is in zoology or botany; human social groups are typically unstable and changeable, not amenable to clear-cut definitions, but sociology cannot hope to make real progress other than through the classification, naming and study of particular social groupings. The weakest point in all sociological systems hitherto, the point at which they have always been attacked, is in the basic social units which they recognise.

There is undoubtedly still a widespread conviction among 'new biologists' that differences between organisms, with which classification is concerned, are superficial, whereas similarities are fundamental—in contradiction to the Aristotelean epigraph to Chapter 5 of this book. Such biologists genuinely feel that the proper aim of scientific study of plants or animals is to 'break through' or sweep aside the apparent differences between organisms, to which systematically inclined botanists and zoologists devote such seemingly excessive attention, and expose the uniform 'mechanism' which these would-be natural philosophers are convinced underlies it all. Such convictions must be regarded as articles of faith resting on no sound evidence. For a long time it was supposed that it was at the chemical level that the supposed 'basic uniformity' of organisms would reveal itself. Present evidence indicates that nearly related species are as likely to differ from each other in details of their chemical functioning as in visible structures. There seems to be no part of the chemical functioning of organisms which is incapable of being changed in the course of evolution. We have no real evidence that the chemical level of explanation is more fundamental than those involving macroscopic functional adaptation; plasticity and change are manifest at all levels of the organisation, and there is no level, above the atomic, at which two different species may be assumed to be identical.

We may quote two recent zoological physiologists as evidence that some of them are beginning to realise the true state of affairs. Hoyle [97] wrote that 'The vertebrate physiologist is no longer justified in complacency regarding the universal applicability of the data obtained from the ordinary muscles of the frog and the cat', and that 'Important new generalisations have emerged from the extension of studies within the phylum Arthropoda and have amply illustrated not only the special advantages of the comparative approach but also the dangers of concentrating on one experimental animal'; a further observation in the same work is 'Of some surprise, first pointed out by Viersma, has been the rapid realisation of the existence of a bewildering variety of response mechanisms defying any simple classification'.

Kerkut [114] expressed a rather similar attitude as follows:

> What then can one conclude about the chemical and physical nature of protoplasm? Simply that we have a very great deal to learn about it. Modern developments are making it abundantly clear that some of our previous concepts are inadequate and that the picture is very much more complex than previously imagined. It would be a great mistake to assume that all is chaos and that there are no general common systems, but it would be a mistake of equal magnitude to assume that everything is simple and that but one system will be found in all protoplasm. From our present viewpoint there would appear to be at least four or five different systems which allow a cell to obtain its energy.

A distinguished biochemist (Davidson, 47a) has recently pointed out that 'It is frequently emphasised that a cell is not merely a bag of molecules but a "system of interacting molecules" whose activities interlock in a most elaborate way, with membranes playing a most important part, and nucleus and cytoplasm supplementing each other in a complicated interrelationship. But it is surprising how often these seemingly obvious considerations are overlooked.'

If botany were limited to those statements which could be made about all plants, and zoology to those which applied to all animals, there could be a satisfactory reduction in the size of textbooks of the two subjects, but also a great loss in their practical utility and in their special appeal to the human mind. The potato and the tomato may be very similar plants to the botanist, but the importance of each rests in features which differentiate it from the other. The vast bulk of both subjects, as at present delimited, consists of statements which apply to certain plants or animals and not to others. Half the content of such a statement is essentially classificatory in character—the specification of which particular plants or animals it applies to. Admittedly, both botany and zoology are at present troubled by vast accumulations of 'floating' statements whose systematic limitation is as yet quite uncertain—they have been shown to apply to one or two species of plants or animals, and possibly not to apply to one or two others, but that is all. This particular evil has grown with the cult of the 'pseudo-natural-philoso-phical' (if the term may be pardoned) outlook among our self-con-sciously 'new' biologists; an enormous amount of systematically-directed research needs to be done before these statements can be given satisfactory form. The role of systematics in these subjects, far from being 'played out', could and should be as great in the future as it ever has been in the past.

Unfortunately, there is probably no university in the world today in which young botanists and zoologists are being trained in such a way as to equip them to achieve the improvements in classification which modern knowledge renders possible. Classification is essentially a cumulative subject; in general, the systematist can only hope to improve on the systems of his predecessors by first understanding and appre-ciating what they did. The value of a classification is more or less pro-portional to the number and variety of the characters used in con-structing it. If you wish to train someone to improve on the existing classification of any group, the first requirement is to lead him to the deepest possible understanding of the existing system and its bases—a task which the universities perform less and less thoroughly every year. Even in those departments which make rather a feature of 'Taxonomy', the emphasis is usually on particular techniques rather than on co-ordinated knowledge—the botany department of one university may specialise in stem anatomy, and another on the extraction and analysis of essential oils, one zoology department may pride itself on its training in numerical taxonomy, while another makes rather a feature of immunology and electrophoresis. The concentration on any one of these technical approaches is harmful to the catholicity of outlook which is the soul of good systematics. The central position in the thinking of a good systematist is always occupied by his organisms. It is perhaps an obscure awareness that, when elaborate and sophisti-cated research techniques are adopted, they are only too likely to usurp the central position in the scientist's thought, which accounts for the notable reluctance of many systematists to become involved with modern and complex modes of investigation.

3 The Species in Biological Systematics

The Oak dies as well as the Lettuce, but its Eternal Image and
Individuality never dies, but renews by its seed.

WILLIAM BLAKE
(annotations to his picture
of the Last Judgement)

Human beings have long been impressed by the permanence and
distinctness of the individual kinds of living things; the saying 'God
made the species, all else is the work of man' expresses an attitude which
is still widely held. Even Darwin [44], by calling his famous book *The
Origin of Species*, showed the influence of this mode of thought. Modern
critics (e.g. Mayr [140a]) have pointed out that nowhere in his work
does Darwin offer a satisfactory definition of the species, nor does he
account adequately for that feature of species-formation which is now
regarded as all-important—the origin of discontinuity. Various recent
systematists have tried to repair the omission of Darwin, and to formulate a set of criteria by reference to which it may be decided whether
a given assemblage of organisms belongs to one or several species. The
fact that this can at least be attempted, whereas none has even pretended
to formulate comparable definitions for higher taxonomic categories
like genera, families and orders, is taken as evidence of the superior
objectivity of the species. These definitions, it must be emphasised, are
concerned only to establish minimal requirements for species-difference,
they offer no assistance in deciding at what point the difference between
two species becomes great enough to entitle them to different genera,
etc. The implications of this state of affairs are considered further in
Chapter 5.

The criteria which have been used in attempts to define the species
may be summarised under five headings:

1. *Museum criteria*

(*a*) Among members of one species, there is normally a limited and continuous variation in characters of structure and pigmentation, whereas a discontinuity in one or both these respects will normally show itself when members of two different species are compared.

(*b*) A species has normally a limited and continuous area of natural distribution, rarely coinciding exactly with that of any other species.

2. *Ecological criteria*

(*c*) Between members of different species, there are normally differences of habits and behaviour, not bridged by transitional forms.

(*d*) It is very rare to find, in nature, matings between members of different species.

3. *Physiological criterion*

(*e*) Within a species, there is normally the same kind of limited and continuous variation in physiological and biochemical characters as there is in structural ones, and the same type of unbridged gap when members of two different species are compared.

4. *Genetical criteria*

(*f*) Sexual crossings between members of one species are normally fully fertile, giving offspring with characters of the same species and themselves fully fertile, whereas interspecific crosses usually yield infertile offspring or none at all.

(*g*) Between members of two different species there are normally differences in large numbers of hereditary factors (genes), usually accompanied by complex chromosomal differences (inversions, translocations, reduplications, etc.); intraspecific variation generally involves far fewer factors, and only simple chromosomal changes (if any) are found in it.

5. *Palaeontological criterion*

(*h*) A species has a limited and continuous range in time.

These eight criteria make quite an impressive list, but rarely (if ever) could they all be applied to one particular case, and each of them is subject to difficulties of application and to exceptions. Each of them will need further critical consideration.

The museum criteria are given pride of place because of the widespread current belief that the proper place for a systematic botanist or zoologist is in a museum, and from the fact that the large majority of described species in both kingdoms are accepted as such purely on the museum criteria. The first thing that needs to be noted about them is that they are reliable only when adequately representative (of the full ecological and geographical range of the species) material is available. Many of the 'species' which have been described as such from the study of scanty museum material have been found subsequently, when more material became available, to grade completely into other species. The museum criteria offer no general means of deciding whether or not two specimens belong to the same species by comparing one with the other; a satisfactory comparison needs to be made with a representative series. The current convention that a single specimen, the Holotype, is the only satisfactory basic criterion for a species would be difficult to justify logically on any theory but that of Special Creation.

This difficulty is particularly manifest in the 'clines' of Huxley [103]. A cline may be defined as a gradual and progressive change in the characters of a species when it is traced along an extensive tract of natural distribution—a number of cases have been described in which the end forms of such a cline look like distinct species and even behave as such when artificially brought together, though they are linked in nature by an unbroken breeding chain. Few museums have adequate representative material from the whole length of such clines; the museum systematist has usually to work with collections in which most of the specimens are taken from only a few, often widely separated, points on a cline. He may well describe them as a series of subspecies, or even full species. Clines will be considered further in the next chapter.

Other difficulties for the museum systematist are presented by species showing what is called 'polymorphism' and by those with 'alternation of generations'. A polymorphic species is one in which two or more distinct phenotypes coexist frequently or constantly in natural populations; the differences between the forms are often sharp and unbridged, and appear to be determined by only one or two genetic factors. Botanical examples are the common Primrose (*Primula vulgaris* L.) with its 'pin-eyed' and 'thrum-eyed' forms, and various 'heterostylic' species of *Limonium*; entomologists will be familiar with the two colour forms of the female in the Clouded Yellow butterfly (*Colias croceus*) and in the Silver-washed Fritillary (*Argynnis paphia*). Other animals manifesting colour polymorphism, such as the snail, *Cepea nemoralis*, and species of *Adalia* among the ladybirds (Coccinellidae), show more or less continuous variation (probably because several pairs of alleles are concerned and some or all of them do not show complete dominance) and present less difficulty to the museum systematist.

Alternation of generations is a term which may be applied to rather diverse phenomena. The succession of haploid and diploid (or game-tophyte and sporophyte) generations is familiar to botanists, but it is probably not very common for systematists to describe the haploid and diploid forms of the same plant as different species, though this may have happened on occasion in the Algae. The 'perfect' and 'imperfect' forms of the same fungus have no doubt at times been ascribed to different genera, but the difference between them is probably not a simple haploid–diploid one.

In animals, alternation of generations is rarely if ever a matter of haploid and diploid individuals; except for such special cases as the males of the Hymenoptera, the haploid stage in animals is represented only by the gametes. Alternation of bisexually-reproducing and uni-sexual (parthenogenetic) forms is not uncommon in some groups of animal, e.g. Insecta, Crustacea, Protozoa; in many groups of 'colonial' animals such as the Coelenterata, Polyzoa, Tunicata, we find 'vegeta-tive' budding alternating with sexual reproduction. There are also cases, at least among the Insecta, in which the alternating generations are both bisexual; a good example is furnished by the central European butterfly *Araschni alevana*. This species has two broods in a year, one with the adults emerging in the spring, the other flying in the later summer; the two forms differ strikingly in colour-pattern, and without breeding investigations a museum systematist could hardly discover that they are not two distinct species.

A further difficulty in the application of the first museum criterion is that differences of this sort are at times exceedingly slight between what are by all other criteria perfectly 'good' species. A well-known example in the animal kingdom is in the genus *Drosophila* (the fruit-flies beloved of geneticists), where *D. pseudoobscura* and *D. persimilis* were recognised as distinct solely as a result of breeding experiments—only after the geneticists had discovered that these forms behaved like distinct species were some exceedingly slight 'museum' differences between them found; it is safe to say that on museum criteria alone none would ever have suspected that two species were present. Two rather similar cases in the animal kingdom were known long before the *Drosophila* one—the Grey Dagger and Dark Dagger moths (*Acronycta psi* L. and *A. tridens* Schiff.), and the Chiffchaff and Willow Warbler (*Phylloscopus collybita* Vieill. and *P. trochilus* L.) among birds. The moth-collectors of the last century were in the habit of breeding their species, and it was thus that the distinctness of the two *Acronycta* was discovered (there is a well-marked colour difference in their larvae); the *Phylloscopus* species were first suspected to be distinct from differences in their behaviour and song, e.g. by Gilbert White [199]. In each of these cases very slight 'museum' differences have subsequently been discovered to separate the species.

In plants, apart from the special case of polyploids which we shall consider later, few instances of what the Americans call 'sibling species' have been recorded. There have indeed often been reports of breeding barriers between plants which do not show evident differences in the herbarium, but such forms have rarely been named as independent species. Botanists commonly refer to such cases as manifesting the phenomena of 'incompatibility' rather than as sibling species. Plant systematists in general seem to rely more exclusively on the museum criteria than do their zoological colleagues; this and other differences are discussed by Turrill *et al.* [193c].

A recent review by Maheshwari (in [215]) entitled 'The Plant Species in an Age of Experiment' distinguishes between 'phenospecies' and 'biospecies', the former being defined on museum criteria, the latter on genetical, ecological and physiological ones. In discussing biospecies he states 'according to this concept, the real species or biological species, as contrasted to the more or less artificial species of the typologists and morphologists, is a natural and non-arbitrary unit of a genetically closed population system that has lost its ability to inbreed with other species'. His final conclusion is 'The species problem is still fluid', and he considers that the phenospecies will retain its practical utility for a long time to come.

The first museum criterion has two great advantages, the ease with which such comparisons can be made, and its applicability to fossils as well as to existing organisms. To compare all the species of a moderate-sized genus in respect of several characters of structure or pigmentation might take a systematist a week, whereas a similar comparison in respect of ecological, physiological or genetical characters might take years of research.

The second criterion expresses the fact that good systematists look at the labels of their specimens. As a principle, it is open to two major objections—first, that there are some species with discontinuous natural distributions, and second, that there are now very numerous species with discontinuous distributions attributable to human influence. Discontinuous natural distributions are found in 'relict' species, whose former continuous areas have become fragmented through extinction by natural (non-human) causes in parts of them. Typical European examples are the more or less sub-arctic forms, which probably had a wide range as far south as the Alps during the last glaciation, but now have widely separated occurrences in the Alps, the Scottish mountains, and in those of Scandinavia. There are also cases where particular species, by some rare non-human means of long-range transport, have established isolated colonies (e.g. on oceanic islands) far outside their normal range. But the impact of human action on the natural plant and animal worlds, which must have been considerable for a long time past in much of the world, is now so enormous and rapidly increasing that, at least in many terrestrial groups, the conception of natural distribution has become an unreal and perhaps unrealisable ideal.

For modern animals, the ecological criteria are probably almost as good in principle as the museum ones, but too few species have been studied from this point of view to make them generally applicable. Habits in plants are largely habits of growth, with resulting differences in general shape etc., and thus become absorbed into the museum criteria. The second ecological criterion may be applied to those lower plants (fungi, algae, etc.) which show conjugation, and could perhaps be extended to cover the development or non-development of pollen-tubes on particular kinds of stigma by the pollen-grains of Angiospermae or Coniferae. A general limitation of the ecological criteria is that they are only properly applicable to species living in the same area—a given species may show habit-differences in different parts of its range, in relation for instance to differing climatic conditions, and organisms living in different areas do not have the opportunity of sexual crossing.

The physiological criterion is in principle of equivalent value to the first museum one, and like it is not limited in its application to forms inhabiting the same area. Even more, however, than the first ecological criterion, it suffers from the lack of the necessary knowledge to make it applicable in most cases. All that can be said is that, where closely related species have been adequately investigated by laboratory physiologists, physiological differences between them have been found. There have been cases among animals in which physiological differences have provided the first indication of the existence of a previously unrecognised species; the mosquito, *Anopheles maculipennis*, provides a well-known example. Investigation of this species as a 'vector' of the malaria parasite, *Plasmodium*, revealed that two forms of the mosquito existed, one breeding mainly in brackish water and capable of spreading the *Plasmodium*, the other restricted to fresh water and not a carrier of malaria. The brackish-water form was first separated as a 'race' *atroparvus*, and then, when it proved to be reproductively isolated from typical *maculipennis*, and distinguishing characters between the two were found in the egg stage, *atroparvus* came to be accepted as a 'good' species.

It is now widely believed that the genetical criteria for species are the most truly fundamental, and many attempts have been made to formulate a new, genetical definition of the species which will make clear its true nature from an evolutionary point of view. In all these attempts, attention is concentrated on the species as a potentially interbreeding unit. Any two non-allelomorphic hereditary factors which occur in different members of the same species may be brought into combination as a result of crossing between the individuals in question or between their progeny; the species may thus be considered to possess a potential 'common gene-pool', a reservoir of hereditary variability on which selection can work. The gene-pools of two different species on the other hand are considered to be irrevocably cut off from one another. On this view the essential feature of species formation is the irrevocable splitting of a previous common gene-pool. This may well be the true explanation of such superior objectivity as the species possesses in comparison with other categories (cf. Dobzhansky [49]), but it will also explain why this objectivity is by no means absolute. A definition of the species on this basis may be satisfactory when applied to forms living at the same time and in the same area, but difficulties may be expected if the attempt is made to apply it to forms living in different islands or continents, or in different geological eras. A gene-pool may be as effectively split by impassable geographical barriers as by behavioural or genetic bars to crossing. On this criterion, specimens of a given species which, by some rare accident, have become established on a remote oceanic island, should immediately acquire the status of an independent species. If the genetical criterion for species is the really fundamental one, then we should expect the objective basis for species-differentiation to disappear when the forms compared are from widely separated areas.

In fact, numerous cases have been recorded, particularly among flowering plants and birds, in which species from Europe and species from the Far East, or species from the Far East and species from North

America, differ strikingly and constantly in phenotypic characters, yet will produce fertile hybrids when artificially brought together. In some groups of birds, notably the ducks, such hybridisable forms may differ so much in appearance that systematists have placed them in different genera. Many well-known horticultural varieties, e.g. *Magnolia soulangeana*, have originated as hybrids of American and Asiatic species.

A further difficulty confronting the genetical definition of the species is mainly botanical, and concerns the treatment of polyploid forms. Most animals have separate sexes and an 'x–y' system of sex-determination (one sex, usually the male, having an odd pair of sex-chromosomes), and for various reasons connected with this system polyploid animals are usually sterile. In the evolution of most groups of animals, polyploidy seems to have occurred rarely or not at all—the exceptions being mainly in those groups which do not have the x–y sex-determining system. In plants, however, this system of sex-determination is comparatively rare, and doubling of the chromosome number is frequently possible without necessarily involving loss of reproductive capacity; the polyploid individuals can either reproduce vegetatively or (being usually hermaphrodite) fertilise themselves. The commonest type of polyploid individual in nature is the tetraploid; these, when crossed with an ordinary diploid produce triploid offspring which are invariably sterile. By doubling the number of chromosomes, we seem to have produced at one bound a breeding barrier like that between two distinct species. This breeding barrier is not, however, really irrevocable; it can be bridged in one direction by producing further tetraploids from the original diploid stock, and cases have been recorded of diploid individuals arising spontaneously from a tetraploid stock.

The genetical criterion for species also breaks down in dealing with those organisms which do not manifest sexual reproduction, e.g. the apomictic *Hieracium* (Compositae), and in such animals as the familiar stick-insect (*Carausius morosus*) and the celebrated Protozoan *Amoeba proteus* (*Chaos chaos* aucct.). In such forms, the genetical criterion would logically require the recognition of every individual as a separate species.

The second genetical criterion (g of the original set) is free from the special objections to the first, but is only applicable where (1) the forms being compared can be crossed to yield fertile offspring, thus violating criterion (f), or (2) detailed chromosome maps have been prepared from linkage studies—these require an enormous amount of work for their preparation, and are only available for a very few species of organisms, or (3) the so-called 'giant chromosomes' occur, as in the salivary gland chromosomes of the insect order Diptera; these have been found in very few groups of organisms.

Support for the belief that the breeding criteria are the really fundamantal ones for the species can be adduced from the fact, well known to zoological systematists, that in those animals in which mating is an active process, species differences are usually most striking in the secondary sexual characters, while in animals with sedentary adults this is not usually the case. In Angiospermae, closely related species often differ most obviously in their flowers, and such differences may be analogous to secondary sexual characters in animals. Some pollinating insects, notably bees, have been shown to have a tendency to go from one flower to another of the same species rather than to one of a different kind—obvious floral differences between species will facilitate this sort of discrimination and thereby increase the chances of successful pollination (cf. [63]).

It is undoubtedly true that, in animals with separate sexes, species differences are apt to be most striking and obvious in the secondary sexual characters, particularly of the males. This observation does not, however, justify the dogma, widely accepted among entomologists in recent times, that the examination of the male genitalia provides a sufficient and infallible guide to the correct discrimination of species. The observations of E. B. Ford [67] concerning the butterfly species *Papilio dardanus* are relevant here:

> Within the region just delimited, the species is subdivided into five races. Two of these, Dardanus and Meseres, comprise a western type. They are larger than the others and the males are less heavily marked with black; moreover, their genital armature differs slightly from that of the eastern and southern forms but not in such a way as to prevent interbreeding with them. This difference is indeed controlled on a single factor basis, subject slightly to the effects of modifiers: the presence of a long spine on the inner surface of the valve in the eastern type is dominant to its absence, which characterizes the western one in which genitalia of the eastern type occur rarely. The point is worth noting, since the anatomy of the genitalia is so often treated uncritically by taxonomists as a criterion of specific differences in Lepidoptera and some other insects. Statements on the structure of these organs are generally made in an unscientific manner: the number of individuals in each species that have been examined in order to establish their characteristics is not recorded nor is any indication given of their variance.

A striking example of the emphasis placed on male sexual characters by some entomological systematists is provided by Coiffait [32a], who not merely erected a new species but even placed it in a separate genus *Sectophilonthus* Coiffait, for a single specimen of a beetle which, while showing all the outward characters of the well-known *Philonthus decorus* Grav., had a strikingly aberrant form of aedeagus (male intro-mittent organ). It seems far more probable that *rossicus* Coiff. represents merely an unusual mutant or teratological form of *P. decorus*.

To these reflections, it may be relevant to add another. The male genitalia of insects are frequently rather complex and very 'three-dimensional' structures, of which two-dimensional pictures are apt to give misleading impressions. The precise appearance of such a picture may depend considerably on the angle from which the specimen is viewed, and on the manner in which the specimen was prepared for observation. The combination of these effects offers a good deal of freedom to the well-known selective powers of the human mind—those powers by which significant shapes are seen in Rohrschach ink-blots. It is often only too easy, for an honest observer with a pre-conceived idea in his head, to persuade himself that he can perceive significant similarities and differences in such comparisons—similarities and differences which are not 'objective' in the sense that they would not be noticed by an observer with different preconceived ideas. It is probable that a good deal of published work, in which insect species are claimed to be distinguishable with dogmatic certitude by reference solely to the male genitalia, is more or less unsound.

The palaeontological criterion (*h*) underlies the use of 'zone fossils' for dating and correlating sedimentary rocks by geologists, and must be taken as a statement of faith rather than as one with really adequate empirical verification. If it is difficult at times to apply many of the criteria for species to forms widely separated in space, the difficulties are even greater when the forms are separated in time. In those cases where we have a full and continuous record of a 'lineage' in fossil animals (few such cases are known in plants), we find a history of gradual and continuous change, rather like that along a cline. Fossils from the bottom of such a lineage may differ from those from the top in a sharp and unbridged way which would satisfy any museum systematist's idea of good species, yet be connected by a perfect series of intermediate forms in intervening layers. Extended lineages of this sort are, fortunately for the palaeontological systematist, not very common in the fossil record.

It is only the gappiness of the record which makes the task of classifying fossils in the accepted fashion practicable.

From direct geological evidence, and perhaps more often from indirect deductions based on geographical distribution (see Chapter 11) it is sometimes possible to estimate the time taken for one species to split into two 'good' ones or for a species to change sufficiently for systematists to regard it as having become a different one. An average figure of a million years would not be an unreasonable guess; however, there are many existing forms treated as species which are undoubtedly much younger than this. It must not be supposed that a law of nature exists according to which after a period of a million years the descendants of a given species must become specifically distinct from their ancestor; it may be quite common for a period of this order to be accompanied by little or no phenotypic change in a particular lineage. This problem will be further discussed in Chapter 9, where classification is considered in relation to phylogeny.

To sum up our conclusions in this chapter, we may assert that the species, though not quite as God-given and objective a category as some have imagined it to be, does have some objective basis, at least when comparisons are being made between sexually reproducing organisms living in the same area and at the same time. It does not, however, seem likely that any really objective basis could be found for species-difference between animals or plants inhabiting different continents or different geological periods. Non-sexually reproducing plants and animals likewise present difficulties for those who work by 'objective' definitions of the species as a category. The species is no exception to the rule that the concepts and categories employed in natural history are never susceptible to precise, rigorous or final definition; any scientist who is not content to operate with more or less vague and inexact basic principles and ideas is temperamentally unsuited to the study of natural history.

4 Classification Below the Species Level

It cannot be taken as accepted by all, nor perhaps even by the majority, of modern botanists and zoologists, that any generally useful purpose is served by the attempt to establish regular classificatory categories below the species level. Nevertheless, it is a fact that more or less extensive variation is to be observed in many natural species, and this variation is by no means always completely continuous and random. Plenty of instances are known of species, within which most specimens can be referred to one or another of a series of sub-types, and intermediates which cannot be so referred are comparatively rare. Where this is the case, and where the sub-types occupy different geographical areas, many systematists would recognise and name *subspecies*. The naming of sub-species is now officially recognised and made subject to rules, in the current codes of botanical and zoological nomenclature. Typical sub-species resemble species in that they are separated from other subspecies by some degree of discontinuity in 'museum' characters, even though the discontinuity is less absolute than in the case of species. It is not easy to cite really good examples of sub-specific differences within the British flora and fauna, though differences at this level are often manifest between British forms and related ones in northern Europe, e.g. the Scottish Red Grouse (*Lagopus lagopus scoticus*) and its Scandinavian cousins (*L. lagopus lagopus*); *Primula scotica* may perhaps stand in a similar relation to *P. scandinavica*. One of the best well-known British examples of sub-specific difference is provided by the Brown Argus butterfly, *Aricia agestis*,★ with the type subspecies (*A. agestis agestis*) practically confined to England and the Scottish examples representing *A. agestis artaxerxes*—the two subspecies scarcely overlap, are phenotypically distinguished in the adults, and usually have different larval

★ The original spelling *agestis*, an obvious misprint for *agrestis*, must be used to comply with the current rules of zoological nomenclature.

food-plants. There is (or was) a zone of overlap in north-east England, in which intermediate forms occurred. Genetically, those subspecies which have been investigated show distinguishing features of similar type to, but less marked than, those of full species—differences are poly-genically determined and not infrequently accompanied by simple chromosomal rearrangements.

In the normal absence of a genetical breeding barrier, the main-tenance of the distinctness of subspecies in nature must result from a relative rarity of interbreeding between them, and perhaps also at times from the less 'adapted' character of the products of such interbreeding. Natural subspecies are, it is generally agreed, potentially at least, a stage in the development of full specific difference. Charles Darwin believed that the evolutionary divergence producing species differences was a completely smooth process, devoid of 'nodes', but it is now generally believed that one or two more or less distinct nodal points can be dis-tinguished in this process—at least in many, it not all, cases. The first nodal point which may be significant is that at which some populations of a species become isolated from each other to such a degree that selection can produce significant adaptive divergence between them—thus providing the basis for the emergence of subspecies. The second important nodal point is that at which the potentiality of genetic inter-change between the diverging forms is finally lost, and they acquire the status of independent species. Sub-specific divergence, it is generally considered, is reversible, two subspecies may come together and merge into a new unity, but this can rarely happen with two distinct species.

On this point there is some difference of opinion between botanists and zoologists. Many botanists consider that interspecific crossing has played an appreciable, even possibly an important, part in plant evolu-tion; we find indeed a tendency to talk about 'reticulate evolution' in some botanical circles. The zoological systematist is inclined to deduce from this that botanists use the term 'species' differently from himself; if they regard interspecific crossing as a frequent occurrence they can hardly work by the definition of the species formulated in the last chapter. There may, however, be a real difference between the higher plants and most animals in relation to speciation [193c]. In their resistant and often long-dormant seeds, the Spermatophyta possess a potential for long-range passive dispersal which few animals could match. It may be comparatively common in plants for geographical isolation to per-mit phenotypic divergence to proceed so far as to establish differences which are sharp enough and constant enough to satisfy any herbarium systematist's idea of species, and for the divergent forms, through some accidents of dispersal, to be brought into contact with each other again before an effective breeding barrier between them had been developed. Such a sequence of events might be much rarer in most animals.

Most biological systematists will admit that, in some cases, objective natural units exist which correspond to the concept of subspecies, and hence are worthy of nomenclatorial recognition. There are many, however, who still have misgivings about the recognition of the subspecies as a category for general application. Once the idea is established that it is a normal and natural thing for a species to be divided into subspecies, it is feared that the floodgates will be opened to permit systematists of 'splitting' proclivity to multiply names beyond any reasonable need. This fear is aggravated by the fact that under the current rules, both botanical and zoological, sub-specific names compete with specific ones in priority and homonymy (see Chapter 20). That such fears are not groundless can be seen from the examples of the butterflies (*Papilionoidea*) and birds in zoology, and the Orchidaceae in botany—in each of these groups, sub-specific naming has been carried to lengths which many regard as excessive. Not only is there the danger of splitters conferring sub-specific names on the very slightest distinguishable geographical variants, they are also liable to try to name and distinguish forms which are not separated by any real discontinuity, and hence not objectively separable.

The most difficult problems for formal sub-specific classification are presented by the clines of J. Huxley [103]. Instances where the phenotypic characters of a species show a gradual and progressive change when it is traced along an extensive tract of natural distribution have long been known in the animal kingdom, and recent taxonomic studies published in the *Flora of Tropical Africa* [21] indicate that similar phenomena occur in the higher plants. That systematists have failed so often in the past to recognise clines is not so surprising when it is remembered that the main bases of their studies are as a rule not natural species but museum or herbarium collections. Such collections have in most cases grown up as aggregates of the collections of various individuals. Thus museum collections rarely comprise adequate representative material from all over the natural distributional range of species, the geographical coverage is as a rule very 'patchy'—a particular species is generally represented by very numerous specimens from certain parts only of its range, and other areas are represented by few or no specimens. If the real geographical variation of such a species is clinal, the collection of it in a large museum is liable to give a spurious impression of discontinuity; most of the specimens coming from certain limited areas in which collectors have been active, they will tend to cluster round a corresponding number of sub-types, which are likely to be distinguished and named as subspecies before it is realised that a cline is present.

No satisfactory method for the designation of clines has yet been put forward; Huxley's own suggestion of naming only the end forms and using some sort of formula to indicate the various intermediate stages has logical attractions but would involve hopeless difficulties with the current codes of nomenclature. If, as is probably the usual case, the 'gradient' of a cline is not constant, but has relatively short 'steep' areas separating larger ones in which the rate of change with distance is relatively low, the most practical solution is probably to distinguish a series of subspecies corresponding to the areas of low rate of change, and to treat the intervening 'steep' tracts as zones of hybridisation between the adjacent subspecies.

Assuming that the evolutionary divergence which produces 'good' subspecies and finally full species is a gradual one, it should be possible to distinguish earlier stages in it than those which would qualify for recognition as a subspecies. The first stage would presumably be a more or less isolated population with some distinct population–genetical characteristics (e.g. unusual frequency of certain recessive factors and rarity of others). The next stage would be a population or group of populations showing some—albeit slight—general phenotypic divergence. Typical examples of this stage would be the populations of one or two species of small mammals on some of the western islands of Scotland. These populations must be of post-glacial origin, i.e. less than 10,000 years old, and there are reasons for believing that at least some of them result from human introductions and are not likely to be more than about 3,000 years old. They have effective geographical isolation and are probably subject to decidedly different selective pressures from their mainland relatives; many of them show very definite phenotypic divergence, and some systematists have treated one or other of them as good subspecies or even full species in the past. A comparable botanical example is provided by the Spotted Orchis (*Orchis maculata* L.; subgenus *Dactylorchis* auctt.); Heslop Harrison [85] described the form of this species occurring on the island of Rum (*Rhum* auctt.) as a distinct subspecies *Dactylorchis fuchsi rhoumensis* H.-Harris. Harrison himself later suggested that the supposed 'species' of *Dactylorchis* (including *fuchsi*) might better be considered as 'eco-species' of the 'coenospecies' *O. maculata* L., and the 'subspecies' (including presumably his own *rhoumensis*) merely as marked ecotypes.

In the matter of sub-specific classification, as in some others, there have been marked divergences between the theoretical utterances of systematists and the practical activities of themselves and their colleagues. In theoretical discussion of the categories, stress is usually laid on the genetically determined characters, but very few systematists make any serious effort to test this in practice. As a result, a curious air of unreality hangs over most of the symposia and colloquia on such subjects which have been held during recent decades. Very numerous different sub-specific categories have been proposed by systematists during the last fifty years, most of them accompanied by carefully worded and apparently objective definitions, yet hardly any of these categories have passed into general use, and few have even been consistently used by their proposers. In practice, the vast majority of systematists continue to work almost exclusively on 'museum criteria'; the question whether to use sub-specific categories, and if so which ones to use, is usually decided by reference to current museum usage in the group in question, not to theoretical criteria. The museum systematists who discovers that specimens of a given species from a particular area can be distinguished from those from elsewhere is likely, if subspecies have been generally named in his group, to describe a new subspecies, otherwise he will probably use some nomenclatorially uncommitted term like form, ecotype, geographical race, etc. If, before publishing anything on the subject, he tries to obtain living specimens for breeding experiments, he will be a very unusual systematist indeed.

The palaeontologists are liable to encounter special difficulties in classification at and below the species level, particularly when they are dealing with a fairly continuous record of a group of organisms over a period of time of the order of a million years or more, as in the so-called lineages. The essential difficulty in dealing with a continuum is not evaded by dividing it into shorter and shorter lengths; this merely has the effect of increasing the number of indefinable boundaries which you have to try to define. Nevertheless, many palaeontologists have tried to define and name subspecies in lineages, as also in respect of 'lateral' (geographical) variation in fossil organisms. The commonest procedure in palaeontology, however, has been to establish species on clearly distinguishable 'type' forms and to refer to divergent or intermediate types by more or less informal procedures, e.g. as 'sp. aff. *typicus*' or sometimes 'sp. cf. *typicus*'. The practical problems of sub-specific classification in palaeontology are usefully discussed by Newell [146a].

Applied biologists, pursuing various practical aims, not infrequently discover that the members of an established species are not all alike in some economically important respect; in a plant species, it might be that certain individuals produce a greater quantity or a preferred quality of a particular essential oil, in an animal, that some specimens will feed preferentially on an economically valuable species, others will not. Such differences may seem to be at least as important as, for example, the slight differences in pigmentation by which ornithologists or lepidopterists are apt to distinguish subspecies; however, unless the applied biologists do so themselves, these 'biological races' are unlikely to be described and published as subspecies. In the case of the fruit-fly genus *Drosophila*, the geneticists and cytologists have seen to it that the sort of non-museum characters in which they are interested are given due classificatory weight—by taking the systematics of *Drosophila* out of the hands of the museum dipterists and into their own.

Drosophila is perhaps the only extensive genus of wild animals in which it might be really practicable to apply the sort of criteria which have theoretically been proposed for sub-specific categories. If the professional systematists have nevertheless largely ignored the *Drosophila* work in their discussions of such matters, this might at least in part be attributed to pique at the intrusion of geneticists into 'their' field. It seems to me that, in any serious consideration of the problems of classification at generic and lower levels, the works of Sturtevant [183], Patterson [150] *et al.* should receive the most serious study. The *Drosophila* workers have generally used the term subspecies for forms which are geographically separated ('allopatric') and show either a marked reduction of fertility in their crosses or a marked phenotypic difference; in cases where a high degree of reproductive isolation has been found to exist between sympatric forms which are more or less indistinguishable phenotypically, these have not usually been named as subspecies. Thus the *Drosophila* workers have tended to follow the precept, if not the practice, of orthodox systematists. A paradoxical result has been that where, as in *D. pseudoobscura*, reproductively isolated sympatric types have been found within a species, these have been known by some non-committal term like 'form' or 'race' unless and until practical phenotypic distinctions between them have been discovered, in which case they are liable to be raised immediately to the status of full species, with no intervening sojourn in the sub-specific category.

With the widespread current conviction that the essence of the scientific method consists in the use of mathematics, it is not surprising that efforts have been made to establish a more 'objective' definition of the subspecies in numerical terms. The oldest and simplest formulation of this kind is the so-called 75 per cent rule. According to this formula, if 75 per cent (for some reason the form $\frac{3}{4}$ is never used in this connection) of the individuals from a given population or area can be distinguished phenotypically (i.e. by museum systematists) from all others of the species, then these specimens represent a 'good' subspecies. This rule has a very attractive simplicity, reflecting perhaps the statistical unsophistication of the average systematist, but is open to many practical objections. Perhaps the most serious one is that the population (or assemblage of populations) will first need to be defined on some other basis. This is usually done on a geographical basis, e.g. all the specimens from Ireland, or from the Pyrenees; most such geographical divisions, however, are botanically and zoologically arbitrary, e.g. that between Scotland and England. In practice, this rule can only be applied satisfactorily where the geographical distribution of the species concerned shows more or less marked discontinuities.

The concept underlying the 75 per cent rule is undoubtedly that if 100 per cent of the specimens exhibited such a phenotypic difference, this would justify a species distinction. If this is accepted, then it seems quite reasonable to use some lesser figure as the criterion for sub-specific rank—though the figure of 75 per cent is obviously an arbitrary one. However, as we have seen in the last chapter, a 'too per cent rule' of this kind would not provide a sufficient criterion for specific rank, so we may legitimately doubt whether any lesser percentage will provide a sound basis for defining subspecies.

Another serious objection to the 75 per cent rule is that it demands a comparison between a representative sample from the population or area under study and *adequate representative material of the whole of the rest of the species*—a requirement which would be very difficult to satisfy in very many cases. This difficulty, however, confronts not merely the 75 per cent rule, but any conceivable objective criteria for classificatory categories below the species; it is discussed by Géry [71].

Recent attempts to improve on the 75 per cent rule have not in the main been concerned to provide it with a more satisfactory definition of a population, or to take into account difficulties similar to those which confront a purely phenotypic definition of the species. The most fashionable current objection to the rule is that the only criterion of distinguishability in it was the 'subjective' judgement of systematists. Many have felt that the scientific status of the rule would be improved if it could be reformulated in such a way as to make its application independent of any prior systematic knowledge of the organisms in question. To this end, attention is concentrated on characters which could be counted or measured (see Chapter 7), and are thus expressible in the form of a mean and a standard deviation. For a given population, such a character can usually be expressed in the form of a histogram approximating in form to the 'curve of random deviation'. If the histograms for the same character for two different populations are plotted on the same piece of graph paper, the area of overlap of the two histograms corresponds to the number of individuals in the two populations which (as far as this character is concerned) could belong to either population. The percentage of either population which falls *outside* this area of overlap could be taken as an index of the divergence of the two populations. Mayr, Linsley and Usinger [141] derived in this way the concept of a 'coefficient of difference' for a measurable character, defined as the difference of the means of the two populations divided by the sum of the standard deviations for the two. A value for this quotient of more than 1·28 is considered by Mayr *et al.* to indicate sub-specific difference between the two populations. This version of the 75 per cent rule looks more 'objective' and scientific than the original one, but in no way overcomes the more serious objections to the original and has the additional drawback of being applicable only to characters which can be counted or measured.

The subspecies is, of course, by no means the only classificatory category below the species which has been proposed or used. An impressive list of such categories was drawn up and circulated by the Systematics Association in 1952. Apart from the subspecies, the oldest and most widely used of such categories are the variety and the form (forma). In the most general usage of the term, a variety is a phenotypically distinguishable, genetically determined variant type which occurs among ordinary examples of the species and does not form distinct populations; varietal characters usually behave as though they were determined by only one or two recessive hereditary factors. A species in whose populations one or more such varietal forms commonly coexist with the 'type' form is described as polymorphic (to be distinguished from a polytypic species, which is divisible into subspecies). The term form is most commonly applied to variant types whose differences are environmentally rather than genetically determined (e.g. the 'phases' of locusts, seasonal forms of certain butterflies, stunted coastal cliff forms of certain trees and shrubs, etc.), but is also used as a non-committal term for variant types of unknown status.

Another term, ecotype, is used by many botanists, to signify a distinguishable form of a species which occurs in a particular area or habitat. It has been shown in some cases, and may be generally true, that the distinguishing features of ecotypes are partly genetical and partly environmental in origin. The same may be true of some of the less mobile types of animals, e.g. the mountain forms of certain rather sedentary and flightless insects. The distinguishing features of these (often small size and dark colour) are rather suggestive of the effects produced in genetically normal individuals when they are reared in cold and generally unfavourable conditions. It is likely, however, that such mountain 'ecotypes' of beetles, etc., usually have a sufficient degree of reproductive isolation to permit more or less considerable genetic divergence.

A recent paper by Alston and Turner (in *Biochemical Systematics* [3]) discusses the possibility of defining sub-specific taxa of plants on bio-chemical characters, and discusses the possible recognition of 'chemo-subspecies'. It seems to me that the subspecies is in essence a genetic phenomenon, indicating a population or group of populations whose interbreeding with other subspecies is in some way restricted so as to preserve a degree of genetic distinctness less than that appropriate to full species. Groups qualified by this definition for the rank of subspecies may be most easily recognised by chemical, structural, behavioural or other features, or by combinations of various types of characters, so that it is not likely to be practicable to distinguish between 'morpho-logical' and 'chemical' subspecies in general.

5 Classification Above the Species Level

Varietates ad Species reducat Tyro. Cum Specialis cognitio fit primum solidae cognitionis.

Species ad Genera reducat Botanicus, cum inde fraterne affinitas plantarum. Genera ad Ordines reducere tentat Veteranus, cum inde prosapiens ex Natura vegetabilium.

LINNAEUS,
Ordines Naturales, 1764

Perhaps the most persistent conflict of opinions concerning the genus and higher categories is whether they are in essence synthetic or analytic constructs. Is a genus, for example, formed synthetically by starting from one species and placing with it all others which show a sufficient degree of similarity to it, or does the systematist start with a large assemblage of diverse organisms and proceed analytically, subdividing it step by step until he comes to what he regards as the generic level? Our Linnaean epigraph seems to imply the first of these approaches. On the other hand, in his earlier (1735) 'Caroli Linnaei, Sveci, Methodus' he sets out his proposed procedure for the systematist in numbered order, putting 'classification as to classes and orders' before the characterisation of genera, and placing the specific *differentia* last. The general system of classes and orders which Linnaeus followed was laid down already in the first edition of his 'Systema Naturae'—and in 1740 he can hardly have thought of himself as '*veteranus*'. If Linnaeus was inconsistent on this issue, so have been nearly all his successors. The two alternative approaches are not necessarily mutually exclusive, and most systematists would probably deny that they pursue one to the exclusion of the other. Inger [106] probably speaks for many of his colleagues among present-day systematists when he asserts that 'Essentially the species is an analytic category whereas the genus is a synthetic one'. For most of us, it is probably truer to say that we adopt sometimes the one approach and sometimes the other, whichever seems more convenient in a particular case, rather than that we successfully synthesise such a pair of Hegelian antitheses.

The distinction, it must be made clear, is not the traditional one of 'lumpers' and 'splitters'—an analytically minded systematist might well be a lumper in that he tended to recognise fewer families, genera, etc., than other specialists in his group, whereas a splitter who multiplies these categories beyond what is customary could well work by the synthetic approach. The artificial classifications of library books, etc., are essentially analytic in character, but the cleavage between the analytical and synthetic approaches in biological systematics is distinct from that between formal and phylogenetic classification. If we were to construct a classification of biological systematists, the formal/phylogenetic, analytical/synthetic and lumper/splitter antitheses would have to be treated as three quite distinct classificatory characters. The 'quantitative systematics' or 'numerical taxonomy' which has recently become fashionable, particularly in the U.S.A., purports to be essentially a synthetic procedure, but as we shall see in Chapter 15, its basic concept is a unit of difference which can only be derived from analysis.

Logically, we might see the problem of supraspecific classification as that of interpolating a certain number of intermediate categories between the commonly agreed upper and lower limits of the kingdom (animal, plant, bacterial or viral) and the species. If the number of intermediate categories could be fixed and not too large (not more than perhaps seven or eight), it might be possible to develop and teach a sort of general classificatory perspective, which might in turn lead to a more uniform use of the categories, both as between different groups at one time and between different times in the same group. In order to maintain this perspective, each group would need to be seen at once synthetically and analytically, as an assemblage of lower-order groups and as a unit in a higher category group.

As we shall see in Chapter 21, practical systematists are almost inevitably specialists in particular groups, and hence have an inherent tendency to see their special groups out of perspective in relation to others. Moreover, they almost invariably find themselves wishing to use more categories than the general system allows them. Both these influences predispose the specialist to raising the previously-established category of his group, whereby he can, so to speak, take in one or two more categories at the top. This has no doubt been a principal cause of the persistent 'inflation' which has affected the application of supra-specific classificatory categories since the time of Linnaeus. A modern genus in most groups of organisms has much the same relation to a Linnaean one as a current shilling bears to one of George III's time.

The tendency to demand more classificatory categories is not wholly to be ascribed to the inflationary effects of specialism; the endeavour to construct a classification corresponding as closely as possible with evolutionary relationships (see Chapter 9) is liable to have similar effects. There is a real conflict of interests here, between phylogenists and specialist systematists on the one hand, and the general body of botanists and zoologists on the other. The most satisfactory solution might be to adopt for general use a system using only a small number of categories, and to provide for the interpolation of additional categories at prescribed points for specialist purposes.

The Linnaean system recognised only three categories between the species and the kingdom—the genus, the order, and the class. Since then, the first and most important additional category to come into general zoological use was the family, interpolated between the genus and the order, and the next was the phylum, between the class and the kingdom. In botany, on the other hand, the tendency until recently has been to depreciate the Linnaean 'order', and to interpolate new categories between it and the class. The accepted zoological system of today recognises, in descending order, sub-kingdoms, phyla, subphyla, classes, subclasses, orders, suborders, superfamilies, families, subfamilies, tribes, genera and subgenera. Additional categories are inserted at various points, by various authors, e.g. between subkingdom and phylum, between subphylum and class, between subclass and order, between suborder and superfamily, between tribe and genus, between subgenus and species. The recent trend among botanists has been to fall more closely in line with the zoologists, though with some differences in the naming of categories; those recognised in the latest botanical code of nomenclature are the following: Divisio, Subdivisio, Classis, Subclassis, Ordo, Subordo, Familia, Subfamilia, Tribus, Subtribus, Genus, Subgenus, Sectio, Subsectio, Series, Subseries, Species. It will be noted that the botanical code recognises no less than five categories, compared with the zoological code's one, between genus and species, but has only two between kingdom and class where the zoologists have three. The retention of Latin (i.e. international) form for the category names, and the relatively elevated status accorded to the genus, are features of the botanical code which may be none the less valuable for being little appreciated at present.

Recent studies in the measurement of the degrees of pairing between different species of single-strand DNA (see Chapter 13) suggest that figures obtained in this way might provide a basis for more or less objective definitions of supra-specific classificatory categories; this was suggested by Goodman in an article on 'Molecular Parameters in the Taxonomic Study of Primates' (in [215]). His figures suggest that an order of Mammalia might be defined as manifesting at least 33 per cent of the 'homologous' degree of pairing when DNA's of any two species of the group are being tested for hybridisation by the Hoyer-Bolton McCarthy technique; the corresponding figure for suborders might be about 50 per cent and for superfamilies 66 per cent and for families at least 83 per cent. The adoption of this criterion for determining the various categories in our system would almost certainly involve considerable changes in our current applications of the categories in some groups, as would the criterion of standard geological ages for common ancestors (see Chapter 19).

It appears to me that, as the available figures for DNA hybridisation within the Vertebrata show a fairly good approximation to inverse proportionality to the ages of the common ancestors as determined from a well-documented fossil record, it might be reasonable to define our categories basically in terms of the ages of the common ancestors of taxa (as is suggested in Chapter 19), and to treat the DNA fibre-pairing figures as measures of such ages unless and until good fossil evidence proves otherwise. The method has the advantage of applicability to groups like the Platyhelminthes for which no significant fossil record exists or is likely to be found; the available figures suggest that there will be an upper limit to its utility and a lower one. DNA's of species of the same genus will probably not manifest a significant departure from 100 per cent of the 'homologous' degree of pairing, and the very low degrees of pairing between DNA's whose common ancestors lie further back than the Carboniferous period may, like 'weak' serological reactions, be unreliable indicators of phylogenetic relationships. It will be necessary to determine far more DNA fibre-pairing figures for groups other than Vertebrata before we are likely to succeed in persuading systematists generally to accept this criterion for categories.

In considering the supra-specific categories, it is perhaps more convenient to proceed upwards from the species, in conformity with our

Linnaean epigraph. In some groups of animals, e.g. birds, and even, despite the botanical code, in some plants (e.g. Graminae) subgenera have been very little used and the genus retains its original position as the next practical category above the species. In such groups, systematists with a synthetic and 'splitting' type of approach are liable to promote any small group of species which seem to be particularly closely related among themselves to the status of an independent genus. This sort of splitting has been carried to extreme lengths among the birds, as also among grasses, orchids and the larger mammals; the analytical approach has been very little cultivated among systematists of these groups. One effect of this has been the discrediting of the genus as a practical category among ornithologists, some of whom have even proposed that the generic names should be dropped from practical use altogether [Cain 25]. This proposal, it may be noted, is not exactly novel, since it was common practice for the moth-collecting amateurs of the last century to refer to their specimens by the specific or 'trivial' names only. No such proposal or practice has to my knowledge been advanced by students of such groups as Coleoptera or Dicotyledons—in both of which determination is commonly by means of analytical keys, so that students soon come to recognise the genera even before the species. On the other hand, identification of birds, as of moths, is rarely by means of keys—species are usually identified directly by comparison with pictures or reference collections, and the average student never learns to attach any particular significance to the generic names. To some extent the same is probably true of grasses and orchids among plants. However understandable the reactions of the lepidopterists and ornithologists may be, systematists working in groups where the generic names really do have a practical use are not likely to agree with Cain's proposal. We cannot help feeling that students of birds and of moths would be better zoologists and better systematists if, despite the difficulties, they seriously tried to observe and appreciate the generic characters in their animals.

If the genus is to retain the practical utility which it should have if Linnaeus' binomial system is to be justified, it will be necessary to eradicate the conception of it as the first classificatory category above the species—in this respect the botanical code of nomenclature can set us a valuable example. The subgenus, it is true, has for a considerable time been an officially recognised and available category in zoology as well as in botany, but it is as yet far from being generally used by specialists. Experience in the larger and better studied genera of animals, e.g. *Drosophila*, indicates that there may be room for at least three categories between genus and species; the work of Sturtevant [183] recognised species-groups as a regular category between subgenus and species, and within some species groups species subgroups were distinguished. Within a species subgroup of *Drosophila*, the breeding barriers are not as a rule absolute, it being possible to obtain first-generation hybrids between many of the species. The revision of *Berberis* by Ahrendt [1] likewise adopts three categories between genus and species. I believe that many of the recently accepted 'genera' of birds are analogous to the species-subgroups of *Drosophila*, and many orchid 'genera' (e.g. *Dactylorchis*) could be compared to subsections of *Berberis*. In *Drosophila* species of the same species-subgroup, it is usually possible to recognise at least some homologous sections, and not rarely whole homologous arms, in the salivary gland chromosomes. The terms 'Artenkreis' [Rensch 156] and 'supraspecies' [Mayr 140a] have been proposed for categories more or less analogous to the species-subgroups of the *Drosophila* workers.

Proceeding up the hierarchy of categories from the genus-group, we come to what the current codes of nomenclature refer to as the family-group, comprising, in ascending order, subtribes, tribes, subfamilies, families and superfamilies. For these, as for all the supraspecific categories, there are at present no objective criteria—whether it might in future be possible to establish any is a question which will be discussed in Chapter 19.

In most groups of organisms, the classificatory work of the normal museum or herbarium systematist extends into the family group of categories, but rarely much above it—these professional systematists are not often concerned with classification in the 'order-class group' of categories. The higher levels of classification have long, by tacit convention, been the preserve of academic botanists and zoologists. This division of function is interestingly reflected in the current codes of nomenclature, which lay down stringent rules for the naming of categories up to the family group (in animals) or orders (in plants), but offer nothing more than 'recommendations' concerning nomenclature of categories higher than these. To borrow a metaphor from former English cricket, it is not felt desirable to subject the 'gentlemen' of the universities to the sort of discipline which is deemed appropriate for the 'players' in museums and herbaria.

With the order-class group of categories (comprising suborders, orders, subclasses, classes, subphyla and phyla in the current zoological practice) we enter the domain of academic biologists rather than museum systematists. According to our Linnaean epigraph, the professional systematist ought not to enter into this field until he has passed retiring age, and to a surprising extent museum taxonomists are faithful to this precept. Academic botanists and zoologists, on the other hand, recognise no such age limits, and not infrequently propose innovations at the order-class level before they reach the age of forty. Museum systematists have a strong tendency to think about and discuss their work as essentially synthetic, though most of them probably practice a good deal more analysis than they are willing to admit. On the other hand, the approach of the academic botanist or zoologist to classification is unequivocally analytic; species to him serve merely as examples of families or orders. This difference of approach is reflected in the current codes of nomenclature. The rules for naming groups up to the family group (in animals) or the order (in plants) are all based on the 'type' principle, which naturally goes with the synthetic approach, whereas no attempt is made to base the higher categories on types in this way. The 'principle of typification' has little justification in relation to categories which are seen essentially as subdivisions of broader ones.

If, as I believe, the ideal systematist should habitually regard his groups, at all levels, at once synthetically and analytically, both the established approaches must be regarded as one-sided and unsatisfactory. To lead the museum workers to an adequate appreciation and application of the analytic outlook in their work may not be easy; to persuade academic biologists to make those extensive studies at lower levels which would enable them to appreciate the synthetic aspect of order-class group categories might well be even more difficult. Some way of achieving both these aims will have to be found if systematic botany and zoology are to make real progress towards stable, uniform and natural systems.

As previously suggested, I think it would be desirable to have only a limited number of categories accepted for general, non-specialist use. A sufficient number of these for most purposes might be provided by phyla (divisions), classes, orders, families and genera. In the interests of long-term stability, it might be a good thing to have at least one type-group in each of these categories designated in each kingdom. Such a proposal might at present seem more practicable for the animal kingdom than for plants; to establish any real equivalence or parallelism between groups of algae and of angiosperms would look to be forbiddingly difficult.

Specialists could insert two, three or even more categories between each category of the general system and the next lower (or higher) one. With three additional categories available between genus and species, generic names might retain some of the breadth and utility which Linnaeus meant them to have; an average figure of twenty or more species per genus might be attained, and it might become practicable to memorise all the valid generic names in quite extensive groups.

Both in their practical work and in their theoretical writings, the attention of most recent systematists has been concentrated on the species problem. This is not surprising, seeing that in ordinary routine determination work it is usually at the species level that the greatest difficulties are encountered—and did not Darwin himself regard the 'Origin of Species' as the prime problem of evolution? All good systematic work, it is commonly argued, must start from species, and higher categories should thus be considered and approached as synthetic agglomerations of species. This attitude, however, does not seem quite so reasonablewhen the systematist is confronted with new and unknown forms. In such cases he does not usually commence by formulating specific diagnoses and then looking round for possibly related forms. The usual procedure is first to 'place' the new form(s) as far as possible in the existing classificatory hierarchy, treating it analytically, and only at the end of such a process to decide on the category to be allocated to the new group and to set about formal definition of it.

It is also possible to approach the classification of a more or less extended group of the plant or animal kingdom analytically from the outset, rather than by the conventionally respectable method of starting with a revision of the species of some genus. The procedure would be to study first the distinguishing features of the group, then those of its major subdivisions, and so on down the heirarchy of categories. In the case of a large group, this sort of analysis may stop well short of the species level. Such an approach is liable to be condemned as fundamentally unsound by the current pundits, but, at least as applied to the insect order Coleoptera in my own case, it seems capable of producing results which may be in some respects superior to those achieved by systematists approaching the same problems in the conventional, species-based manner. It may be worth while to consider further advantages and disadvantages of such an analytical approach.

The first advantage I would claim for it is that of following naturally from the way in which university students normally learn about the plant and animal kingdoms. The young graduate, first starting on species systematics, can hardly help feeling that this is something utterly different from the 'systematic' botany or zoology he learned as an undergraduate, something for which his university training was not a very useful preparation. A second advantage is that of encouraging a broad, synoptic outlook and the capacity to see various groups in perspective in relation to each other—a capacity which is by no means conspicuous in many current exponents of 'species systematics'. When the study of a major group is approached in a piecemeal synthetic manner, there is always the danger of aberrant and interesting types being 'lost' or omitted from consideration altogether; this is far less likely when the classification of the group is approached analytically. Finally, systematists with an analytical approach may perceive significant phylogenetic analogies and links which would probably be missed by those with the synthetic approach.

There are, of course, disadvantages in the analytic attitude to systematics. The most obvious one is that it encourages generalisation on bases which are liable to be inadequate. Characters may be predicated for extensive groups without having been confirmed in anything like all the species within them; further investigations not infrequently reveal exceptions to generalisations of this kind. Another is that for ecological, behavioural, physiological and genetical purposes, species differences are of prime importance, and by neglecting them the analytically minded systematist may alienate himself from sources of direct information about living organisms. A third factor which may reasonably deter an aspiring young systematist from adopting the analytical approach today is that is is unfashionable. The power of fashion in the modern scientific world can hardly be over-estimated; an unknown young scientist who adopts an unfashionable approach is liable to find himself in a similar position to that of a young musician whose artistic conscience impels him to compose in the style of Delius or Rachmaninov. However, in science as in the arts, fashions are essentially temporary, and the theorists of species-systematics have already enjoyed a considerable period of ascendancy.

In the opinion of the present writer, there is no evidence to prove that those who have the most experience in species determination in a group are also the most successful in dealing with problems of higher classification within that group. Admittedly, it is as a rule the established 'specialists' in a group who compile the catalogues and handbooks which are followed by most other systematists, and by non-specialists dealing with the group. The fact that it is the classifications published by the museum taxonomists which are usually followed does not, however, establish that these classifications are, in fact, the most natural (in any sense of that term), any more than the fact that newspapers and advertisements have more influence than anything else in the gradual changing of our accepted standards of language proves that journalists should be the supreme authorities in linguistics.

The practical utility of supra-specific classification (except as a step towards species determination) is often questioned or denied today; the only useful function of a systematist, it is frequently asserted, is to name species. Such assertions are, I think, flatly untrue, and reflect ignorance or misconception in those who make them. In any natural classification, as we have seen, an organism may be referred to a particular group by reference to comparatively small numbers of observed characters, but this reference will permit the deduction that it possesses, or is very likely to possess, many other characters, some of which may not be observable from the material at hand. The extent to which such deductions can be made will depend of course on how thoroughly the group has been studied from the point of view of supra-specific classification. Deductions of this sort play a vitally important part in our practical dealings with plants and animals.

6 The Classification of Fossils

Systematists dealing with fossils tend to give insufficient consideration to the peculiar nature and difficulties of their task. There are two radically different types of approach to it open to them—either to restrict themselves to classifying fossils as such, or to attempt to place the organisms, from which the fossils were derived, in relation to the system of modern animals and plants. There may even be a case (practically if not grammatically) for a third alternative—to erect for the organisms from which fossils were derived a system distinct from though related to that of modern organisms. Systematists adopting the first-mentioned, strictly 'geological' approach, would not require any great systematic knowledge of modern plants or animals; their practical requirements would consist mainly in an adequate training in stratigraphical geology and a good knowledge of previous work on their particular group of fossils. Such training could quite well be carried out in geological departments; their professional work would normally be in specifically geological institutions, and its results would be published in geological journals.

On the other hand, anyone aspiring to 'place' the organisms of the geological past in relation to the systems of modern plants or animals would require at the outset a thorough knowledge of the recent forms, a knowledge embracing not only the dominant 'important' types but also those rare and little-known forms which may be 'living fossils'; all this in addition to the basic requirements of the 'pure' palaeontologist. He would require at least as much training in botany or zoology departments as in geological ones, a requirement which recent trends towards 'biological' integration in our universities are likely to make even harder to satisfy in the future. And in later life he can expect to be continually subject to the tiresome necessity of transgressing the deeply entrenched boundary lines separating geological institutions (and journals) from botanical or zoological ones.

In practice, the study of fossils has been, until now, largely left in the hands of specialist palaeontologists. The systematist who turns from classifying modern plants or animals to the attempt to do the same for the fossils of his group is apt to be deterred at the outset. He finds himself expected to describe, name and place specimens which are almost always vastly inferior to the most defective examples of modern organisms with which he had had to deal. He will probably say 'I cannot observe in these specimens the characters which would be needed in order to place them in the system; in the circumstances I should feel quite unjustified in attaching definite names and systematic positions to them.' The more thorough and conscientious he is as a systematist of modern organisms, the more likely it is that he will take this attitude to the fossils; the converse unfortunately applies too—if a systematist of modern organisms devotes much attention to fossils, it is only too likely that his systematic work will not be of the highest thoroughness and accuracy.

So, by default of the neontologists, classifying fossils is usually left to the palaeontologists. These lack the inhibitions of the systematists of modern plants and animals, and are as a rule content to classify the fossils as such. In adopting this practical approach, palaeontologists have been enabled to perform very important services to geology, by distinguishing the 'zone fossils' which are so indispensable to the stratigrapher. The cleavage of attitude, usually accompanied by segregation in different institutions, between systematists of Recent and of past organisms has however been unfortunate for botany and zoology. If we are successfully to unravel the mysteries of phylogeny, it will be necessary to include both fossil and recent forms in a single study. The problem arises—who is to do this?

There are, of course, many 'pure' palaeontologists who have written extensively about phylogeny, in fact, few of them can resist the temptation to construct 'lineages' from the stratigraphical succession of their fossils. In view of the importance often assigned to these lineages, as direct records of evolutionary change, it is unfortunate that their objective basis is not always as secure as the unwary may be led to suppose. There is no serious reason to doubt that the described successions in Jurassic *Gryphaea* oysters, or Cretaceous *Micraster* sea-urchins, are what they purport to be; on the other hand, doubts about Simpson's [169] lineage from *Hyracotherium* to *Equus*, spread over fifty million years and three continents, are quite legitimate. It is perfectly possible that another specialist, approaching the same material with a background different from Simpson's, might construct from it a different yet equally plausible story.

When dealing with entirely extinct groups like Pteridospermae or Trilobita, the pure 'palaeontological' approach might seem not merely the most practical but the only logical attitude. However, understanding of Pteridospermae is likely to be much deepened by a knowledge of analogous vegetative types among modern ferns, and of the reproductive arrangements of Cycads and heterosporous modern Pteridophyta. Similarly, anyone seriously concerned to understand Trilobita would benefit from a knowledge of the structure and mode of life of such modern forms as Xiphosura, Notostraca, Mystacocarida, Branchiura and Cephalocarida. Admittedly, some types of fossils which can have real stratigraphic value may be totally inadequate for 'whole organism' classification, e.g. Conodonts and spores—for practical purposes, the erection of purely *ad hoc* classifications of such objects may be justified (see discussion of the 'Parataxa' proposals in Chapter 17).

As compared with palaeontologists, serious systematists of modern organisms are as a rule more cautious and inhibited in phylogenetic theorising. Anyone with a really extensive and detailed knowledge of some particular group of plants or animals can hardly fail to be conscious of the difficulty of constructing a phylogenetic scheme which will be consistent with all the manifold similarities and differences with which he is familiar. How much less formidable the task would look if one knew only a tenth or a hundredth as much! Just this, he may suspect, is the position of the palaeontologist. To this imputation the palaeontologist may reasonably retort that, however fragmentary his knowledge of past organisms may be, he is directly acquainted with the succession of forms in time, a line of phylogenetic evidence which is at least as valuable as any of those available to the neontologist.

If we are called on to arbitrate in this dispute over who is to be the authority on phylogeny, I think we shall be compelled to decide against either of the main contestants. The neontologist, in refusing to accept responsibility for the fossils of 'his' group, was doubly to blame —for one thing, he thereby threw away the chance of acquainting himself with the most direct evidence of phylogeny available to us, and furthermore, he ensured that the study of fossils would be left in hands less well qualified than his own. If, as a result, the study of fossils has not contributed as much as it might to our knowledge of evolutionary history, it is hardly fair to blame the palaeontologists.

Some of the difficulties which may arise when the attempt is made to use the work of the palaeontologists as evidence for the phylogeny of modern animals may be illustrated by reference to Simons's [167] review of the Tertiary fossil record of the mammalian group Primates. The oldest fossils attributed to this group are some from the North American Palaeocene; the two known skulls of these show no trace of such a typical feature of the Primates as the post-orbital bar, their incisor teeth are rodent-like, and their dental formula is not that (2.1.3.3.) which would be expected in an ancestor of modern Primates. The only reason for referring these skulls to the Primates is the character of their cheek-teeth. It seems perfectly possible that these Palaeocene fossils are ancestral to modern Rodentia rather than Primates; the differences between their cheek-teeth and those of typical rodents parallel those between cheek-teeth of pigs and those of higher Artiodactyla. Up to the present, the classification of fossil Primates has been based almost entirely on the cheek-teeth; a tacit and unchallenged dogma has grown up that the cheek-teeth are the one and infallible guide to the recognition and classification of this group. This is, of course, a very convenient theory for the palaeontologists, as by far the majority of the described fossils of the group consist of cheek-teeth, with or without attached fragments of skull or jaw. Whether it is scientifically justifiable to make far-reaching deductions about evolution and phylogenetic relationships on such grounds is, however, very questionable. Fortunately, there are fossils belonging unmistakably to Primates and showing considerably more than the cheek-teeth; for example, some more or less complete skulls from the Eocene (*sensu Americano*) are unmistakably of lemuroid-like Primates with quite well-developed brains—these have the 2.1.3.3 dental formula and in the character of their cheek-teeth somewhat resemble New World monkeys (Ceboidea). Numerous Oligocene fossils from North Africa, and a few from Asia, have been attributed to the Primates, but none of them represent anything like complete skulls, and they contribute little to our knowledge of the phylogeny of the order. According to the cheek-teeth classifications, some of the Oligocene forms have been attributed to the anthropoid group, and others resemble New World monkeys, but none unequivocally belong to the tailed Old World monkeys (Cercopithecoidea). Unless and until more complete fossils are found, I do not think any definite conclusions are justified by the Oligocene fossil Primates. From Miocene deposits of Europe and Africa there are, however, several more or less complete skulls of Primates and even one or two almost complete skeletons. All of these have been attributed—

by the criteria of cheek-teeth systematics—to the Anthropoidea, though the skull, at least in the celebrated *Proconsul* is more like that of a cercopithecoid monkey than of any modern anthropoid in its general characters. Even the more or less complete skeletons do not establish whether a tail was present or not. Unquestioned cercopithecoid monkeys, close to existing types, are known from Pliocene deposits in Europe and elsewhere in the Old World.

The systematist of modern Primates, who takes his information about fossils of the group from such standard works of reference as Romer's *Vertebrate Palaeontology*, will learn thence that fossils of lemuroids are known from the Palaeocene and of anthropoids from the Lower Oligocene, but will probably not realise that in both cases the attributions are based on the form of the cheek-teeth unsupported by any other evidence whatever. Another dogma which has influenced classificatory placings of fossil Primates, especially by American palaeontologists, is that of 'the permanence of the continents and oceans' (cf. Chapter 11); thus none of the fossils from the Old World has ever been attributed to the New World monkeys (Platyrrhina), though on known characters such an attribution would be at least as reasonable for some of the fossils as the ones that have actually been published. It is generally admitted that, among modern Primates, the Lemuroidea and the Old World monkeys are rather widely separated, with the New World monkeys occupying a rather intermediate position. Presumably the ancestry of Old World monkeys must at some time have passed through a stage with characters similar to those of modern New World monkeys, and such a stage might reasonably be expected to be represented among the Oligocene fossils.

A special difficulty confronting the systematist dealing with fossils is that of unreliable labelling, extending at times even to deliberate fraud. A very large number of the fossils in museum collections have passed through the hands of amateur collectors; very often the original finders were unable to ascertain the precise stratum from which their specimens came, or at least they neglected to do so, and more or less conjectural stratigraphic origins have been added subsequently to the labels. The possibilities of deliberate deception were dramatically illustrated in the celebrated 'Piltdown skull' case; a somewhat analogous but much less publicised instance cocurred in Australia in the 1920's. The eminent 'palaeo-entomologist', R. J. Tillyard, was confronted with an insect wing embedded in a crystal of gypsum; he proceeded to name it as a new (extinct) genus and species of the order Orthoptera. It was later revealed that the wing belonged to a modern Australian grass-hopper, and that its inclusion in the crystal of gypsum had been cleverly faked. The story illustrates the weaknesses of palaeontologists —how many of them would be more successful than Tillyard in recognising a fragment of a modern organism presented in the guise of a fossil?

One of the more serious instances in which deliberate deception may have played a part in spreading misconceptions about the life of the past is that of the fauna of the celebrated Baltic amber [116]. This occurs *in situ* in early Oligocene sediments in the former East Prussia; many geologists consider that it is of the nature of a 'derived fossil' in these Oligocene deposits and has been washed out of somewhat older (? late Eocene) strata. Since prehistoric times this amber has been the basis of a flourishing and profitable trade; the techniques of its manipulation have been jealously guarded trade secrets of skilled German craftsmen. The amber is itself a fossil resin, derived supposedly from an extinct species of Coniferae, *Pinus succinifera*; it can be distinguished chemically from other known fossil (and Recent) resins by its high content of succinic acid—it is also distinguishable physically by its infra-red absorption spectrum. The main scientific interest of the Baltic amber lies, however, in its inclusions, and more particularly, in the insects and plant remains embedded in it. Though the amber plant fossils are extremely fragmentary, the insects are complete and often so well preserved that it is possible to see in them almost as much structure as can be made out without dissection in dry specimens of modern insects in museum collections. In fact, the Baltic amber insects have been studied to a considerable extent by specialists on modern insects, e.g. Wheeler [198] on the ants, Alexander [2] on the Tipulidae and Schedl [163] on the bark-beetles. The Baltic amber insects constitute by far the most important evidence we have of the insects of the lower Tertiary era, and perhaps the most extensive body of evidence we have for any form of life at a particular time in the geological past. Their interest is by no means only for entomologists; they may be important also from the points of view of zoo-geography (see Chapter 11) and palaeo-climatology, and given the occurrence of specialised herbivores like Chrysomelidae and Curculionidae, and of vertebrate parasites such as fleas, the amber insects may also provide evidences of the flora and of the vertebrate fauna of the period. The Amber flora has been treated by Conwentz [33a].

As soon as those in the amber trade realised that 'flies in amber' were

not mere imperfections but highly saleable commodities, such specimens were carefully sought out and specially marketed; the vast numbers of amber insects which have been collected and recorded are the
result of this careful and systematic searching. Most of this material
went to private collectors, mainly German. These collectors seem to
have been willing to pay quite high prices for 'good' specimens—a
good specimen from their point of view being one containing a
relatively large and distinctive-looking insect which was very clearly
visible. It is not surprising that attempts were often made to pass off
specimens of insects in much more recent resins (e.g. East African
copal) as amber fossils, and many amber collections contained a proportion of specimens in copal and other more recent resins. These can
usually be detected even without chemical analysis, by their lower
hardness and specific gravity as compared with genuine Baltic amber;
the insects in them usually prove to be identical with modern tropical
species.

A much more serious question is whether it is possible to fake inclusions in the Baltic amber itself. According to the eminent amber
collector, Klebs [116], faked inclusions were current at the time,
where specimens had been enclosed between two carefully prepared
pieces of amber which were glued together by means of another
natural resin-like substance, gum mastic. These fakes could generally
be detected by application of heat or suitable solvents. It is also known
that amber technologists had a technique whereby two pieces of amber
could be welded indistinguishably together by application of heat and
pressure, possibly assisted by certain solvents. It has not to my knowledge been revealed whether it would be possible in this way to produce an insect inclusion. If it were possible, the prices offered for 'good'
amber fossils might have made it worth someone's while to indulge in
such practices.

Some of the peculiar, apparently contradictory features of the
described Baltic amber fauna might become explicable on some such
basis. For example, among the Baltic amber ants described by Wheeler
[198], we find practically all the dominant modern European genera
recorded, side by side with a series of primitive extinct genera whose
closest living relatives are almost invariably extra-European and usually
tropical or subtropical. The described amber representatives of modern
European genera like *Lasius* and *Formica* are, however, apparently
almost indistinguishable from Recent species. On the other hand, a
specialised ant-parasitic group of beetles, the Paussidae (see Chapter 11),

is represented in the amber solely by primitive types whose living relatives are tropical or subtropical, and often Australian—and there is no question of any of the amber Paussidae being indistinguishable from recent species; the only paussid types represented in the European fauna have no relatives in the amber. Likewise, the bark-beetles (Scolytidae) of the amber, studied by Schedl, comprised only Hylastinae (mostly extinct genera) and a few primitive types of Ipinae; the Scolytinae, Platypodinae, and more advanced Ipinae (*Ips, Orthotomicus, Pityogenes, Pityophthorus,* etc.) were entirely lacking in the material seen by Schedl. Rather similar considerations apply to the amber Syrphidae, studied by Hull [100].

We may note that the ants, unlike the other insect groups just mentioned, represent a type of insect which is familiar to the non-specialist and to which a great deal of popular lore attaches. If one imagines some enterprising German with the idea of helping to satisfy the inordinate demand for 'insects in amber', what more natural group would there be for him to experiment with than ants?

The peculiar composition of the amber ant fauna described by him led Wheeler [198] to the conclusion that the group had reached the full limits of its evolutionary capacity by the early Tertiary period, and had stagnated ever since. This remarkable behaviour in a group which has often been considered to offer the closest non-human parallels to our own social evolution, seems to have had a strong influence on the evolutionary thinking of Julian Huxley and others. If, however, we exclude from consideration all the forms which appear to be identical or nearly so with modern European species, the residual Baltic amber ants would constitute a picture not very unlike that given by the other insect groups mentioned—an assemblage of more or less primitive types, often of extinct genera, whose living relatives are usually in warm climates and often in the southern hemisphere. If this is the true picture, the evolutionary history of ants would not support the sort of conclusions which Huxley has tried to draw from it.

Particularly in micropalaeontology, serious mistakes may arise from the accidental contamination of rock specimens with modern organic material. The most generally recognised danger of this concerns pollen and other spores, and it offers an additional argument in favour of the principle suggested elsewhere in this book (Chapter 17), that evidence from fossil pollen and spores should not be accepted as establishing the existence of groups of plants in rocks older than any in which probable macro-fossils of the groups concerned are known. A zoological example of a somewhat similar nature may be provided by certain insects and mites described from the celebrated Rhynie Chert—a Lower Devonian deposit with beautifully silicified fossils of very primitive vascular plants. The study of this by palaeo-botanists has been mainly by way of examining rock sections. The insects and mites which have been described from this deposit have all been found in rock sections already prepared by palaeo-botanists. They include a mite (Arachnida-Acarina) *Protacarus crani* Hirst [94a], and a spring-tail (Insecta-Collembola) *Rhyniella praecursor* Hirst and Maulik. From Hirst's figure, *Protacarus* looks like an absolutely normal modern mite. Material of *Rhyniella* has recently been restudied by Debouteville and Massoud [48a], who draw attention to its complete similarity to modern Collembola of the family Neanuridae.

It needs to be emphasised that the Acarina have been generally regarded by comparative anatomists as the most 'advanced' group of the Arachnida, or at least the group which shows greatest structural modification in comparison with the assumed ancestral forms. The mites, for example, have completely tracheal respiration, with no trace of the 'lung-books' of primitive Arachnida. Apart from scorpion-like (and possibly aquatic) forms from the late Silurian, the oldest fossils of terrestrial Arachnida come from the Upper Carboniferous deposits, whence a considerable diversity of primitive-looking types have been described, many of them attributed to extinct groups and none of them closely similar to any of the dominant modern types. *Protacarus*, as a Devonian organism, cannot be fitted into a rational scheme of arachnid phylogeny which is consistent with the other evidence available to us.

The position of *Rhyniella* is very much the same. It belongs to a group of Collembola which are characterised, among other things, by the complete loss of the 'spring tail' or furcula, which represents the appendages of abdominal segment 4 and is probably subject to Dollo's Law (see Chapter 9). Various stages in the reduction and loss of this organ may be seen in related modern forms. Apart from *Rhyniella*, the oldest fossils of primitive wingless insects (Apterygota) are forms rather like modern bristle-tails (Thysanura) from Upper Carboniferous and Lower Permian deposits; these, however, differ from the modern forms in having only one 'tail' instead of three and are placed in an extinct order Monura (see Rohdendorf 1962 [161a]).

The well-known palaeobotanist, Professor J. Walton, some years ago showed me a rock-section of Rynie Chert containing what he thought to be an insect. After some study, I was able to recognise it as a nymphal thysanopteran, representing quite an advanced group of winged insects. It seemed fairly evident to me that the insect was a modern contaminant of the rock specimen. Rhynie Chert is a rock tending to contain cracks and crevices into which small insects might crawl, and it seems probable to me that both *Protacurus* and *Rhyniella* represent modern organisms accidentally contaminating rock specimens.

A particular difficulty confronting the palaeobotanist is that, unlike the students of many groups of fossil animals, he can hardly ever hope to find a 'complete' fossil of one of his organisms. A plant fossil may comprise a stem, a root, a leaf, a cone, a seed, pollen grains etc., but rarely two or three of these things in association and practically never all of them together. In these circumstances, it is hardly surprising that different parts of the same kind of fossil plant have very often been described under quite different generic names; the process of integration of such fragments into more or less complete plants has been and still is a very slow one, depending on very occasional discoveries of fossils in which the association of different parts can actually be seen.

The overall organisation of plants would appear to be in some way 'looser' than that of most animals, the parts being less intimately bound up with each other and there being consequently less possibility of deducing the whole organism from knowledge of one part of it; a fossil molar tooth of a mammal is probably more reliably classifiable than is a leaf of an angiosperm. The fossil record of Angiospermae is indeed very largely composed of leaves, many of these fossils being very beautifully preserved and offering considerable complexity of observable detail, certainly as much as is present in the average molar tooth. One cannot help feeling that it *ought* to be possible to place such fossils as *Credneria* fairly reliably in the classificatory system. If previous placings of fossil leaves have too often been mistaken, this is probably in considerable degree attributable to inadequate study of modern leaves as such. A really comprehensive collection of dried and pressed leaves, accurately named, and classified according to their externally observable characters—somewhat in the manner of the palynologists' artificial classifications of pollen grains—would be a very desirable basis for systematic work on fossil Angiospermae.

For those who attempt to fit fossils into classifications, whether phylogenetic or Aristotelean, devised for modern organisms, the difficulty of observing the requisite characters is by no means the only, perhaps ultimately not even the most serious, of the difficulties to be faced. Among existing organisms there are real and evident discontinuities which provide an objective basis for the divisions of our hierarchy —but if past as well as Recent organisms are included, these discontinuities will vanish. Such discontinuities as there are among fossils must be attributed to the 'gappiness' of the fossil record [Newell 146], for which palaeontological systematists should be duly grateful. Simpson [169] himself pointed out that *Hyracotherium*, included by him in the horse family (Equidae) would serve equally well as an ancestor of the rhinoceros (Rhinocerotidae)—but he does not consider adequately the problem that this will present for those who wish to place *Hyracotherium* in the system of mammals. He further points out that *Hyracotherium* does not differ from its contemporary ancestors of Artiodactyla or even of Carnivora to a degree anything like commensurate with the placing of these forms in different orders. How this state of affairs is to be reconciled with and expressed in the sort of formal classification he favours Simpson does not explain, though he quotes instances of this sort in arguing against the possibility of consistent phylogenetic classification.

There appears to be only one possible logical way of avoiding this difficulty—adopting the third of our suggested alternative possibilities for palaeontology. Instead of trying to fit fossils into classifications devised for Recent plants and animals, we should have to construct a separate classification for each era of the geological past. For practical purposes, it would probably be necessary, at least at the outset, to adopt the conventionally recognised geological periods for this purpose—this would involve separate classifications for Cambrian, Ordovician, Silurian, Devonian, Carboniferous (the Americans would no doubt prefer separate classifications for the Mississippian and the Pennsylvanian), Permian, Triassic, Jurassic, Cretaceous, Palaeogene (= older Tertiary) and Neogene (= younger Tertiary) organisms. The classifications for succeeding periods could be systematically related to each other, as suggested in Chapter 19 of this work. The resulting overall system, however strange and cumbersome it might appear, would in effect have gained a new dimension and would partially escape the insoluble contradictions which confront the attempt to incorporate fossils and modern organisms in a single system.

7 On Classificatory Characters, and Their Correlation

Character ergo non constituat unquam genus, sed sedulo secundum genus naturae conficiendum est.

LINNAEUS, *Systema Naturae*, Tomus II,
pars I, Regnum vegetabile, Introduction

A character, in systematics, may be defined as any feature which may be used to distinguish one taxon from another—cf. Mayr, Linsley and Usinger [141]. According to this definition, there seems to be no reason for us to expect the 'character' to have any special significance as a natural, objective unit; it will probably be something more or less arbitrarily selected to suit the convenience of human systematists. Whether it might be possible to define a unit character in some other way, according to which it might come to have some objective reality as a unit, is a question we shall consider later.

Systematic characters, as currently used in analytical keys and definitions of taxa, are very diverse; some are evidently complex and intrinsically capable of subdivision, for example, the presence or absence of angiosperm-type carpels in vascular plants, or of pentadactyl limbs as against fins in Vertebrata. Others are of the nature of a more or less arbitrary line drawn across something which is in principle a continuum, for example, 'outer involucral bracts longer than the inner ones' as against 'inner involucral bracts as long as or longer than the outer ones'.

Linnaeus, in the well-known passage quoted at the head of this chapter, asserted that it was not the character which made the genus, but the genus which made the character—thereby expressing a truth about the procedure of good systematists which is still not as widely understood as it should be. It is a common experience for the systematist revising a particular group to reach the conclusion that a particular assemblage of species constitute a natural sub-group, before he has discovered any single character by which this sub-group could be defined. The conclusion usually follows from the discovery that the various subtypes within the subgroup are connected to each other by transitional forms. When eventually he does discover a character by which to distinguish his new subgroup, it may well be one which *a priori* he would not have expected to have such classificatory importance.

Many systematists have maintained and, despite the denials by Inger [106], Wagenitz [195] and others, many probably still believe, that it is in general possible to distinguish different grades of classificatory characters, some suitable for distinguishing species, others appropriate for the definition of genera, still others for families, and so on. There are three possible connotations for this supposed principle. It might be merely an arbitrary procedural rule, descriptive of the behaviour of the systematists concerned, it might imply that some characters are by their nature more 'essential' in the Aristotelean sense than are others, or it might reflect some differences in the modes of evolution of groups at different levels. The first of these possible interpretations of the rule would imply that such systematists are liable to define genera, families, etc., which would not otherwise be justifiable. The third possible implication is likewise rarely drawn, because systematists who claim to be able to distinguish specific, generic, familial, etc. characters *a priori* rarely profess to be constructing phylogenetic systems.

As already mentioned, some 'single classificatory characters' are obviously complex, like the angiospermatous carpels or pentadactyl limbs, these manifestly represent the results of relatively lengthy bouts of adaptation, whose intermediate stages have disappeared entirely; the groups distinguished by them are 'isolated' in the sense explained in Chapter 16. Such features are clearly of high classificatory importance, but there is little justification for treating them as unit characters.

Despite Wagenitz's (loc. cit.) assertion 'Fur den "systematischen Wert" der Merkmale gibt es eine in über 200 Jahren immer wieder bestätigte Erfahrung der Taxonomie: Die einzige Regel ist, das es keine Regeln gibt oder doch nur solche mit vielen Ausnahmen. Das heisst, es gebt keine Merkmale, die a priori und in den verschiedensten Gruppen taxonomisch höher zu bewerten sind als andere', botanical systematists usually agree that the gamopetalous condition of the corolla in dicotyledons is a character of familial or ordinal importance, whereas flower colour is regarded as of not more than specific value. Yet in some families (e.g. Ericaceae and Cucurbitaceae) the corolla may be gamopetalous or polypetalous in related genera, and in others blue flowers (e.g. in Campanulaceae) or white ones (e.g. in Umbelliferae) may be almost universal. Furthermore, it has recently been shown that an unusual type of flower pigment (the betacyanins) distinguishes a whole group of families in the Centrospermae—see Hegnauer [88], Harborne [84] and Mabry [132].

Innumerable examples could be quoted from the animal kingdom of characters which are constant in extensive groups, yet at times differ between related genera or even species. A striking example concerns the cervical vertebrae of mammals. These are almost invariably seven in number, from the Monotremata upwards, but in the sloths (Order Xenarthra, family Bradypodidae) the number varies from six to nine in different genera, and the number six recurs in some genera (e.g. *Manatus*) of Sirenia. Analogous phenomena are shown by the tarsal formula in beetles; the number of tarsal segments is usually a 'good' familial character, but sometimes differs between related genera (e.g. *Apion* and *Heterapion* in the Curculionoidea, or *Lyprops* and *Heterotarsus* among the Heteromera) or even within a genus (e.g. *Agathidium*); the formula sometimes differs between the sexes of one species. Unlike the mammalian cervical vertebrae, however, the tarsal formula in beetles never exhibits an increase from the basic number (5).

It will be noticed that the characters I have quoted are of the type known as 'meristic', concerning the numbers of repeated structures. It is probable that such characters really do behave as 'unit characters' in many instances, i.e. the transition between one number and another may take place in a single step rather than by gradual loss or acquisition of individual members. Where the number seems to have increased, as in the cervical vertebrae of some sloths, it is difficult to imagine the additional members appearing first as minute vestiges which progressively increase in size until they become complete vertebrae. Comparable botanical examples may be the occasional appearance among the Monocotyledonae, with almost universally trimerous flowers, of forms like *Paris* or *Aspidistra* with floral parts in fours, or the trilocular carpels which often occur among cultivated forms of naturally bilocular Solanaceae.

In Chapter 16 it is suggested that active evolution is liable to affect only one character at a time, and that this may offer a basis for an objective definition of a unit character. We might define it as the amount of phenotypic change which could be brought about by a single continuous bout of natural selection. However, even if the basic idea in this definition is sound, it does not follow that it will be in general easy to recognise such unit characters among the present-day products of evolution. Doubtless there are numerous cases in which important intermediate stages are entirely unrepresented among existing organisms, and we are left to guess how many 'unit characters' went to the making of some of the major differences. One thing which can be asserted on the basis of present-day knowledge is that it is rare for two distinct classificatory characters to be 100 per cent correlated in their incidence, and where two characters appear to show such a complete correlation, they are usually very closely related functionally (e.g. the cribellum and calamistrum in spiders).

Our definition of a classificatory character, it will be noticed, would not restrict the term so as to include only the characters which have generally been used by systematists—the words 'may be' rather than 'has been' were carefully chosen. Differences in internal anatomy, development, behaviour, physiology, biochemistry could all be used to distinguish classificatory groups (see Chapter 8), though few professional systematists have made much use of them hitherto. Undoubtedly, there is a great deal to be said in practice for the definition of groups by readily visible structural characters; unfortunately it is by no means always possible to construct even a satisfactory Aristotelean classification on this basis, let alone a phylogenetic one. We can, however, retain the practical convenience of externally visible structures by using them in making artificial keys for determination, while basing our actual classification on a very much wider range of evidence.

Physiological characters, such as the degree to which low atmospheric humidities may be endured, rate of growth at a given temperature, ability to digest a particular type of substance, ability to form adventitious roots or to survive in either fresh or salt water, have undoubtedly been of great importance in evolution, and should therefore receive the attention of phylogenetic systematists, though there may be no compelling reason for their utilisation in an Aristotelean system. Physiological differences between species may, however, be both practically important and useful for determination. The same applies to behavioural characters in animals; the distinction between physiology and ethology (or ecology) is in any case artificial and subjective.

In so far as the characters in which organisms differ can be counted or measured, the systematist using such characters may claim to be a practitioner of the mysteries of biometry. As a rule, in any character which can be counted or measured (e.g. number of stamens, number of scales along the lateral line in a fish, length of leaves, total body weight), members of a species show a limited and continuous (though meristic characters may have only whole-number values) variation, usually expressible as a histogram similar in form to the 'normal distribution' curve of the statisticians. This sort of variation is adequately expressed by stating the mean value and the 'standard deviation'; from these two figures, curves can be drawn which will agree closely with histograms constructed from large samples of actual organisms. Different species, as a rule, show more or less different values for the mean and standard deviation in histograms relating to the corresponding character, though often the two curves show more or less extensive areas of overlap, corresponding to individuals which as far as this character is concerned might belong to either species. A histogram of this sort, if it is to be regarded as a characteristic of the species, needs to be constructed from very widely representative material. Particular populations of the species may give markedly different histograms for the same character —usually in connection with differences in mean body-size; thus populations from unfavourable environments often consist mainly of very small individuals, with correspondingly aberrant values for most of their measurable 'parameters'.

A histogram for a particular character in a given species may at times show a double peak, what is known as 'bimodal distribution', instead of a normal distribution curve; this results as a rule from sex dimorphism or from what is called 'polymorphism' in the species. The common primrose (*Primula vulgaris*) shows 'heterostyly' or polymorphism in the length of the style; histograms of style-length in the flowers would show a markedly bimodal distribution; a familiar example of sex dimorphism is shown by the crest on the top of the head in the common fowl (*Gallus domesticus*).

Where two species are mixed together in a sample, a histogram for a particular character in the sample will not necessarily be bimodal, even when the means and standard deviations differ in the two species —for clear bimodality to appear in the curve, the difference between the two means must nearly equal the sum of the standard deviations, at least. Thus a histogram of this sort cannot always be depended on to reveal the presence of two species in a sample.

A different, on the whole more systematically valuable, procedure is to plot individual values for one parameter against the values for another in the same individuals. Each measured specimen will then provide a point, and a numerous sample will provide a large number of discrete points. If a sufficiently large number of such points are plotted, they will normally appear to be clustered round a definite 'line of maximum density'. Quite often this line appears to be more or less straight, and if prolonged would pass through the 'origin' of the graph; in this case the value of one parameter is approximately proportional to that of the other, and the equation connecting them will be of the form $y = kx$. In many cases, however, the line will appear to be more or less curved; theoretical considerations about the process of growth, and practical experience, both indicate that the curves thus obtained correspond to equations of the form $y = kx^p$.

Instead of drawing a free-hand line through the estimated maximum density of points on such a 'scatter diagram', it is possible to calculate the values of k and p from a simple table of the corresponding values of x and y. Methods of doing this are explained, e.g. in *Quantitative Zoology* [168a] chs. 13–14. In actuality, there is another variable, besides k and p, involved in such a scatter diagram—the degree of divergence of the actual points from the line corresponding to the equation $y = kx^p$, to which a figure analogous to a standard deviation may be attached.

If scatter diagrams for the same two parameters are made for two different species, their 'lines of maximum density' are liable not to coincide; values of k are liable to differ between species and where p is not equal to one (i.e. where heterogony occurs), this constant too is liable to differ from one species to another. A species is often better characterised by values for k and p for a given pair of parameters than it is by the means and standard deviation of either of these parameters by itself.

As already indicated (Chapter 2), a very important feature of the classificatory characters of organisms is that they are not, in general, randomly distributed in relation to each other. This has been statistically demonstrated for a number of groups in recent years, for example, for the families of Dicotyledonae by Sporne [178] and Stebbins [179], for the species of *Drosophila* by Sturtevant [183], and I have drawn up a similar table for the families of Coleoptera-Polyphaga from which it is evident that several pairs of characters show significant correlations in this group. Thus in the 154 families analysed, 42 may be described as having the hind coxae normally more or less excavate, the other 112 as having them mainly flat; several other characters show more or less strong correlations with this one, for example, the form of the Malpighian tubules and presence or absence of spiracles in abdominal segment 8 of the adult, the presence or absence of a spiracular closing apparatus and of separate galea and lacinia to the maxilla, in the larvae. As an example, 74 of the families with 'flat' hind coxae have the ends of the Malpighian tubules closely attached to the hind gut and only 35 are known to have them free (in three families the condition is unknown); among the families with excavate hind coxae, 32 have free ends to the Malpighian tubules while only 8 have them closely attached to the hind gut. This, treated as a 2 × 2 contingency table, gives a highly significant value for chi squared.

The large majority of the families of Polyphaga with flat hind coxae and attached ends of the Malpighian tubules belong however to a single section, Cucujiformia, of the suborder; if these are excluded from consideration, there is left only one family (Bostrychidae) known to have this combination of characters, the contingency table then takes on a very different appearance, indicating if anything a negative correlation between flat hind coxae and attached ends of the Malpighian tubules. This I believe illustrates a general principle: if two characters which are not closely linked functionally, show a marked correlation in their incidence within a group, this is because a particular combination of them characterises a large subgroup. In the case of the dicotyledonous families analysed by Sporne and Stebbins, this principle is less evident than in the Polyphaga. However, among the positively correlated characters listed by Sporne are leaves stipulate, petals free, and stamens pleiomerous; there are reasons for supposing that a large group of families, the 'Gamopetalae', in which all three characters are normally lacking, represent a natural subgroup of Dicotyledonae, and this may be sufficient to account for the positive correlations found by Sporne and Stebbins.

It is evident from the studies of Sturtevant, of Sporne, of Stebbins and from my own observations on Coleoptera-Polyphaga, that correlations between particular pairs of characters may vary from strongly positive through zero (i.e. random association) to strongly negative, and that some characters show very significant correlation (positive or negative) with many different ones, while others have very few such significant correlations. This frequent 'correlatedness' between classificatory characters is a very significant fact in nature, and one which is ignored in the standard procedures prescribed by the 'numerical taxonomists' (see Chapter 15). It is however the principle underlying traditional or 'Aristotelean' classification. The characters used in defining major groups in traditional classifications have been those which show strong correlations with others in their incidence within the higher group in question.

Correlations between particular pairs of characters may of course be attributable to functional connections—thus, for example, in animals, carnivorous adaptations of the mouth and alimentary canal tend to be correlated with adaptations of the sense-organs and locomotor apparatus, and among Dicotyledonae the positive correlation between the 'woody' habit and leaves with secretory cells is probably due to the functional correlation of both characters with long-persistent leaves (which place a premium on the possession of some sort of chemical defences). Aristotelean classification made no distinction between functional correlations and such apparently non-functional ones as those between alternate leaves and arillate seeds, or between flat hind coxae (in Coleoptera-Polyphaga) and Malpighian tubules with their ends attached to the hind gut. As we shall see in Chapter 9, the distinction between the two types of correlation is important for the purposes of phylogenetic classification.

Looked at from the Aristotelean point of view, the characters of higher systematic importance are more 'essential', they are ones which change less readily in response to changes in the mode of life of the organisms concerned. These characters are sometimes called 'non-adaptive', to distinguish them from the 'adaptive' features which are readily changed in the course of evolution. This distinction, however, reflects a misconception; the 'deep', 'phyletic', unchanging characters must also have been adaptive in order to have arisen at all; if such characters are really liable to persist after they have lost any positive adaptive value or have even become disadvantageous, the explanation is probably to be sought in their genetic basis. A character which retains a high selective advantage for a long time is liable to be subject to 'developmental canalisation' or 'buffering', as a result of which it becomes increasingly immunised against the effects of casual mutation.

From the point of view of Aristotelean classification, an important character is easily defined as one, similarities or differences in respect of which are correlated with similarities or differences in respect of many other characters, but this definition is not always a satisfactory measure of importance from the point of view of phylogenetic classification. For the phylogenist, an important character would better be defined as one, a change in respect of which characterised an important step in evolutionary history. As is explained in Chapter 19, the new form of the character, resulting from such a change, may not be preserved in all or even most of the posterity of the changed stock, and the changed form of the character may not, among modern organisms, show a very high degree of correlation with other characters in its incidence. For example, many botanists believe that the angiosperm type of flower was developed in relation to pollination by beetles, against whose strong jaws and protein-hungry appetites the closed carpel served to protect the ovules. Existing Magnoliaceae and a few other supposedly primitive types of Angiospermae are reported, in their natural habitats, to be pollinated largely by beetles, and it is strongly suspected that this characteristic has been preserved by the Magnoliaceae line from the very beginning of its history. If so, a phylogenist would certainly regard it as an 'important' character, but it would hardly qualify for a high ranking on purely Aristotelean principles. Similarly, small size, nocturnal activity and carnivorous habits are believed to have characterised the first Mammalia, but all three features have been lost on many separate lines among their descendants.

8 The Classificatory Use of Non-Structural Characters

La seule route féconde que puisse suivre la biochimie comparée dans son domain taxonomique, est celle qui trouve à chaque pas ses fondations dans l'acquit actuel de la systématique animale.

MARCEL FLORKIN

The systematist dealing with Bacteria (see Chapter 17) is obliged, from the extreme simplicity of their observable structure, to make extensive use of physiological and biochemical characters; such characters undoubtedly have classificatory value in higher animals and plants also. Naturally, the professional systematist in his museum is often reluctant to rely much on them—to him, a proper systematic character is one which he can observe and compare in preserved specimens. Yet a field ornithologist will rely mainly on features of behaviour and song to separate the Chiffchaff (*Phylloscopus collybita* Vieill.) from the Willow Warbler (*P. trochilus* L.) and may well distinguish the two species as reliably as (or even more reliably than) the museum systematist working with the dried skins; for a comparable botanical example we need only cite the field mycologist, tasting his *Russula* species before naming them. Many more such examples could be quoted.

As we have seen in the last chapter, long experience of the classificatory use of structural characters has firmly established two important principles (1) that the classificatory value of a character is not to be established *a priori*, in relation to its apparent functional importance, (2) that no definite classificatory 'weight' can be assigned to a particular character in general. A feature which in one group is constant through whole families or orders (e.g. the dentition in Rodentia) may elsewhere differ between related genera or even species (e.g. dentition in ceboid monkeys). The only way in which the classificatory importance of a character can be assessed is by studying it in conjunction with as many other characters as possible; this is implicit in the epigraph to this chapter, and should apply alike to structural, physiological, biochemical or behavioural characters. At present, far more complete and extensive information is available from the field broadly describable as comparative morphology than from any other, so that, for higher organisms at least, the best basis for estimating the classificatory value of a non-structural character will be to study it in conjunction with structural ones.

In this chapter, we shall consider briefly the present and the potential classificatory value of biochemical, physiological, and behavioural characters, in that order; comparative serology and kindred techniques are considered in Chapter 13, and evidence from the field broadly known as 'cytogenetics' in Chapter 12. Other characters which might be considered here but are treated in separate chapters are geographical distribution (Chapter 11) and host-parasite relations (Chapter 10).

The works of Florkin [65, 66], Baldwin [9] and others suggest that chemical characters could be used to define groups in the animal kingdom at all levels, from phyla downwards, and a similar principle is in the process of being demonstrated in botany, notably in the work of Hegnauer [88]. The evidence so far supports the idea that chemical and physiological characters are potentially equal in classificatory value to structural ones. The extent to which such characters have actually been used hitherto is affected by two main factors—first, the relative ease or difficulty of observing them, and second, the predilections of individual systematists. A believer in traditional, formal classification, seeing his task as that of building a practically useful pigeon-hole system for the species of his group, will see no reason why he should pay attention to non-structural characters if the species can be distinguished and grouped without their use; as a rule, he would only appeal to chemical or physiological characters as a last resort, in order to distinguish species (e.g. the *Phylloscopus* species already mentioned) whose structural differences are very slight and obscure. The phylogenetic systematist, on the other hand, is (or ought to be) always greedy for more comparative information about the forms he is studying; for him, the present era should offer exciting opportunities. The new chemical and physiological information that is becoming available for both plants and animals should, if only it could be made more systematically representative, open the way for major advances in our understanding of phylogeny.

Unfortunately, much of the new non-structural information is only obtainable by the use of complex and difficult techniques and apparatus; those who are masters of such techniques have not often had a thorough systematic training. If they are at all disposed to undertake comparative work, their probable procedure will be to apply the technique, perfected in studies of some 'laboratory' species, to as varied as possible an assortment of other species. Having done this, they are likely to suggest a new classification of the group to which these species belong, based solely on their own results. Now the limitations of this kind of approach were strikingly demonstrated a generation or two ago, when very similar things were done in the name of 'comparative morphology'. A young graduate would set out to make a comparative study of some particular piece of structure in as many species as possible of an extended group, and would proceed from this to propose a new classification of the group, based (ostensibly, at least) entirely on his own results. These new classifications were invariably unsatisfactory; established systematists, reading through publications of this kind, usually drew in their turn an unjustifiable conclusion—that the new information could safely be ignored for the purposes of serious classification.

The dividing lines between the different types of non-structural information which may have classificatory value are, of course, arbitrary, cutting through the unity of the organism. For practical purposes we may reasonably include under the heading of biochemistry all those substances (up to the level, perhaps, of peptide chains) for which it is reasonably possible to establish precise structural formulae; this domain was named 'micromolecular systematics' by Turner (in [215]). The study of proteins themselves is a border-line case, considered both in this chapter and in Chapter 13. The term physiology might perhaps cover things like functions of sense organs, effectors, digestive system, transpiration, osmotic relations with the environment, temperature adaptations, modes of growth, etc.; for animals, behaviour (studied by ethologists) may be considered as a separate category of evidence. The study of DNA base sequences in the chromosomes is a new and undoubtedly very important field, considered in Chapter 13.

The work of Hegnauer [88] presents a systematically arranged* conspectus of the known chemical constituents of plants, and is thus complementary to the numerous volumes in which the same information is presented from the point of view of chemistry rather than

---

* A regrettable feature of Hegnauer's work is the alphabetical rather than systematic arrangement of the families within groups such as Dicotyledonae.

botany. Each group in turn is given as full as possible a chemical characterisation. In some groups, e.g. the Gymnospermae, the characterisation is taken down to the level of genera and even species; other groups are treated in less detail, and the fungi receive rather cursory treatment. Hegnauer does not, unfortunately, discuss the chemical evidence in relation to other available indications of relationship. From the evidence he presents, however, it is clear that anyone aspiring to make a phylogenetic system of the Gymnospermae should consider very seriously their chemical characters, and there are many instances among Dicotyledonae and Monocotyledonae where the chemical evidence he presents is at least highly suggestive. Merxmuller [141*a*] provides a critical review of the work of Hegnauer.

A biochemical character which is evidently of considerable classificatory value in Angiospermae is the presence of nitrogen-containing pigments of the type known as the betacyanin group; a useful summary of current knowledge about the betacyanins is provided by Mabry [132]. There seems to be good reason for believing that all the betacyanin-containing dicotyledons come from a common ancestor which possessed this type of pigment. Mabry himself suggested that the order Centrospermae should be redefined so as to include all the dicotyledonous families in which betacyanins have been found, and no others. Thus defined the order will include, according to available information, the families Chenopodiaceae, Didieraceae, Amaranthaceae, Nyctaginaceae, Stegnospermaceae, Phytolaccaceae, Ficoidaceae, Cactaceae, Portulacaceae and Basellaceae; it will exclude notably the Caryophyllaceae, Elatinaceae and Cynocrambaceae. As Mabry points out, the evidence of the betacyanins conflicts strongly with the classificatory view of Hutchinson [101], but agrees rather well with those of Hallier [82], Engler and Prantl [59] and Wettstein [197]. Acceptance of Mabry's views on the monophyletic origin of betacyanins does not, however, oblige us to define Centrospermae in the way he suggests. For one thing, it seems possible that some forms descended from the betacyanin-containing ancestor may have lost this type of pigment, just as the betacyanin-containing families appear to have lost ordinary anthocyanins. Also, it is possible that betacyanins did not appear at the very beginning of the line to the present-day Centrospermae; there may be forms whose closest relationship is to this line but which split off from it before betacyanins had evolved.

Besides the betacyanins, the normal flower pigments known as anthocyanins may also provide useful phylogenetic indications. There are three main widespread types of these, delphinidin (and its methylated derivative petunidin), cyanidin and pelargonidin—each of these may occur in combination with various sugars or in metal complexes, such combinations being liable to modify the colour of the pigment. Delphinidin and its derivatives are responsible for most of the blue to purple colours; cyanidin, whose basic molecule has one less –OH group than delphinidin, similarly produces red to purple tints, while pelargonidin, with one less –OH than cyanidin, gives orange and red colours. Two or more of these may occur in combinations, and any or all of them may coexist with various lipid-soluble yellow pigments. Horticultural experience indicates that it is usually possible, starting from delphinidin-pigmented flowers, to produce mutants in which the pigment is replaced by cyanidin or pelargonidin, but there are few authenticated cases of delphinidin being produced in mutants from flowers which naturally contain only cyanidin or pelargonidin, such as roses. It is, however, possible to have blue flowers without delphinidin derivatives, for example, in the usually delphinidin-less Compositae, the blue of the Cornflower (*Centaurea cyanus*) is produced by a metal complex of cyanidin. See Harbone [217].

The widespread occurrence of delphinidin or its derivatives in both Monocotyledonae and Dicotyledonae strongly suggests that this type of pigment was present in the flowers of ancestral Angiospermae; in the course of phylogeny, whenever selection tends to shift the flower colour away from blue tints, it seems that delphinidin is liable to be lost. This must have happened on numerous phyletic lines, very few of which are likely to have later regained the lost delphinidin. It is interesting that those blue-flowered Campanulaceae which have been investigated contain delphinidin, which has not been proved to exist in any Compositae; if the Compositae arose from Campanulaceous ancestors, one of the basic changes involved must have been a change of flower colour involving the loss of delphinidin. Such a change in flower colour (cf. Verne Grant [80]) in turn probably reflects a change in the types of insect mainly responsible for pollination, e.g. from bees to flies.

As a third botanical example, we may consider briefly the evidence, as published in the Symposium 'Chemical Plant Taxonomy' [185], concerning the relationships of the family Scrophulariaceae; this may be compared with the serological indications discussed in Chapter 13, and with the evidence of the herbivorous insects considered in Chapter 10. According to the percentages of various fatty acids in the seeds,

Shorland (loc. cit., ch. 10) places Scrophulariaceae in the same group
as Oleaceae, Plantaginaceae, Pedaliaceae, Polemoniaceae and Solanaceae but in a different sub-group of it from Boraginaceae and Labiatae;
Bignoniaceae are placed in a group of their own; Buddleiaceae,
Martyniaceae and Rubiaceae are in another group, and Verbenaceae
with the Convolvulaceae in yet another. The presence of substantial
quantities of mannitol in most species of Scrophulariaceae links the
family particularly with Oleaceae (Plouvier, loc. cit., ch. 11). Aucubin
is reported by Bate-Smith and Swain [10] to be particularly prevalent
in *Buddleia*, Scrophulariaceae, Lentibulariaceae, Orobanchaceae, Globulariaceae and Plantaginaceae among Gamopetalae. The presence of
orobanchin is recorded by Harborne [84a] in *Buddleia*, Oleaceae,
Verbenaceae, Scrophulariaceae, Bignoniaceae, Orobanchaceae, Gesneriaceae and Acanthaceae.

In another work, Gibbs [72] reported that nearly all Scrophulariaceae
gave a marked positive reaction with the Ehrlich test, as do Bignoniaceae, Pedaliaceae, Acanthaceae, Orobanchaceae, Selaginaceae, Globulariaceae, Myoporaceae, Martyniaceae, Verbenaceae, Labiatae, but not
Solanaceae, Nolanaceae, Convolvulaceae, Boraginaceae etc. No report
is given of the reaction to this test of *Plantago*, *Buddleia* or Oleaceae.

The sum total of this evidence to a large extent supports traditional
anatomically-based views on the relationships of Scrophulariaceae.
However, there are chemical indications of relationships to Oleaceae,
Plantaginaceae, and *Buddleia* (Loganiaceae) which many systematists
have not recognised but which receive support also from serological
studies, from the host-selection of herbivorous insects (see Chapter 10),
and from some recent anatomical studies, e.g. that of Hartl [86]. It is
interesting that the chemical evidence, like the insects, offers little or
no indication of a particular affinity between Scrophulariaceae and
Solanaceae (or Convolvulaceae).

Chemical evidence may also be used to clarify relationships within
particular families of Angiospermae; Hegnauer [89a] reviews, for
example, the system of subfamilies and tribes in the Ranunculaceae in
relation to the distribution of alkaloids, etc. Sorensen [185] considered
acetylenic compounds as evidences for relationships within Compositae,
drawing attention to the apparent absence of such substances in Cicho-
riae, which distinguishes members of this tribe from all other Compositae investigated.

Relationships within and between genera have also been investigated
on a basis of chemical evidence, for example, species of *Lathyrus* and

Vicia have been investigated by Bell [125] for their peculiar non-protein amino-acids. The essential oils of *Eucalyptus* have long attracted the attention of phytochemists, and the results are reviewed by Gibbs [185] from a systematic point of view. The distribution of various terpenoid substances has provided similarly valuable evidence for relationships within the genus *Pinus*. At the opposite end of the classificatory hierarchy, Goodwin (in [3]) shows that characteristic types of carotenoids and xanthophylls can be used to distinguish the major divisions of the plant kingdom.

As yet no comprehensive work, comparable to that of Hegnauer, has been produced on the chemical classification of animals, though the subject receives more or less cursory consideration in most textbooks of biochemistry; see however Florkin and Mason [66a]. Nevertheless, chemical characters have long been used to define certain groups in the animal kingdom. The nature of the skeleton—whether calcareous, siliceous, or horny (organic), has traditionally been used as a fundamental character in Porifera and Protozoa-Rhizopoda. The presence of uric acid in the excreta has long been known to distinguish birds and reptiles from other groups of modern Vertebrata. The presence of cantharidin appears to be common, and perhaps universal, in beetles of the family Meloidae. The final products of nitrogenous excretion seems often to be constant throughout extensive classificatory groups (e.g. the guanine of spiders, the urea of Amphibia and Mammalia, the substituted amines of fishes, etc.). The nature of the 'phosphagens' involved in the 'Krebs cycle' in muscles and other tissues has received a good deal of recent attention, and may be of classificatory value at quite high levels, even phyla at times. An interesting paper by Watts and Watts [194a] suggests that the study of the enzymic reactions involved in the production of phosphagens can provide bases for conclusions about their phylogeny; they conclude 'all other phosphagens are formed from the transamination of arginine', and imply that this is true phylogenetically as well as in the individual modern organism. Among Vertebrata, creatine seems to be the universal and sole phosphagen in the muscles; in arthropod muscles, the same role seems to be played exclusively by arginine. As is shown in recent papers by van Thoai and Roche [189], quite a number of different guanidine derivatives may function as phosphagens in other invertebrates, and quite a number of animals are reported as possessing two or more different phosphagens (generally in different tissues). Among the non-vertebrate Chordata and in the Echinodermata, but in no other animals so far studied, creatine and arginine may occur together. Lombricine has been found in all the Oligochaeta studied, and is recorded also for a few sedentary Polychaeta and for the only echiuroid studied—it thus appears to be confined to the Annelida. The conclusion of van Thoai and Roche—'Finally, the distribution of phosphagens seems to defy any existing classification of the animal kingdom'— embodies the usual experimentalists' misapprehension of the nature of systematics. It is apparent that the nature of the phosphagens would not alone provide a satisfactory basis for an overall classification of the animal kingdom—but nor for that matter would any other comparable character. The nature of the phosphagens seems, however, to be constant at times throughout a whole phylum, and in other instances may be useful at lower levels; in the case of the Echinodermata and Chordata they may even provide evidence for interphylum relationships.

Current knowledge of the ionic components of the blood plasma of a considerable number of species of Insecta and other Arthropoda is summarised in a paper by Sutcliffe [184]; sodium, potassium, calcium, magnesium, chloride and phosphate ions are recorded, also amino-acids. The plasmas of nearly all the higher insects (Endopterygota) studied prove to contain notably high concentrations of amino-acids and low values for the chloride ion; in the Exopterygota, and even more in the apterygotan, *Petrobius*, in Myriapoda and in Crustacea, also in spiders, sodium and chloride ions are in relatively high concentration and amino-acids are low. There seems to be a tendency for aquatic forms to have higher values for sodium and chloride ions than are found in related terrestrial species, and for herbivores to have higher potassium values than are found in related carnivores. The general picture that emerges from the data tabulated by Sutcliffe is of a progression from plasma resembling sea-water to something more similar to cell-sap in the higher insects.

As is so often the case with characters of this kind, the data come from the results of many different investigators, working at different times and places and not always using identical techniques; furthermore, some of them studied immature forms and some studied adults, and we have evidence that in some cases the composition of the plasma may be different in different developmental stages of the same species. From such causes, discrepancies are likely to arise which may partly obscure the systematic value of the characters in question.

The theory underlying the technique known as chromatography (or partition chromatography) is still imperfectly understood, and in any case is too complex to be explained here; for the non-specialist an adequate account is given, e.g. in *Chambers' Encyclopedia*. The form of the technique which has been most used in systematic studies is called paper chromatography (or 'fingerprinting' in the U.S.A.); it is adapted for the recognition of amino-acids and similar diffusible substances in extracts of whole animals or plants, or of parts of them. The method produces a spatial separation, on a sheet of prepared filter-paper, of patches of different diffusible substances; these patches may be made visible by staining techniques such as the ninhydrin reaction for amino-acids, or by the application of ultra-violet radiation, in which many amino-acids fluoresce in characteristic colours; specific tests for particular substances may be applied to individual patches. The work of Thompson et al. [190] offers a detailed and comprehensive review of these techniques as applied to plant material, but gives little consideration to the systematic value of the results.

Paper chromatography may be 'one-dimensional' or 'two-dimensional'; if the diffusible substances are separated along a narrow strip of paper the result is a linear 'chromatogram', but if different gradients are established parallel to the two axes of a rectangular sheet of paper and diffusion starts near one corner of it, the result will be a two-dimensional chromatogram. Two-dimensional chromatograms are more laborious to prepare, but they are correspondingly more useful for systematic purposes.

In order to be properly comparable, chromatograms need to be made from samples drawn from exactly the same parts and developmental stages of the organisms, and the preparation of the samples needs to be carefully standardised. Examples of the results achieved by one-dimensional chromatography may be seen in the papers of Micks [144] on insects and of Riley and Hopkins [157] on species of the liliaeceous genus *Haworthia*. Both in mosquitoes and in *Haworthia*, clear differences were demonstrated between nearly related species. The value of such chromatograms for classification at higher levels seems doubtful; in Mick's results, two species of the hemipterous genus *Triatoma* gave chromatograms which differed from each other at least as much as did those of any two of the three widely separated genera of cockroaches (Blattoidea) which they studied; likewise, three closely related species of the mosquito genus, *Culex*, gave chromatograms which were not notably more similar to each other than to one of a species of the quite distinct genus *Aedes*.

Results using two-dimensional chromatography are rather more promising. One of the most impressive demonstrations of the possible systematic value of it is provided by the work of Hill, Buettner-Janusch and Buettner-Janusch [93]. They studied haemolysed blood of various species of the mammalian order Primates, treated with trypsin. The patterns obtained were not merely distinctive for the species, but provided indications of generic and familial groupings. They were able to deduce the number of polypeptide chains in the haemoglobins of different species—two in the lower Primates, four in the Anthropoidea, etc.; actual amino-acid sequences were worked out for particular polypeptide chains in a few species. In general, the results agreed moderately well with the immunological studies of Paluska and Korinek [148] (see Chapter 13), and with general ideas on the relationships within the Primates, but it appears that the results of this work are not quite equal in systematic value to those of good comparative serology.

Another modern method, somewhat analogous to chromatography, is electrophoresis; an electric current is used to produce a spatial separation of electrically charged molecules. The method is in itself one-dimensional but may be superior in some respects to one-dimensional chromatography; it can, however, be combined with chromatographic separation at right angles to itself, to produce a two-dimensional pattern, perhaps superior in classificatory value to a two-dimensional chromatogram. Zuckerkandl, Jones and Pauling [213] made a study of tryptic peptides of haemoglobin from various vertebrates in this way; their material included four species of Primates, two Ungulata, a bony fish, a lung-fish, a shark, a myxinoid, and a haemoglobin-containing invertebrate (the echiuroid *Urexis*).

Comparative physiology has been a fairly popular denomination among zoologists and (less often) botanists for a long time, but very few of those who called themselves comparative physiologists have made any significant contribution to the systematic ordering of the sort of information in which they deal. In the case of zoology, at least, the epithet 'comparative' usually implies no more than that they are liable to experiment on organisms other than placental mammals. For a mere systematist, the attempt to establish some order among the masses of data provided by the various journals of physiology is a forbidding task. Nevertheless it is already clear that some of the characters which are liable to be reported in these journals possess considerable classificatory value. For example, the presence of full trichromatic colour-vision seems to be confined to, and universal in, the order Primates among the Mammalia, and among insects, the ability to digest true cellulose seems to be confined to the termites (Isoptera) and the closely related cryptocercine cockroaches (cf. McKittrick [133]), all of which also harbour trichonymphid Protozoa. The deciduous habit is a good classificatory character at species level among various dictyledonous trees, as are features like 'long-day' and 'short-day' reactions and flowering seasons in flowering plants generally. The halophytic tendencies of the Chenopodiaceae may be a basic feature of the family.

Florkin [66] suggests that, if chemical characters are to be used as evidences for phylogeny, we need first to establish evidence for homology. The presence of a particular chemical substance in two different taxa, he says, should not be accorded much weight as an indicator of a relationship between the two unless it has been shown that in both groups the substance in question is produced by the same series of chemical steps and under the influence of a strictly homologous series of enzymes. Theoretically, this is undoubtedly a sound principle, but its practical adoption would probably have an unfortunate effect, in discouraging descriptive chemical work on organisms. A similar principle could be asserted in respect of structural characters—that we should not rely on them as evidences of relationship until their ontogenetic development had been thoroughly studied and shown to be the result of a homologous series of factors. If the descriptive anatomists of the last 200 years had been inhibited by such a principle as this, we could hardly have achieved our present level of systematic and anatomical knowledge of the plant and animal kingdoms. If biochemical knowledge of organisms is to take its proper place in systematics it will be necessary for numerous and extensive studies to be made involving a simple comparative and descriptive approach. The sophisticated phylogenetic principles of Florkin are unlikely to stimulate this. In the evolution of our subjects, the systematic collection and ordering of data should precede strictly phylogenetic research.

A general difficulty in trying to estimate the classificatory value of comparative physiology is that of distinguishing it from biochemistry, from ecology, and sometimes even from morphology. If particular cells in the leaves of certain plants regularly contain globules of some 'essential oil' or characteristically shaped crystals of some specific chemical compound, is this a structural, a physiological, or a biochemical character? If the leaves of a particular group of plants are always folded in a special way in the buds, is this a structural or a

D

physiological character? Are the 'innate' and characteristic songs of some birds to be considered as 'ethological' or physiological (of the syrinx) characters? Physiology, like most of the currently fashionable 'disciplines' of biology, has no objectively definable boundaries, a circumstance which does not favour the development of any stable organisation within it.

It may be appropriate at this point to consider briefly the classificatory value of enzymes—being proteinaceous substances, they also fall within the scope of 'macromolecular systematics' of Turner [215] considered in Chapter 13, though, of course, biochemists are very much concerned with the reactions which enzymes catalyse. Up to the present, the study of enzymes has been very much hampered by the lack of a satisfactory classification of them. They are normally named and considered in relation to the reactions which they catalyse—as 'lactic dehydrogenases', 'lipases' etc.—and in relation to the temperature and pH optima of these enzymic reactions. For a satisfactory classification we shall need far more information than we yet have on the constitution and origins (both ontogenetic and phylogenetic) of particular enzymes; if and when such information becomes available it should become possible to distinguish 'homologous' and 'analogous' enzymes, and to use them as bases for far-reaching phylogenetic conclusions.

It is often supposed that something like Dollo's Law (see Chapter 9) should apply to enzymes—that particular enzymes are liable to be lost but not regained in the course of evolution. This may well be true to an extent comparable to that within which Dollo's Law is applicable to structural characters; where an enzyme catalysing a particular reaction has been lost in the course of evolution, if an enzyme affecting the same reaction is re-acquired during subsequent evolution, it is not likely to be identical with the original one. In recent work [64a] on the genetics of moulds (*Aspergillus*, etc.), most of the mutant strains recognised are distinguished from the wild type by their inability to grow on nutrient media lacking specific amino-acids; this inability is attributed to the lack of some enzyme required for the synthesis of that particular amino-acid. However, there have been many reports of 'wild-type' moulds arising from such mutant cultures, presumably as a result of mutation. Where such types have been genetically analysed, the reversion to wild-type appears to result, not from true 'back-mutation', but from mutation at a new locus—so the new 'wild-type' will not be genetically identical with the old. It would be interesting in such cases to determine the precise structures of the relevant enzymes. Such instances, however, offer a poor parallel for natural evolution; the loss of a particular enzyme in phylogeny would normally occur in association with a whole series of other adaptive changes, involving large changes in the genotype.

It seems probable that, in the course of evolution, enzymes may at times undergo changes whereby they come to catalyse what are really new reactions, and thus organisms may come to possess what we would consider to be new enzymes. Some such process has presumably been involved, for example, in the development in at least two different insect groups of the ability to digest keratin. It may well be that in most cases the acquisition of a 'new' enzyme is paid for by the loss of an old one; there is little evidence, for example, that higher plants and animals possess more enzymes than do the simpler ones.

A very interesting paper by Wilson and Kaplan [206] discusses enzyme structure and its relation to taxonomy. As these authors point out, there are several ways in which enzymes may be compared. They assume that enzymes are homologous when they catalyse the same reaction and the reaction in question is a universal one within the group studied. This assumption may not always be justified; it is quite conceivable that, in the evolution of a particular lineage, the function of a particular enzyme may come to be first supplemented and then entirely replaced by that of another, non-homologous one. However, in the examples quoted by Wilson and Kaplan, their assumptions seem to lead to reasonable and consistent results.

To quote Wilson and Kaplan: 'The ideal approach is to determine the amino-acid sequences. Because of the difficulty of determining sequences, only a few results of taxonomic interest are yet available. These results already imply that enzyme structure varies in relatively good accord with taxonomic relationships.' They suggest that 'enzymes from different families of mammals differ in sequence by about 2 per cent, enzymes from different cohorts of mammals by about 12–20 per cent, and enzymes from different kingdoms by more than 90 per cent'. Another mode of comparison they discuss is the so-called 'fingerprinting', the combined chromatographic-electrophoretic analysis of peptides of the enzyme in question; their conclusion about this is 'It is reasonable to conclude from this and from the myoglobin case that fingerprinting is an unreliable guide to the degree of structural resemblance between homologous proteins' . . .

The third Wilson–Kaplan method of comparing homologous

enzymes involves the study of differences in their catalytic behaviour, manifest particularly in their reactions with co-enzymes differing from the 'normal' one, referred to by these authors as 'analogs'. They report the results of an extensive study of this sort on lactic dehydrogenases in both vertebrate and invertebrate animals. Their results are certainly suggestive, e.g. in showing that the lactic dehydrogenase of higher Crustacea is much more similar in its reactions to those of Myriapoda and lower Insecta than it is to the corresponding enzymes in *Limulus* and Arachnida.

The fourth method which Wilson and Kaplan discuss for comparing enzymes is serology or immunology. In the present work, taxonomic serology is considered specially in Chapter 13. Wilson and Kaplan noted that the results of comparative serological studies of enzymes agreed for the most part very well with those of their first and third methods of comparison. The method had two principal variants— either to test the amount of a given anti-serum required to inactivate a given enzyme-preparation, or to use the various methods of quantifying the precipitin reaction itself. One of these precipitin methods, that of 'complement-fixation', was extensively used by Wilson and Kaplan. They concluded that 'the complement-fixation results are probably more reliable guides to the degree of overall structural resemblance between these enzymes and the chicken enzymes' as compared with their enzyme-inactivation results.

Human interest in the ways in which animals behave is a very old thing, but the establishment of 'ethology' as a specialist 'discipline' (the implication being that it is desirable that the study of 'behaviour' should be segregated from the study of animals from other points of view), with its own journals, organisations and technical jargon, is quite modern. It is far from self-evident that it is more important for someone studying the behaviour of (for example) rats to be well primed in 'ethological' literature concerning sea-gulls, bees, sticklebacks or flat-worms, than it is to know a great deal about his animals from other points of view, ecological, physiological, phylogenetic or genetical. However this may be, the ethologists are now with us and they are publishing increasing amounts of information about detailed behaviour of diverse kinds of animals. Much of this information consists of laborious and elaborate statistical demonstrations of things which ordinary naturalists knew, or thought they knew, a long time ago, but a significant part of it is effectively new.

Even on a purely Aristotelean approach to systematics, there is no reason why we should not use such behavioural characters as seem important to us for classificatory purposes—especially as the characteristic behaviour of a species of animal may be, from the human point of view, the most important thing about it. The song of the European Cuckoo (*Cuculus canorus* L.) provides perhaps the simplest character by which to recognise the species, and the habit of laying eggs in nests of other birds is a perfectly good one by which to distinguish the group to which it belongs. Among the Insecta, the habit of running on the surface of water is a perfectly 'good' classificatory character for the group Amphibicorisae of the Hemiptera.

The classificatory and phylogenetic value of behavioural characters of termites have been reviewed by Emerson [57].

From the point of view of the phylogenetic systematist, information about behaviour may be even more important, in that it can help us to an understanding of the functional significance of structural and physiological characters, which in turn may lead to a sounder estimate of the likelihood of various postulated evolutionary sequences. Conversely, some understanding of the evolutionary history of a particular type of animal may help the understanding of its behaviour. If we as systematists can do something to break down the Maginot Line of technical vocabulary behind which some of the ethologists are trying to entrench themselves, we may benefit them as well as ourselves.

In all too many twentieth century works on animal evolution, the process has been discussed in terms of a mysterious active something called the Environment, which is represented as directing and controlling evolution with the arbitrary omnipotence of the human plant or animal breeder. In this picture, the only important function attributed to the animals is that of producing, involuntarily and at random, mutations. To anyone who has seriously observed, or thought at all deeply about, animals in the wild, such a conception is bound to appear fantastic. For normal, free-living animals, it is surely truer to say that they select their environment than that the environment selects them. The behaviour of the animals is of crucial importance in this connection, in determining the sort of environment in which it lives, the type of food it eats, its reactions to other species, and so on. Before a population can undergo selection for structural adaptations to a particular habit, the animals must first have the habit.

That animal behaviour is a profoundly important factor in evolution is a fact which has been recognised by a number of recent ethologists, notably by Tinbergen [192], who also points out that behavioural characters may be used to characterise classificatory groups at various levels. At the same time, he draws attention to a weakness in many behavioural characters, considered from a systematic point of view; 'Behaviour is known to be subject to phenotypic change to a much greater extent than are morphological characters. Learning processes may even keep changing the behaviour throughout the life of the individual. It is therefore of special importance, when dealing with a behaviour character, to investigate whether observed differences are genetically determined, or merely reflect differences in the environment.' Tinbergen goes on to point out that even when a particular feature of behaviour is nearly or quite universal among members of a wild population, it may not be, in the strict sense, genetically determined; examples are quoted of species of birds in which a standard type of song is, in fact, learned from the parents, and in which young birds, brought up under foster-parents of another species, learn a quite different song from the normal one of their own species. In some other species the song seems to be strictly inborn, and various intermediate conditions also occur. There are also reports that in certain species of insects the selection of a host (animal or plant) on which to lay eggs may be influenced by the female's experience during the larval stage.

The prospect of our being able to account for all the evolutionarily significant differences in behaviour among animals in terms of genes and their mutations appears remote. The comparatively recent development, among British urban Great Tits (*Parus major* L.), of the habit of pecking the tops off milk bottles, might well exemplify a type of change which could lead to important evolutionary consequences; if both milk-bottles and great tits persist for long enough as ingredients of our urban scene, adaptive modifications of the feet and bills of the birds may be expected to occur. I doubt whether many serious students of birds would be content to regard the origin of this particular habit as sufficiently explained by the word 'mutation'.

Tinbergen asserts that 'compared with the morphologist, the ethologist aiming at evolutionary interpretation has a very narrow and restricted inductive basis to operate from'—meaning, I take it, that comparative information on animal behaviour is as yet very far from being as detailed and comprehensive as is that on structure, also that, as Tinbergen points out, we have nothing analogous to a fossil record of behaviour. However, the 'ethologist aiming at evolutionary interpretation' is not obliged to confine himself to behavioural evidence, and Tinbergen himself says 'It is fortunate that the classification of many groups has already been worked out satisfactorily'. On the actual relations between classification and evolution, the author in question is somewhat obscure; he seems not to have considered the possibility of a strictly phylogenetic classification and writes: 'Unawareness of the difference in aims between classifiers and evolutionists could give rise to endless disputes between lumpers and splitters.'

In a special consideration of the behavioural differences between kittiwakes and other gulls, Tinbergen writes:

> The classifier is fully entitled to use, e.g., the tameness of kittiwakes, their nest-building behaviour, the black neck-band of their young, and their nidicolous habits as four separate characters. But the evolutionist is not entitled to treat them as four independent characters. To him, the correct description of the characteristics of the species would be in terms of adapted systems, such as (1) cliff breeding, (2) pelagic feeding, (3) orange inside the mouth and related habits of posturing; and a few other characters such as the shape of the black wing-tip and the underdeveloped fourth toe. The character 'cliff-nesting' characterises the list of 25 separate characters mentioned separately above. The classifier can use all 25 characters to show how different the kittiwake is from other gulls; the evolutionist concentrates them into a few systems.

There are certain difficulties in this passage. For example, is Tinbergen suggesting or implying that all the characters composing one of his 'adapted systems' will have originated simultaneously, together, that there was never an ancestral form possessing some but not all of them? The analogy of the Non-congruence Principle as applied to structural characters (see Chapter 16) suggests that such a series of behavioural characters, however closely they may be linked in function, will probably have originated one by one rather than concurrently—in which case the phylogenetic systematist is entitled to consider them as separate characters.

A particular difficulty in using ethological or ecological similarities as evidences for phylogenetic relationship may be illustrated by an imaginary example. Let us suppose that two taxa, A and B, are believed by some to be related to each other, on account of certain structural similarities—but that other systematists are more impressed by structural differences between A and B, and on account of these do not believe that the two taxa can be nearly related. Then some new ethological or ecological studies reveal previously unknown similarities in mode of life between members of A and B. The advocates of a near relationship between A and B will be likely to seize on the new evidence to re-inforce their case. In reality, however, if the structural similarities between A and B can be considered as adaptations to the mode of life which has been found to be similar in the two taxa, then the new evidence does not really strengthen the case for a phylogenetic relationship between the two; it may indeed actually weaken it. One might even say that structural similarity despite differences in mode of life is a stronger indicator of phylogenetic relationship than structural similarity correlated with similarities in mode of life. Likewise, structural differences despite similarities in mode of life are entitled to more classificatory 'weight' than structural differences correlated with differences in mode of life. The difference in structure between the wing of a bat and that of a bird is more significant than the difference between the wing of a bat and the hand of a man.

9 Phylogeny as the Basis of Classification

'Our classifications will come to be, as far as they can be so made, genealogies'

C. DARWIN, 1859

We have already considered, in Chapter 2, the Aristotelean definition of a natural classification, and in Chapter 15 another possible definition, which I have called the statistical one, is discussed; neither of these definitions, it will be noted, contains any specific reference to natural history, and the statistical definition in particular could equally well be applied to classifications of pebbles, books, postage stamps, motor cars, etc. In the present chapter we shall be concerned with a third possible definition of a natural classification, one which will be applicable only to the products of organic evolution. A natural group, above the level of a species, could be defined as one in which all members share a common ancestral species which is not ancestral to any form outside the group. Classifications based on this principle can properly and strictly be called phylogenetic, in accordance with the usage of the term by Hennig [91]—though Simpson [170] and other American authors have used the term in much less stringently defined ways. In a phylogenetic system, the various categories could be defined in relation to the various degrees of remoteness of this common ancestral species, measured in years or in generations; it is, in fact, the only type of classification which offers the possibility of really objective criteria for supra-specific categories. The successive divisions in the classificatory hierarchy will then correspond to successive forkings in the 'family tree', realising the idea embodied in the Darwinian quotation at the head of this chapter.

This phylogenetic definition of a natural classification is quite distinct from the Aristotelean and statistical ones; we may well ask, to what extent will a 'phylogenetic' classification be natural according to the other two definitions? The strict answer is, of course, that we do not know, having no classifications which we can be sure are perfectly natural according to any of the three definitions. Probably the majority of zoologists and a minority of botanists share the comfortable faith that in practice all three definitions will lead to the same system; the zoological minority, and the much more considerable botanical fraction, who doubt this are themselves divided on the question—if different definitions of a natural classification lead to different results, which of them should we follow? The botanists could justly point out that the zoological majority, while professing to believe that a natural classification is the same as a phylogenetic one, accept without demur in textbooks etc., classifications which are at variance with the family trees depicted in the same works. The most obvious example concerns the classes of the Vertebrata. Thus in Romer [160], the 'dendrograms' in figures 31–32 show the birds sharing a common ancestor with the crocodiles more recently than the latter group do with snakes and lizards, yet in that work Crocodilia are placed together with Squamata in a class Reptilia while birds form a separate class Aves. In the same work, figure 22 shows the lung-fishes (Dipnoi) sharing a common ancestor with the Amphibia more recently than they do with ordinary fishes (Actinopterygii), yet Dipnoi are placed by Romer with Actinopterygii in the class Osteichthyes, while Amphibia are treated as an independent class.

The points at issue in these cases are worthy of closer examination. Assuming that the phylogenists are right, and that the Australian lung-fish, *Ceratodus* (*Epiceratodus*, *Neoceratodus* auctt.), shares a common ancestor with a frog (or a reptile, or ourselves for that matter) more recently than it does with actinopterygian fishes like the Gar Pike (*Lepisosteus*) or Bow Fin (*Amia*), can we say that it shows similar relationships either in its 'Aristotelean essence' or by statistical preponderance of its characters? If it does, the fact has escaped the attention of all systematists hitherto—none of them have yet, on non-phylogenetic grounds, placed *Ceratodus* with Tetrapoda and apart from Pisces. Similar considerations apply in the case of crocodiles in relation to lizards and birds. It seems fairly evident that, at least in some instances, Aristotelean or 'statistical' classifications may differ considerably from phylogenetic ones.

Ceratodus has always been classified in Pisces because it possesses a whole series of correlated characters—e.g. the scales, gills, operculum, fins and lateral line system—which together compose the accepted

'Aristotelean essence' of fishes. A frog possesses a corresponding series of correlated 'Tetrapod' characters. The leading characters of fishes, it is generally agreed, are adaptive for fully aquatic life, and those of Tetrapoda are correspondingly suited for life on land. We can say, in fact, that the correlation of the characters composing either of these 'essences' is at root functional; an evolutionary change from aquatic to terrestrial life might be expected to lead to the relatively rapid loss of one set of characters and the acquisition of the other. The phylogenist will, in general, attach much less importance to characters which seem to be thus functionally correlated, and more to coincidences in characters with no evident functional connection. The differences separating lobe-finned fishes (including Dipnoi) from Actinopterygii are not at all obviously connected with differences of habitat or mode of life; for this very reason the phylogenetic systematist might suspect them of being deeper-rooted than the distinctions between lobe-finned fishes and tetrapods. The same is true of the differences between lizards and crocodilians, as compared with those between the latter group and birds. Both these cases may be considered as somewhat untypical, in that they show phylogenetic and Aristotelean classifications in particularly sharp conflict; it would be easy to quote others in which the two approaches would lead to more consonant results.

The concept of a 'mode of life' to which an organism is 'adapted' structurally and physiologically, is undoubtedly more difficult for us at present to apply to plants than to animals. Animals live faster, their behaviour and reactions can be directly observed by human beings, the relationship between particular features of behaviour and specific structural characters is often self-evident. It is far more difficult as a rule to form a mental concept of the mode of life of a particular kind of plant. This difficulty might be mitigated to some extent by the more frequent preparation and study of speeded-up films showing the growth and reactions of plants in as natural conditions as possible. Different phenomena in the life of plants would, of course, call for different speeds and techniques; things like pollination and seed-dispersal would no doubt be filmed at their natural speeds, and the opening and closing of flowers would with advantage be filmed much faster than general vegetative growth. Whatever the difficulties in their preparation, films of this type would surely be worth making, and their widespread use in teaching might revolutionise the general attitudes of botanists—almost certainly, adaptive and phylogenetic thinking would be greatly stimulated thereby.

It is none the less true for having been pointed out *ad nauseam*, that in practice we cannot as a rule base our classification directly on phylogeny, since the latter cannot be immediately known or observed, but has itself to be deduced. A would-be phylogenetic system can be founded only on evidences for phylogeny. The best known and most obvious of such evidences is 'comparative morphology', which is the basis also of Aristotelean and other so-called 'phenetic' classifications. Thus the phylogenetic systematist will normally treat established 'morphological' classifications as first approximations to the desired phylogenetic system, and will be ready to modify them in accordance with such further evidences of phylogeny as may become available. This is the procedure I have adopted in the study of Coleoptera, and I cannot see that it is in any way inconsistent or illogical. Opponents of phylogenetic systematics will, of course, argue that, in the absence of direct knowledge of phylogeny, we are in no position to assess the degrees of reliability of different types of phylogenetic evidence, particularly when these seem to conflict with one another. The same could, of course, be said about the evidences for human history. The ultimate criterion, in both human and natural history, is expressed in William Blake's [15] aphorism 'Truth can never be told so as to be understood, and not be believed.' We must proceed on the assumption that the true story will possess a degree of internal consistency, of conformity with laws, and of agreement with the most diverse kinds of evidence, which no false theory could equal. This is, of course, an article of faith, and an aspect of the general belief in the rationality of the universe which is fundamental to the scientific attitude. There is no foreseeable prospect of an 'objective' method for measuring the degree of internal consistency, conformity with laws etc. of any historical hypothesis. Any such 'measuring' can only be a function of the higher, non-mechanisable, powers of the human mind—the same powers no doubt which apprehend the qualities of works of art. Numerical taxonomists may look forward to their own ultimate replacement by computers, but historians and phylogenists have no need to fear any such fate, even though Wilson [207] has suggested a way of testing phylogenetic hypotheses for consistency, criticised by Colless [33].

Three main problems confront those who would pursue the Darwinian ideal of phylogenetic classification. The first, most interesting and doubtless most difficult one is that of discovering the evolutionary ancestry of plants or animals; the second is how to convert a family tree into a formal classification. The third, perhaps least serious, problem is that of defining phylogenetically natural groups. Evolutionary history can only be reconstructed from evidence available in the present, and by arguing from effects to causes, thereby reversing the usual sequence of scientific reasoning. The characteristic attitude of mind of natural philosophy includes a conviction that the past is in some way unreal, or at any rate of diminished scientific validity; most physicists like to imagine their world as created afresh at every instant. A serious natural historian must embrace the contrary attitude to this— for him, the past must be eminently real and discoverable, and in a sense more significant than the increasingly human-dominated present. Naturally, those biologists who like to think of themselves as natural philosophers are not as a rule very interested in phylogenetic classification. The present chapter is devoted to the first problem of phylogenetic systematics, the second and third being considered in Chapter 19.

In the vertebrate groups so far considered in this chapter, knowledge of phylogeny is based on a relatively good fossil record. Fossils, as the most direct source of information about organisms of the past, are potentially the most valuable of all evidences for evolutionary ancestry. In those organisms for which a representative fossil record is available (see Chapter 6), and in which the fossils preserve sufficient structure to provide a satisfactory basis for classification, palaeontology indeed provides the best phylogenetic evidences. It is very rare, however, for the fossil record by itself to establish unequivocally the family tree of an extended group of plants or animals; almost invariably, the fossil evidence could be reconciled with more than one phylogenetic theory. We need to consider the fossils in the light of knowledge gained from the study of modern organisms.

In the animal kingdom, the fossil record is of major importance in establishing the evolutionary history of mammals, of reptiles and of bony fishes, but is of much less value in the Amphibia, Aves, Elasmobranchii and Cyclostomata. Of the invertebrates many of the groups most richly represented in the fossil record, such as Mollusca and Foraminifera, are represented only by shells or tests which are much inferior as a basis for classification to vertebrate skeletons. There is one group of invertebrates—the Arthropoda—in which the skeleton is at least equal in classificatory value to that of the Vertebrata, but complete and well-preserved skeletons of Arthropoda, are, unfortunately, rare as fossils. Echinodermata also have a relatively complex skeleton, and are well represented through most of the fossil record; of all invertebrate groups, this one offers perhaps the best possibilities for reconstructing phylogeny from fossil evidence alone. Brachiopoda also appear fairly promising from this point of view.

Despite the abundance in many parts of the world and of the geological record of fossils of vascular plants, and the considerable amount of structure which can be elucidated in many of the fossils by modern palaeobotanical techniques, the fossil record has failed hitherto to provide a very secure basis for conclusions about the phylogeny of plants. Some 'petrifactions' indeed preserve minute details of cellular structure of stems, leaves, seeds, etc. in a way scarcely to be matched in fossil animals; taking all things together, the fossil record of vascular plants must be at least as representative as that of land vertebrates. The explanation of this seeming paradox may lie in the looser organisation of even the higher plants as compared with that of the higher animals —a fossil plant is less likely to be preserved as a whole, and there is less possibility of deducing the whole from a knowledge only of some of the parts. Furthermore, the lower present level of our functional-adaptive understanding of plants as compared with Vertebrata hampers phylogenetic interpretation of the fossil evidence.

From fossil evidence, it can at least be said that the sequence of appearance in time of the main groups of plants is in accordance with ideas about phylogeny based on other evidence. For example, Silurian alga-like forms (e.g. *Parka*, *Prototaxites*) are older than any well authenticated Archegoniatae; the homosporous Devonian Psilophytineae antedate any known heterosporous types; pteridophytic types in the early Carboniferous (Mississipian) deposits appear at a considerably lower level than do any true Gymnospermae; Coniferae, Ginkgoales and similar gymnospermatous types appear well before any definite Angiospermae, and within the latter group, fossils of ranalian and similar types are known from older deposits than any certain Gamopetalae.

The likelihood of fossilisation depends a great deal on the habits or habitat of the organism, and we must reckon with the possibility that when a major change of habits occurs in a phyletic line, we may find as a result that a group previously well represented in the fossil record suddenly becomes rare or absent—or *vice versa*. The sudden disappearance of representatives of a particular line above a certain level in the fossil record does not necessarily prove the extinction of the line, just as the sudden appearance at a particular horizon of a type without evident fossil ancestors does not prove special creation.

Opponents of the principle of phylogenetic classification commonly argue that, in most groups of plants and animals, the fossil record is a quite inadequate basis for conclusions about ancestry. In practice, they say, your conclusions about phylogeny are based on the comparative study of modern organisms, i.e. on the same data as are used by ordinary systematists. Your 'phylogeny', they triumphantly conclude, is a theory based on the facts of classification, and not *vice versa*. The short answer to this is that phylogeny, even in the total absence of a fossil record, is not deduced solely from the facts of classification. We can also employ empirical generalisations about the process of evolution, such as Dollo's Law (see below). Furthermore, we can bring into consideration numerous lines of evidence, such as the nature of an organism's parasites (see Chapter 10), its protein and DNA structures, and immunological reactions of its body proteins (see Chapter 13), which are not used in ordinary classification.

Dollo [50] himself, with the typical post-Darwinian addiction to grand generalisations, expressed his law as 'The Principle of Irreversibility of Evolution', thereby making it sound like a biological equivalent of the Second Law of Thermodynamics (which underlies the irreversibility of many physical phenomena). In this form his 'law', whether true or not, is hardly useful to the phylogenetic systematist. In a much more restricted form, e.g. 'complex, or phylogenetically ancient, characters, once lost in the course of evolution, are not regained in the same form', the rule is, however, of prime value in deducing phylogenetic relationships. The main practical difficulty lies in deciding how complex, or how ancient, characters need to be in order to come within its scope. In the absence of any simple, objective criteria, this has to be decided more or less intuitively; those who consider that no function which could not in principle be performed by an electronic computer has a legitimate place in science, will conclude that Dollo's Law is not scientific.

A simple example of the application of this rule may be drawn from the class Reptilia. There is no doubt that snakes (Ophidia) and lizards (Lacertilia) are nearly related to each other; both are normally placed in one class Squamata and some more or less transitional forms between them are known. From these facts alone, the answer to the question 'did snakes arise from lizards, or lizards from snakes?' is not, however, evident; the transitional series (of modern forms) could be 'read' either way. Other, specifically phylogenetic, considerations are needed in order to settle the question. The pentadactyl limbs of lizards are certainly complex, and phylogenetically ancient, enough to come within the scope of Dollo's Law. We thus conclude that snakes might arise from lizards, but not lizards from snakes.

It is not easy to find an equally clear and generally accepted botanical illustration of Dollo's Law, a fact which may be connected with the current unpopularity of phylogenetic classification among botanists. Plants in general are simpler in organisation than the higher animals, and less prone to the evolutionary loss of complex structures. Possible examples might be quoted from the structure of the seedlings of Angiospermae, where the reduction of the cotyledons to one in Monocotyledonae seems to be an irreversible process; no plants are known which are monocotyledons in other respects but have two cotyledons, though there are a number of Dicotyledonae whose seedlings have only one cotyledon.

It is rather paradoxical that the claims of Bailey and his fellow workers (see Eames [52]) to have deduced, with dogmatic certitude, the phylogeny of xylem elements, on principles akin to Dollo's Law, have been rather uncritically accepted by many botanists, even among those who are otherwise highly sceptical about phylogenetic theories. Attempts to do the same sort of thing in respect of particular pieces of structure have often been made by zoologists, and have frequently led to highly controversial conclusions. Thus Jarvik [110], basing his conclusions on certain small details of skull structure, has postulated a polyphyletic origin of Amphibia (and Tetrapoda generally) from different groups of lobe-finned fishes—these conclusions being very pertinently criticised by Thomson [191]. Hennig [92] has similarly criticised phylogenetic conclusions about Diptera by G. H. Hardy, based on a too-rigid application of the Dollo principle to specific veins in the wings.

It is commonly assumed that 'nomomeristic' structures (meristically repeated ones with a fixed number) are liable to undergo reduction, but not increase, in number in the course of evolution. There is evidently a large measure of truth in this assertion, but like practically all rules in natural history, it seems to be subject to occasional exceptions. The tarsal segments of the higher insects have a basic number of five, subject to reduction in many lines, but no clear examples of an increase in the number of tarsomeres are known; similarly the corolla in Scrophulariaceae and numerous other families of Gamopetalae has a basic number of five lobes, subject to reduction in many groups but rarely, if ever, to increase. On the other hand, in the insect order

Coleoptera the antennae have a basic number of eleven segments, but in a few widely scattered groups (e.g. Lampyridae, Rhipiceridae, Oedemeridae, Cerambycidae) the number may be increased to twelve or even more, though reductions are very much commoner. Seven cervical vertebrae seems to be a basic feature of the true Mammalia, but in some of the sloths (*Xenarthra*, family Bradypodidae) the number is increased to eight or nine. A similar exceptional process seems to have been responsible for increases in the numbers of legs within the Pycnogonida (sea-spiders). Another probable zoological exception to the rule that nomomeristic structures are liable only to reduction in number in the course of evolution is furnished by the gill-slits and branchial arteries of sharks and their allies (Elasmobranchii); here the basic number seems to be five or perhaps six, but in the existing genus *Heptanchus* the number is increased to seven. Comparable botanical exceptions are: the stamens of Rosaceae, where the primitive number is probably ten, but many modern forms have much higher numbers; the trilocular condition commonly found in the fruits of our cultivated tomatoes which has presumably arisen from the bilocular carpel which is characteristic of all the more primitive Solanaceae; in the Monocotyledonae, the genus *Paris*, belonging to a family (Trilliaceae or Liliaceae) in which the floral parts are usually and no doubt primitively in threes, has all the floral parts in fours or fives. All these examples are quoted as exceptions to the rule about nomemeristic structures; a vastly longer list could be made of instances in which the rule of reduction applies.

There is a widespread, but I believe mistaken, idea that an antithesis can be drawn between 'conservative', 'non-adaptive', 'phyletic' characters which are valuable for higher level classification, and 'adaptive' characters which are liable to differ between nearly related forms. If a character persists for a very long time in a phyletic line, many would argue that the character must be unaffected by natural selection. To me it seems much more probable that such characters are adaptive at a very deep level, and that casual variations in them would be very unlikely to be advantageous within the broad mode of life represented by the line in question—the long term action of natural selection in eliminating any such variations would have had the effect of 'stabilising' the character developmentally. Really non-adaptive characters, such as wings in flightless insects, eyes in truly cavernicolous animals, or leaves in completely parasitic plants, are liable to be extremely unstable, and to vary greatly between nearly related forms.

The available evidence indicates that the evolutionary reduction, leading often to complete disappearance, of structures which have lost any adaptive value, is liable to be quite rapid. Vestigial structures do not, as a rule, persist for very long in phyletic lines, they are not usually of value in indicating relationships at much above the generic level. The main reason for the rapid degeneration of non-functional structures is, no doubt, that suggested by Sewall Wright [209]—the 'pleiotropic' effects of other adaptive genetic changes—rather than pure mutation pressure as suggested by Brace [18]. In some cases, no doubt, 'physiological economy' may be an important factor, for example, in the reduction of the flight muscles of flightless insects. The large amount of high-grade protein incorporated in these can advantageously be diverted to the service of the reproductive system—such diversion has been shown to occur in the individual development of some insects, e.g. bark-beetles [28]. Phylogenetically, the flight muscles seem to be the first things to go when the function of flight is lost in a phyletic line; the reduction of the wings themselves is a much slower process, for which the main cause is, no doubt, that suggested by Sewall Wright.

A general and major problem facing phylogenetic systematists is that of determining which, of two alternative conditions of a given character in a particular group, is the primitive, ancestral one. If one of the two conditions is established as primitive (plesiomorph of Hennig), the correct epithet for the alternative one is derivative (apomorph of Hennig). It has, unfortunately, become common practice in the present century to use 'specialised' as the antonym of primitive. The correct antonym for specialised is, of course, 'generalised'. In this connection, generalised means, or should mean 'not closely adapted to some specific function or mode of use' whereas specialised should mean 'closely adapted to a specific function or mode of use'. The confusion between the two pairs of terms derives from a once prevalent theory that ancestral forms must be 'generalised'. For example, modern mammals include forms specialised for feeding on other mammals, fish-eaters (seals, otters, etc.), insectivores, browsers, grazers, etc.; many would argue that an ancestral mammal would have eaten all these things indiscriminately and lacked special adaptations for feeding on any of them. Fossil evidence, however, indicates that, as far as can be judged from their dentition, the ancestors of mammals were quite specialised carnivores.

It seems very probable that, at any particular time in the geological past, the large majority of species of animals and plants had quite specific habits or modes of life, to which they were structurally and physiologically adapted, just as is the case today. There are some modern organisms, and doubtless similar ones existed in previous eras, which are unusually 'generalised' in their modes of life, seeming to rely more on adaptibility than on specialisation for their continued survival. There is, however, little or no evidence that it is only such 'generalised' types which have given rise to new evolutionary developments. An interesting example among the insects is that of the cockroach group (Dictyoptera). The living species of true cockroaches (Blattodea) are noted on the whole for their 'generalised' modes of life and their lack of structural specialisations. Exceedingly similar forms are known as fossils, going right back to the Pennsylvanian (Upper Carboniferous) deposits—they furnish a better text for a sermon on the persistence of conservative, slow-evolving types than for one on the immense evolutionary potentialities of the generalised. It is true, however, that during the immense span of geological time since the coal-measures, the Blattodea have given rise to two other important groups—but in each case the evolutionary efflorescence followed the development of un-cockroach-like specialisation. The first such development involved a change to specifically predaceous (Blattodea are nearly all omnivorous) habits, with great modifications of the front legs for seizing prey. Thus arise the great group Mantodea (praying insects), probably quite early in the Mesozoic era, and a highly successful group at the present day. Considerably later, a group of cockroaches adapted to digest wood and cellulose with the aid of symbiotic micro-organisms; this led to social habits and the origin of the termites (Isoptera), one of the dominant groups of social insects of the modern world (cf. McKittrick and McKerras [133]).

If a zoologist had lived and studied fishes during the Devonian period, he would probably have considered the lung-fishes, with their special adaptations to living in water bodies which are liable to be poorly aerated or to dry up, as one of the most specialised types of the group—yet it was this group which made the great evolutionary advance leading to the Tetrapoda. I am not arguing that the conventional attitude should be 'stood on its head' so to speak; it would be no more justifiable to assume that an ancestral form must be specialised than to conclude that it must be generalised—evidence suggests that it might be either. Admittedly, this mode of thought is as yet rather difficult to apply to plants; if it is difficult to form a clear mental concept of the 'mode of life' of a particular type of plant, it is even harder to distinguish between specialised and generalised modes of life in the vegetable kingdom. One might consider as 'generalised' plants which are able to grow and reproduce successfully in a great variety of plant assemblages, and as 'specialised' those which are narrowly restricted in this respect. In many cases, it has been shown that such specialisation is the result of competition; if competitive pressures are relaxed, many species will reproduce successfully in conditions quite different from those in which they naturally occur.

If the specialised–generalised antithesis cannot be relied on to distinguish primitive from derivative characters in a group, there are other criteria to which we can turn. One is that a condition which occurs in related forms outside the group is probably primitive within it. The chief qualification to this principle is that the related forms in question should not be actually derivable from members of the first group. Thus termites (Isoptera) have commonly been given the status of an order independent of the Blattodea, but a conclusion that the ability to digest wood with the aid of symbiotic micro-organisms was a primitive character in cockroaches, because it occurs in the related group Isoptera, would be unjustified—Isoptera being presumably derived from true cockroaches [133]. Similarly, it would be wrong to postulate that fertilisation by means of a pollen-tube was a primitive feature in Gymnospermae on the grounds that it is universal in the related Angiospermae. Of course, if you already possessed a phylogenetic classification of the groups concerned, you could assume that any groups outside a given one were not derivable from it—but if you have a phylogenetic classification of a group, you would already have discovered which characters were primitive within it.

Another qualification to the principle that a condition which occurs in related groups is primitive in a given group, is that some derivative characters are polyphyletic, i.e. have arisen independently on more than one line. The extent to which such polyphyly has actually occurred in classificatory characters is a highly controversial one in many groups of organisms. I doubt whether there is any extended group of plants or animals for which it would be possible to formulate a theoretical phylogeny which did not involve polyphyletic development in at least a few characters; no serious phylogenist would deny that particular characters may at times arise polyphyletically. Many botanists and zoologists argue that any character which possesses an evident adaptive value is *a priori* likely to have arisen polyphyletically; not a few carry this argument beyond individual 'characters' to suggest that some or even many established classificatory groups of organisms are of poly-phyletic origin. Others, like myself, consider that polyphyletic origins of characters are 'entities' subject to the law of Occam's Razor; i.e. those phylogenetic theories are most likely to be correct which postulate polyphyly in the fewest and simplest characters.

We do not, alas!, possess any certain and fully documented instances of actual evolution which could be regarded as test-cases in choosing between these opposing attitudes. Nearly all the cases in the fossil record which have been interpreted in terms of polyphyletic develop-ment (e.g. that of the mammalian type of lower jaw suspension) could be interpreted in other ways; the predilections of particular palae-ontologists have entered into their theory-making. It is often suggested that where a character arises polyphyletically, it is most likely to do so in related evolutionary lines—as indeed Simpson and others suggest was the case for the mammalian type of lower jaw suspension. In such cases, the common possession of a derivative character might be an indicator of phylogenetic relationship even though the character in question was not actually present in the common ancestor. As a matter of scientific principle, I consider that 'sophisticated' explanations of this kind are to be avoided unless there is strong evidence in support of them.

A further difficulty confronts the principle that we have been discussing—that derivative conditions may be subject to secondary loss. Within a given group you may find some forms preserving a genuine primitive condition 'a', others with a derivative condition 'A', and still others in which 'A' has been replaced by a secondary condition approximating to the original 'a'. This might appear to be a breach of Dollo's Law in its more ambitious guise, as the Principle of Irrever-sibility of Evolution, but is not necessarily, or even usually, a breach of our more restricted form of the law. A botanical example can be drawn from the seed-plants, in which Cycadaceae preserve a primitive unisexual and perianth-less type of flower; bisexual flowers with a perianth arose in an ancestor of the Angiospermae, but some of the latter have reverted to unisexual and often perianth-less flowers, e.g. Salicaceae. Characters liable to secondary loss are those involving the gain of new structures, or the elaboration of old ones; our version of Dollo's Law indicates that where the derivative character involves the loss of something, it is not usually reversible.

The botanist, K. R. Sporne [178], has suggested that primitive characters should show a positive statistical correlation with each other in their incidence in a particular group—so that it would only be necessary to establish that one character was primitive in a given group, and it could then be deduced that any other characters whose incidence showed a positive correlation with this one would be primitive too. As suggested elsewhere in this work (Chapter 7), I believe that significant correlations in the incidence of different classificatory characters are in the ultimate analysis functionally determined; if a series of primitive characters tend to occur together, this is because all of them are adaptive to a definite, presumably primitive, mode of life. In the animal kingdom, at least, it is clear that characters which manifest positive correlations with primitive characters in their incidence are not necessarily primitive; for example, the presence of lungs (as opposed to air-bladders) in Vertebrata would be strongly correlated with that of tetrapod limbs, and the presence of the reptile-like 'dromaeognathous' palate in birds would be correlated with reduced functionless wings. Among dicotyledons, the correlation demonstrated by Sporne between the woody habit and unisexual perianth-less flowers is probably comparable to these zoological examples. Another botanist, Stebbins [179] has reached conclusions differing from Sporne's and more similar to my own, from essentially the same data. In general, I think Sporne may be right in supposing that primitive characters have a tendency to occur together—an organism which can be shown to be primitive in one respect will be rather more than usually likely to be primitive in others. This 'rule' will, however, be subject to many exceptions; once again it is manifest that 'rule of thumb' methods are not reliable in natural history.

The already mentioned doctrines of I. W. Bailey and his co-workers concerning the phylogeny of xylem elements call for some critical consideration here. No doubt the impressive-looking amount of experimental and observational work carried out by Bailey's school has won for its theories a degree of acceptance which is hardly justified —very much as the important studies of S. M. Manton on the loco-motor apparatus in Arthropoda have ensured for her phylogenetic theories an over-respectful hearing from the scientific community at large.

The Bailey school claim to have established three phylogenetic principles: (i) that evolutionary advances are manifested first in the secondary xylem, then in the metaxylem, and finally in the protoxylem —thus they expect a sort of Haeckelian recapitulation in the ontogenetic development of the xylem; (ii) that, in the terminal joins between conducting elements (tracheids, vessels, and perhaps also sieve-tubes), the primitive condition is always the presence of a very oblique scalariformly-perforated partition, from which condition more or less transverse partitions with a simple round perforation are always ultimately derivable, while scalariform perforations can never be derived from the round ones—perhaps a special case of Dollo's Law; (iii) that the general evolutionary trend in Angiospermae is towards shorter and shorter individual xylem elements. Zoological experience suggests that generalisations of this kind will be subject to various qualifications, but there may be a good case for the provisional acceptance of these ones. However, provisional assent to Bailey's principles does not necessitate acceptance of all the phylogenetic theories which his followers have advanced.

The first of these we may quote from Eames [52], 'Vessels in Angiosperms arose without question from scalariformly-pitted tracheids. Such an origin makes impossible the derivation of Angiosperms from Gnetales (*sensu lato*) or from other higher gymnosperms that have tracheids with only round-bordered pits.' The critical systematist will note that this generalisation about the Gnetales is based only on the three recent genera, fossil woods of the group not being as yet satisfactorily recognised. Existing Gnetales (*sensu lato*) comprise the monotypic *Welwitschia*—which could be described as an enormous, lignified, neotenic seedling—the few species, mainly woody climbers, of the genus *Gnetum*, and *Ephedra* with a number of species of shrubs of the 'switch plant' type. The three genera differ widely from each other, and each evidently represents a highly specialised type. A Mesozoic ancestor of the group might well have been a tree of more or less normal habit and could perfectly well have had scalariformly-pitted tracheids (and vessels) like its presumed ancestors among the Bennetitales. Those angiosperms which most closely approach *Gnetum* and *Ephedra* in habit (there is nothing really comparable to *Welwitschia*), e.g. *Lardizabala* and Tamaricaceae, likewise do not have scalariformly pitted partitions.

A second doctrine of the Bailey school is that vessels have developed polyphyletically within the Dicotyledonae. The reason for this assertion is that a few modern dicotyledonous trees have 'homoxylous' wood without vessels, and that these trees nearly all show other features considered to be primitive. Among more 'advanced' dicotyledons, homoxylous wood is practically confined to specialised aquatic or parasitic types. If we assume that the ancestral dicotyledons had vessels appearing only in the secondary xylem, whereas in the higher types they are present already in the metaxylem, secondary loss of the vessels might be expected to occur more readily in the more primitive types than in the advanced ones. Putting it in another way, in the early dicotyledons, the basic features of the group were less developmentally 'canalised' (to use Waddington's term) or 'stabilised' (to quote Schmalhausen) and so more liable to modification than they are in higher types. The systematist will note also that acceptance of the theory that homoxylous wood is a primary feature in the group does not help at all to clarify the classification of Dicotyledonae, on the contrary, it seems to make confusion worse confounded. It is quite possible, however, to consider homoxylous wood to be secondary in angiosperms without rejecting any of Bailey's basic principles.

A third Baileyan dogma, is that vessels have evolved independently, and probably polyphyletically, in Monocotyledonae. Writers of this school (e.g. Eames, loc. cit.) argue that resemblances between such forms as Alismataceae or Butomaceae and Ranunculaceae or Nymphaeaceae merely concern adaptive features of aquatic herbs, and are not phylogenetically significant—yet these same authors cite Alismataceae and Butomaceae as examples of primitive Monocotyledonae. We see here a clear example of the false antithesis between 'adaptive' and 'phylogenetic' characters. There seems to be no great difficulty in supposing that original monocotyledons were aquatic herbs with characters more or less similar to Butomaceae, and that they were derived from dicotyledonous ancestors of more or less Ranalian type. Very possibly, as the Bailey school suggests, ancestral monocotyledons had vessels only in the roots, the later extension of them into the stems being no doubt connected with the development into larger and terrestrial plants.

It is common in both plants and animals for a particular classificatory character to be distributed through a group in such a way that it is hardly possible to postulate for all the forms possessing it a common ancestor which is not common to any of the forms lacking it. This would apply, for example, to such characters as pinnate leaves or absence of stipules in Dicotyledonae, or to the reduction of the toes on the front leg from five to four in tetrapod vertebrates. In such cases, there may be two distinct types of possible explanation, which are not necessarily mutually exclusive. The first, and currently most popular, explanation of such 'patchy' distributions of derivative characters is the appeal to *polyphyly*—to the independent development of the same character on more than one distinct evolutionary line. There can be little doubt that this is, in many cases, the true explanation—for example, for the peculiar distribution of zygomorphic flowers in Dicotyledonae, of hooves in living and fossil mammals, and for the reduction of the toes on the front limbs of tetrapods. In cases like that of the compound leaves, however, the question is more complex, and it may be necessary to postulate the secondary loss of the character in some cases (e.g. in *Hydrocotyle* among Umbellifera and *Hedera* among Araliaceae). Where a systematic character is constituted essentially by the loss of something, e.g. of stipules, or of one of the toes on the front legs, Dollo's Law may be cited against the likelihood of the recovery of the lost structure. Where the new character is essentially the gain of some new structure, e.g. of thorns on the stem or leaves, or of adhesive lobes on the tarsal segments of insects, I know of no *a priori* reason for supposing that its secondary loss is any less likely than its polyphyletic origin. Indeed, there are good reasons for supposing that the 'facilis est descensus ad averno, sed retrograre gradum' principle applies to organic evolution, and that the loss of structures occurs much more frequently and readily than their gain in the course of evolution.

The proper application of Dollo's Law in such cases is at present a highly controversial matter. There are some zoologists, and many more botanists, who either ignore the 'law' altogether, or consider it to be subject to so many exceptions as be of little practical use; on the other hand, quite a number of zoologists are unwilling to admit any real exceptions to it at all. Both these opposing attitudes stem from a misconception of the nature of generalisations in natural history, a desire to put them on the same basis as the laws of natural philosophy. If and when an exception to a physical law is discovered, physicists normally consider that the 'law' in question must be either abandoned, or reformulated in such a way as to eliminate the exception. If the term 'science' is restricted to cover only those branches of knowledge which can be dealt with entirely in terms of laws of the natural-philosophical type, then natural history in general, and systematic botany and zoology, in particular, are not 'science'. If science is defined in such a way to exclude natural history, the latter is not thereby abolished; whether it is 'science' or not, particular knowledge about plants and animals will retain its importance in relation to human activities, and it will remain necessary to organise this knowledge in some way. Generalisations such as Dollo's Law play an important part in this organisation. In principle, it is always possible to render such generalisations exception-free if sufficient restrictions are embodied in their formulation, but this almost invariably results in making the applicability of the law so narrow as largely to deprive it of practical utility.

Practically all zoologists will admit that the loss of such complex characters as pentadactyl limbs in vertebrates, and wings in pterygote insects is an irreversible process; many botanists might be prepared to concede as much for the loss of the corolla in angiosperm flowers, or of true leaves in Cactaceae or Euphorbiacaeae. Most zoologists would also admit the applicability of Dollo's Law to less complex organs like particular teeth in mammals, or reductions of the tarsal segmentation in insects, fewer of them would assert with any confidence the irreversibility of the loss of a particular seta on the thorax of *Drosophila*, or the reduction of the number of segments in a beetle antenna from eleven to ten. From my own systematic studies in the very large and diversified insect order Coleoptera, I feel fairly sure that in the evolution of this group, changes have at times occurred which might be considered as violations of Dollo's Law. There are indications in several different families of the development of additional veins in the hind wings, for example, in opposition to the normal evolutionary trend to reduction in the wing-venation. Similar phenomena are reported from the Diptera by Hennig [92]. Also, it is difficult to devise a phylogenetic scheme for the families broadly related to the Elateridae which is consistent with other known facts and which does not involve the secondary redevelopment of a closing apparatus (otherwise lacking in the entire group) in the larval spiracles of Buprestidae and of Eucnemidae–Trixagidae. In such cases, the organs concerned are not of a very high degree of complexity, though at least the closing apparatus of the larval spiracles is of considerable phylogenetic antiquity—it probably goes back to the very beginning of the Pterygota stock of insects, perhaps in the Mississipian (Lower Carboniferous) period.

Broadly speaking, it may be said that any positive character (i.e. not the loss of something) which is acquired as an adaptation to a specific mode of life will be liable to secondary loss if that particular mode of life is abandoned in the course of subsequent evolution. Although most recent writers on organic evolution would accept this generalisation, I believe that all too many of them have failed to apply it adequately in practice. Thus, nearly all botanists would agree that a specialised non-green perianth represents an adaptation for insect pollination of flowers, yet numerous systematists, following Engler and Prantl, have been very loath to admit that the wind-pollinated 'catkin' type of flowers found in Amentiferae, etc., may result from the reduction of insect-pollinated flowers. Those who have discussed the evolution of aposematic and similar appearances and displays in insects (e.g. Blest [16]) have generally attempted to explain their 'patchy' distribution in systematic groups in terms of polyphyletic evolution rather than of secondary loss.

An interesting recent review by Stebbins (in [49a]) considers the question of reversibility and irreversibility in structural evolution of higher plants. He presents evidence indicating that the numbers in meristic characters have at times been subject to evolutionary increase as well as decrease, for example, in numbers of leaflets of compound leaves, numbers of floral parts, etc. Experience in the animal kingdom indicates that, while increases in number in meristic characters have occurred occasionally, decreases have been much more frequent, and I suspect that this will prove ultimately to have been true in plants as well. Stebbins suggests that the connation of parts (e.g. in gamopetalous corollas) and the development of zygomorphy represent evolutionary changes which are rarely if ever reversed. In discussing zygomorphy, Stebbins refers to *Plantago* as an example of actinomorphic flowers arranged in a spike—but this instance might well tell against his general argument, as Hallier long ago suggested that Plantaginaceae might be a direct offshoot from the zygomorphic Scrophulariaceae, and a good deal of recent chemical evidence (see Chapter 8) supports this conclusion.

Stebbins also supports the Baileyan doctrines, discussed previously, concerning supposedly irreversible changes in the evolution of xylem elements, which I think are by no means convincingly proved by empirical evidence.

Evidences for phylogeny, and thus for phylogenetic classification, in addition to the comparative anatomical ones considered in this chapter, will be discussed in subsequent chapters. The fossil evidence has already been considered in Chapter 6, in connection with the classification of fossils; as our knowledge of fossils is almost entirely structural, the main principles involved in drawing evolutionary conclusions from them are those developed in this chapter. When comparative anatomy is supplemented by the study of geographical distribution (Chapter 11), of host–parasite relations (Chapter 10), of developmental stages (Chapter 14), of proteins and serology (Chapter 13), and biochemistry and behaviour (Chapter 8), when evidence from all these fields is collated and functionally interpreted, I believe we can hope to achieve quite a high degree of understanding of evolutionary history, even in the absence of a fossil record.

10 Hosts, Parasites and Classification

As a rule, parasites are smaller and less complex organisms than their hosts, so that the natural relationships of the hosts might be expected to be better understood than those of their parasites—in which case we should look to the classification of the hosts to clarify the affinities of the parasites rather than vice versa. In practice, systematists working with parasitic organisms usually treat the hosts of a particular group as a character of more or less systematic importance; they are apt to err at times through attributing too much weight to such evidence. At the species level, parasitologists are only too likely, when they find a parasite of a particular genus in a host from which this parasitic genus has not previously been reported, to describe the parasite as a new species; they are almost certain to do so if the parasite in question shows the slightest phenotypic difference from known related forms. This is despite the fact that phenotypic characters in some species of parasites have been shown to be influenced by the species of host (or even the precise part of the host's body) in which they have developed; known examples of this are the Liver Flukes (*Fasciola hepatica*) reared in ungulates and in rodents, and various parasitic Hymenoptera from different insect hosts. Parasitologists are also liable to be influenced by information about the hosts, to a greater degree than they are often willing to admit, in their placing of species in genera and higher groups.

In recent times, much publicity has been given to parasites as indicators of the relationships of their hosts (see, for example, [155]). In justification of this, it is frequently suggested that parasites evolve more slowly than their hosts, so that relations which have been obscured by recent rapid evolution in the hosts may still be apparent in their parasites. The host-selection of parasites is undoubtedly apt to be influenced by chemical–physiological features of the hosts which are not used as data for ordinary classification; it is often, and almost surely wrongly, assumed that such characters are more 'fundamental' and evolutionarily conservative than are visible structures. This is but one manifestation of a currently widespread misconception, that chemical phenomena in organisms are somehow more fundamental than structural adaptations or habits.

Evidently, if the established classifications of parasites are to be used as evidence for the relationships of their hosts, there is serious danger of a circular argument. In particular, vertebrate zoologists have, as a rule, too little systematic knowledge of the parasites to do anything but accept uncritically the current classifications devised by helmintho-logists, entomologists, etc.

Various generalisations about systematic relations between hosts and parasites have been published, for example, Fahrenholz's rule, Szidat's rule and Eichler's rule [see 178*a*], all of them based mainly on specialist work on particular groups. For none of these rules has general applicability been adequately demonstrated, and it does not seem worth while to discuss any of them in detail here. The soundest procedure is probably to consider the hosts and evolution of any particular group of parasites as a unique and particular problem—'rule of thumb' methods should be as suspect in natural history as they are among the best historians of humanity.

Most parasites, among both plants and animals, are more or less restricted in their choice of hosts—a given species of parasite occurring naturally only in members of a particular genus, family or some such classificatory group. Furthermore, it is common for a related species of parasite to affect a related group of hosts—to the extent that in certain cases the classificatory hierarchy of the parasites more or less closely reflects that of their hosts. There are two distinct, though not mutually exclusive, possible explanations of this state of affairs. One is that the mode of transmission of the parasites is such that they will almost inevitably be transferred from a given host to another of the same species. The other possible determining factor is a close behavioural, physiological or structural adaptation of the parasite to its host, with the effect that it becomes unable to survive on hosts which, in certain critical respects, differ notably from the usual one.

Where the mode of transmission is such that the parasites are almost inevitably passed from one host to another of the same species, the evolution of the parasites may be expected to parallel that of their hosts, and the natural classifications of parasites and hosts should be largely 'congruent'—the system of either could be used as a check on that of the other. Where the host-limitation of the parasites is attributable to physiological-structural adaptations, the problem is likely to be more complicated—the host-records of the parasites may still be a useful indicator of similarities, but hardly an infallible guide to evolutionary relationships.

There is probably no large group of parasites whose evolution has completely paralleled that of their hosts; in the long run of phylogeny *all* types of parasites are liable to occasional transmission to radically new kinds of host. There is evidence, for example, that predators may acquire lice (Anoplura) from their prey; the lice of rodents seem to have been a major source of infestation among other groups of mammals. Cuckoos seem to have acquired Mallophaga from their hosts, and there are indications that lice may at times be transferred between quite unrelated animals which may share a common burrow, or otherwise be brought into unusually close contact.

In most parasitic plants, for example fungal groups like Uredineae and Ustilagineae, and seed plants such as Loranthaceae and Orobanchaceae, host specificity is probably attributable in most cases to physiological adaptations; the methods of distribution of spores, conidia or seeds are not usually such as to impose a narrow restriction on possible hosts. Exceptions to this generalisation might be found among those fungi whose spores are carried by specific oligophagous insects, e.g. the notorious *Ceratomostella ulmi* transported by the Elm bark beetle, *Scolytus destructor*. Unselective means of distribution, comparable to those of most plants, are found in many groups of parasitic animals, such as the Platyhelminthes and Nematoda. In all such groups, eggs, seeds or spores are produced in very large numbers, and only very few of them normally succeed in infecting the right species of host. For such parasites, the natural hosts must be similar to each other in certain specific, critical features (structural or physiological)—such similarity being often, but by no means always, an indicator of phylogenetic relationship.

There is at least one group of parasitic fungi—the Laboulbeniaceous ectoparasites of insects—in which the mode of transmission is probably analogous to that in the previously mentioned lice of birds and mammals; but in the present [see Thaxter 188*a*] state of knowledge of the Laboulbeniaceae it is not yet possible to say to what extent the natural system of these parasites reflects that of their hosts. In the case of the biting lice (Mallophaga) of birds and mammals, as in the sucking lice

(Siphunculata) of mammals, both the hosts and the parasites have been studied fairly intensively, and these groups provide some of the best examples of the possible classificatory value of host-parasite relations.

The rusts (Uredinales) are a group of parasitic fungi whose host-relations may have considerable systematic interest. The life-cycle in them commonly, and probably primitively, involves alternations of two distinct hosts and two different modes of reproduction, phenomena perhaps analogous to those displayed by plant-lice (Aphididae) in the animal kingdom. The 'aecial' stage of a rust produces male and female sexual cells and usually manifests self-sterility, while the 'telial' stage produces basidia and teliospores. In his valuable review of the genera of the group, Cummins [38] states that 'Host restriction may, in heteroecious species, apply to both phases of the life-cycle or only to one phase. . . . . While such restrictions may be taxonomically useful, because of the short-cuts to identity which they provide, there is nothing inviolable about them. Whether or not exceptions are known, there is always the possibility that such occur'. Cummins defines and distinguishes his genera purely on structural features of the reproductive organs; if this system does not agree fully with groupings based on the aecial or telial hosts, this does not prove that the hosts have no value as a systematic character—non-congruence (see Chapter 16) would be expected between such different characters. I think it is likely that, for example, serological studies will show that in some cases the nature of the host-plants will prove to indicate relationships which are not evident in the structures used diagnostically by Cummins. The rust fungi may also prove to have value as indicators of relationships among their hosts.

As a first detailed example we may consider the lice affecting mammals of the order Primates, an account of which may be found in the work of Hopkins [96a]. *Homo sapiens*, as is well known, supported—and still supports in many parts of the world—two species of Siphunculata of the family Pediculidae—*Pediculus humanus* (including the sub-species *corporis* and *capitis*) and *Phthirus pubis*. With the exception of species of *Pediculus*, exceedingly similar to *humanus*, found on one or two New World monkeys in captivity, all known species of Pediculidae have been found exclusively on Old World monkeys (Catarrhina). Hopkins suggests, very plausibly, that the *Pediculus* found on New World monkeys represent secondary infestations derived from *Homo sapiens*. Apart from this, the species of *Pediculus* have been found only on Anthropoidea (chimpanzees, gibbons, etc.), as is also the case for *Phthirus*. The tailed Old World monkeys (Cercopithecidae) support sucking lice of a distinct genus, *Pedicinus* which has certain similarities to *Pediculus* and has been placed by most authorities (though not by Ferris) in the Pediculidae. The relations of Pediculidae are rather uncertain; the genus *Pedicinus* has been considered by Ferris to show affinities to the Hoplopleuridae, which nearly all parasitise rodents. There is no evident connection between the lice of the Catarrhina and those affecting New World monkeys, lemuroids or tupaioids; the three last-mentioned groups do, however, seem to be remotely linked by their ecto-parasites.

New World monkeys (Platyrrhina or Ceboidea) significantly resemble other old Neotropical groups or mammals in having no endemic sucking lice of their own; they do, however, have distinctive Mallophaga, the most interesting of which is perhaps the genus *Cebidicola*, with one recorded species on howler monkeys and another from a species of spider monkey. *Cebidicola* is considered to show definite affinities to *Lorisicola*, with a single recorded species from an Asiatic *Loris* (Lemuroidea). The lemuroids of Africa and Madagascar have specific Siphunculata—*Phthirpediculus* from the true lemurs of Madagascar, and *Lemurphthirus* from African Galagos; Madagascan lemurs carry also a peculiar and isolated mallophagan genus *Tricho-philopterus*. The relationships of *Phthirpediculus* and *Lemurphthirus* are somewhat obscure; they are placed by Ferris in the Hoplopleuridae-Polyplacinae and do not appear to have any particular connection with *Pedicinus* (placed by Ferris in an independent subfamily of Hoplo-pleuridae) or other Pediculidae. *Phthirpediculus*, however, has some similarity to *Docophthirus*, reported from an Asiatic tree-shrew (Tupaii-dae), a representative of a group often considered to be remotely kindred to lemuroids; *Hamophthirus*, from the 'Flying Lemur' *Galeo-pithecus*, may also be related to these.

The Pediculidae (including *Pedicinus*), it appears, may have evolved on and with the Old World monkeys since the latter group lost effec-tive contact with its New World cousins, i.e. probably not before the Oligocene period. The data strongly suggest that the original lemuri-form ancestors of higher Primates had no sucking lice (Siphunculata) though they may have carried Mallophaga akin to *Cebidicola* and *Lorisicola*. The relatively deep-rooted (possibly Oligocene) divergence between the great apes (Anthropoidea) and the tailed monkeys (Cercopithecoidea) may be reflected in the considerable differences

E

between *Pedicinus* and *Pediculus* (or *Phthirus*). Mallophaga are now generally considered to be older than, and ancestral to, the sucking lice; the geographical and host distribution of *Cebidicola* and *Lorisicola* would suggest that the common ancestor of the two was probably in the early Eocene period, when sucking lice may not have been in existence. The absence of any records of Trichodectidae (to which family *Cebidicola* and *Lorisicola* belong) from African or Madagascan lemuroids is, however, not easily explained on this theory. *Trichophilopterus* is a peculiar type, possibly independently derived from a bird-mallophagan and not directly related to the Trichodectidae (all of which parasitise mammals). If the Siphunculata are as recent in origin as we have suggested, the common ancestry of *Phthirpediculus, Lemurphthirus, Docophthirus* and *Hamophthirus* must be more recent than that of their hosts. It is notable that *Phthirpediculus* seems to be the only member of the Siphunculata recorded from any endemic Madagascan mammal.

The supposedly recent origin of Siphunculata is supported also by the absence of any members of this group specific to marsupials or monotremes, or to the older types of Neotropical mammals, and by the difficulties it presents to the systematist—genera and higher groups being not very well defined and classificatory characters markedly labile. The marsupials, incidentally, have specific mallophagan parasites, but these, both in Australia and America, belong to a distinct major group (Amblycera) which is probably of older origin than the Trichodectidae (or *Trichophilopterus*); no specific anopluran parasites have yet been reported from the Monotremata, whose separate ancestry certainly goes far back into the Mesozoic period and may well antedate that of Anoplura altogether. From these and similar considerations, it appears that the family-tree of lice proposed by Symmons [186] is more plausible than most others, and that the generally accepted classifications of the group are by no means phylogenetic in the sense of Hennig.

It will be noted that in this case our relatively well-based knowledge of mammalian phylogeny is used to elucidate the evolution of lice rather than *vice versa*. Many ornithologists have recently looked to the classification of Mallophaga (Philopteridae) to elucidate the relations of their bird hosts, thus betraying a touching faith in the superior classificatory understanding of their entomological colleagues. As an entomologist myself, I have regretfully to confess that this faith is not very well founded—it being very doubtful whether current classifications of Philopteridae are any more 'natural' than are those of their hosts. If anything, I should say that it is the relationships of the Mallophaga which are likely to be clarified by further studies of the birds rather than *vice versa*. The host-distribution of avian Mallophaga has been well discussed in various papers by Clay [31]. Undoubtedly, the data presented in her papers will be of some value for systematic ornithology, especially if considered in conjunction with other lines of evidence such as comparative serology.

In the case of the Anoplura, it appears that the unduly high classificatory status hitherto given to the group by most entomologists has been a source of error in attempts to use the group as evidence for relationships among its hosts. The lice have generally been treated as an independent order (Anoplura or Phthiraptera) or even two separate orders in systems of the Insecta. Orders of insects are usually old groups, at least as old as Aves or Mammalia in Vertebrata; there has thus been a tendency to assume that the lice must have originated far back in the Mesozoic (or even in the Palaeozoic) era, and have evolved *pari passu* with their mammalian and avian hosts practically from the beginning of those groups. If this were the case, we might expect to find the most isolated and primitive of existing lice on ratite birds (or tinamous) and on Monotremata; in fact, there is no indication that the Mallophaga found on Ratitae or Tinamiformes are in any way more primitive than those of Gallinaceae for example, and no specific lice at all have been found on Monotremata.

If, however, as suggested by Symmons, the lice represent merely a comparatively recent parasitic development from the Psocoptera, they may well be younger than Mammalia (or even Aves); a quite late Cretaceous origin for the lice seems perfectly conceivable. As we have seen, the Mallophaga are doubtless older than the Siphunculata, and within Mallophaga the Amblycera are presumably older than the Ischnocera. If the Siphunculata did not appear till late Eocene or Oligocene times, as we have suggested, they would be of later origin than most of the orders of mammals, and could hardly be expected to provide valuable evidence for mammalian relationships at ordinal or subordinal levels.

The Mallophaga would doubtless be somewhat younger than the Aves, and probably younger than the separate ancestry of Ratitae, so that existing ratites have probably acquired their Mallophaga at some stage from more normal flying birds. This group of parasites should thus not be expected to provide conclusive evidence on such questions as the relationships of the New Zealand Kiwi (*Apteryx*).

The fleas (Siphonaptera or Aphaniptera) are fairly certainly an older group than the Anoplura; for various reasons the view once held, that they arose as a parasitic development from true Diptera, is now considered to be untenable. It appears rather that fleas are phylogenetically entitled to the status of a distinct order, whose origin must have lain somewhere among the pre-dipterous types of the 'Panorpoid complex' in the Triassic period. On this assumption, the beginnings of the Siphonaptera must have been at least very early in the Jurassic, and more probably in the Triassic period, probably more or less synchronous with the appearance of mammals which were no doubt the ancestral hosts of fleas. Unlike the lice, fleas have a fossil record, one or two representatives of the group having been found in the Baltic Amber, of early Oligocene or late Eocene age. These amber fleas are attributed to extinct species but appear to be close to some existing types parasitising Insectivora. They do not seem to be notably more primitive than existing Siphonaptera, and support the assumption that the origins of Siphonaptera lie very much further back.

Larval fleas do not occur on the body of the host at all (though usually living in its nest) and the adults of most species are highly mobile and readily transferred from one host to another, unlike the lice. Many of the species do not seem to be closely adapted to particular hosts. In the circumstances, a close correspondence of the classification of fleas with that of their hosts would hardly be expected, nor do we find it. In considering the relationships of fleas, geographical distribution is frequently as valuable a guide as the nature of the hosts. It is noteworthy that most of the fleas affecting marsupials belong to groups which are confined to the Neotropical and Australian regions, e.g. Stephanocircidae and Pygiopsyllidae. The Stephanocircidae, with two subfamilies, one in each of these two regions, provide a striking parallel to the mallophagan Amblycera—the Australian Stephanocircinae corresponding to Boopidae, and the Craneopsyllinae to Gyropidae, both in distribution and in hosts. The family Ischnopsyllidae, as far as is known exclusively parasitic on bats, has two subfamilies, Thaumapsyllinae on Megachiroptera and Ischnopsyllinae on Microchiroptera. This suggests that the Ischnopsyllidae are of similar (i.e. probably early Eocene) age to the Chiroptera and have probably evolved with them.

Little attention has so far been paid to the host-specificity of parasitic Crustacea, though in some cases, e.g. the Copepoda and Cirripedia-

Rhizocephala available data suggest that it may have some systematic value. For example, the rhizocephalan family Peltogastridae appears to be exclusively attached to hermit-crabs, whereas Sacculinidae are parasites of true crabs; the somewhat controversial relations between these great groups of Decapoda may in some degree be clarified by further knowledge of their rhizocephalan parasites, even though the greatly simplified structure of the latter does not offer a very good basis for natural classification. Conversely, of course, improved understanding of decapod phylogeny may throw valuable light on the evolution of Rhizocephala.

Among the parasitic helminths—Nematoda, Acanthocephala, Trematoda, Cestoda, etc.—host specificity is no doubt mainly attributable to physiological and structural adaptations. In almost all cases, the relationships and phylogenetic history are better understood for the hosts (very largely vertebrates) than they are for the helminths; the study of host specificity in such groups is thus likely to be of more systematic value for helminthology than it is for vertebrate zoology. We have as yet very few useful clues to the antiquity of any of the major groups of parasitic helminths. The problem is discussed however by Llewellyn [187]. It is reasonable to assume that groups which are almost exclusively parasitic as adults on true Vertebrata—for example, the Trematoda and Cestoda—can hardly be older than these, which would establish Silurian or late Ordovician as maximum possible ages for them; the general diversification of structure and hosts in these two groups is such that origins within the Tertiary era are hardly thinkable. The time between the end of the Ordovician and the beginning of the Tertiary is, however, an exceedingly long one. One possible, if slight, indication is that what is believed by some to be the most primitive living type of Cestoda s.str. (i.e. excluding Cestodaria), the genus *Haplobothrium* (representing the monotypic family Haplobothriidae) is a parasite of the fish *Amia*. The latter is the sole surviving type of a group (Holostei) of bony fishes which was numerous in the Mesozoic era and whose ancestry appears to have been separate from that of any other living fishes since the Jurassic period. The forms most resembling *Haplobothrium* among other living Cestoda parasitise more modern types of bony fishes (Teleostei). This suggests that the Cestoda proper may have originated in the Jurassic period, as parasites of the Ganoid precursors of Teleostei.

Among the Trematoda, the typical digenetic flukes (i.e. excluding Aspidogastra) may be of comparable age to the Cestoda proper, if diversity in structure and hosts is a reliable guide. The Digenea probably stand to the Aspidogastra in much the same relation as that of Cestoda s.str. to Cestodaria. Cestoda and Digenea are both richly represented among the parasites of higher vertebrates, birds and mammals, while Cestodaria and Aspidogastra, like Monogenea, are largely confined to the lower aquatic types, fishes and Amphibia. Both Cestodaria and Monogenea have peculiar and isolated members parasitising Holocephali—unquestionably a very ancient and isolated group of cartilaginous fishes, whose separate ancestry can certainly be traced back into the Palaeozoic era. It is reasonable to conclude that the origins of both Cestodaria and Monogenea are likely to have been Palaeozoic; they may plausibly be associated with the great proliferation of aquatic fish-like vertebrates which was at its height in the Devonian period. Aspidogastra may well have appeared rather later, from some monogenean ancestor; most of the existing species are attached to chelonians, a group whose origins are believed to be Triassic—though some Aspidogastra have been found in elasmobranchs, including Holocephali.

The general simplicity of structure of parasitic Protozoa (and a fortiori, of Bacteria and Viruses) makes them rather difficult subjects for natural classification. Parasitologists dealing with Protozoa consequently rely heavily (often too heavily) on the relationships of the hosts as a guide to the classification of the parasites. A few of the more complex and host-specific protozoan parasites, such as the trichonymphoids occurring in termites and cockroaches, and the Protociliata in Amphibia, may, however, be of some interest from the point of view of the classification of their hosts.

The term parasitism is often not considered to include the relationship between a phytophagous insect and its food-plant, but in fact there are considerable analogies between this relation and that of fleas, lice, helminths etc. to their vertebrate hosts. The main difference is that insects are usually much more mobile than their host plants, whereas vertebrate hosts are normally far more mobile than their parasites. Host specificity in phytophagous insects is almost entirely attributable to behavioural, physiological or structural adaptations, and there is normally no means of 'automatic' transmission to another host of the same species; insects actively seek and select host plants. As might be expected, the value of phytophagous insects for plant systematics is more comparable with that of helminths than with that of lice for vertebrate classification.

It is very common for herbivorous insects of a given species, living in a given area, to feed only on one species, or a few closely related species, of plants; the insects, if brought into contact with them (by introduction either of new plants into the area or of the insects into new areas) will often eat other related plant species which do not occur naturally in the given area. There are, however, a number of instances known where a given insect species has two alternative food-plants which are systematically remote from each other—a good instance of this is the Brown Argus butterfly (*Aricia agrestis**) whose larvae eat either Rock-Rose (*Helianthemum*—Cistaceae) or Stork's Bill (*Erodium* —Geraniaceae). In such instances it is to be presumed that the insect selects its host plants by some single, often by no means obvious, presumably chemical, feature in which they resemble each other; the subject is discussed more fully by Schoonhoven [164a].

Herbivorous insects often seem to be 'good botanists' in the sense that they 'recognise' relationships which are by no means obvious to the uninitiated, e.g. that of Ash (*Fraxinus*) to such Oleaceae as Privet (*Ligustrum*) and Lilac (*Syringa*). In a number of cases the relation between alternative host-plants of a particular type of insect is problematic, for example, that between Cruciferae and the genus *Tropaeolum* (Tropaeolaceae) which, as many gardening readers will know too well, is an alternative food-plant for some of the 'Cabbage butterflies' (*Pieris* spp.), as well as for the crucifer-eating weevil *Ceuthorrhynchus contractus*. Up till now, most botanical systematists have placed *Tropaeolum* in or near the Geraniaceae rather than in Rhoeadales with Cruciferae, but increasing uncertainty about the natural position of the genus has been manifest in recent works. It may be added that larvae, not only of *Pieris* spp. but also of the large majority of Pierine butterflies, feed either on Cruciferae or on the undoubtedly closely allied Capparidaceae. Caterpillars of the 'Bath White' (*Pontia daplidice*) have been reported as eating *Reseda* as well as various Cruciferae, a relationship which is supported by *Phyllotreta* flea-beetles and by the beetle genus *Bruchela* (*Urodon*—Anthribidae)—the larvae of which live in the seed capsules and all of whose North African species reported on by Peyerimhoff [152] are attached either to *Reseda* spp. or to Cruciferae. In floral characters Resedaceae and Capparidaceae show much more similarity to *Tropaeolum* than is apparent in Cruciferae themselves.

* The species named appeared as *agestis* in the original definition, an obvious misprint which under the current Zoological code of nomenclature acquires the force of law.

The butterflies may in this instance be testifying to a genuine phyletic relation which has been largely overlooked by botanists. In the case of *Pieris* species there is experimental evidence that egg-laying by the female and normal feeding by the larvae are stimulated by mustard oil; the butterflies will lay their eggs, and the caterpillars will feed, on plants of many orders if their leaves are brushed with mustard oil. Mustard oil, and the correlated enzyme myrosin, have been found in Cruciferae, Capparidaceae, Resedaceae, one or two small related exotic groups, and in the genus *Tropaeolum*.

Another instance in which the evidence of phytophagous insects may be systematically important concerns the family Scrophulariaceae. Most field botanists in Europe will have seen plants of *Scrophularia* more or less severely eaten by slimy larvae—these will be of species of *Cionus* or *Cleopus*, belonging to the tribe Cionini of weevils (Coleoptera-Curculionidae). The natural food-plants of all the European, North African, and Canadian species of *Cionus* and *Cleopus*, as far as recorded, are species of *Scrophularia*, *Verbascum* and *Celsia*, in many cases two or all three of these genera may be attacked by the same species. *Cionus scrophulariae* has been reported also as feeding on the South African *Phygelius* (Scrophulariaceae) in gardens. No less than three different species of *Cionus* have, however, been reported as sporadically attacking cultivated species of *Buddleia*—exotic shrubs usually classified in Loganiaceae. Beeson [11] reported an Indian species of *Cionus* as feeding on wild *Buddleia*, and another on *Dolichandrone* (Bignoniaceae); *Cionus radermacheri*, in Indonesia, is said to breed normally on *Radermachera* (Bignoniaceae). The reported food-plants of other European genera of Cionini are as follows: *Stereonychus* has a species feeding on foliage of Ash and *Phillyrea* (Oleaceae), another in Europe and a third in the Canary Islands both attached to *Globularia* (Globulariaceae); *Cionellus* has a species feeding on *Phillyrea* (Oleaceae). The known food-plants of Cionini agree only partially with the established classification of the tribe but are very suggestive botanically.

Not very distant from Cionini in the established system of Curculionidae are the Mecinini, with three European genera, *Mecinus*, *Gymnetron* and *Miarus*; in important structural characters the first two of these are nearer to each other than either is to *Miarus*, and in at least one of these respects (connate tarsal claws) they resemble Cionini. The known food-plants of European *Miarus* spp. are all species of *Campanula*; a Japanese species of the genus is reported from *Lobelia*. In both *Mecinus* and *Gymnetron* there are some species living on *Linaria* and allied Scrophulariaceae, and some feeding on *Plantago* species, a few *Gymnetron* being attached to *Veronica*, *Verbascum*, *Scrophularia*, etc. The Mecinini clearly indicate some special similarity between Scrophulariaceae and Plantaginaceae, and possibly a more distant link with Campanulaceae.

The caterpillars of moths are, as a rule, less selective in their food-plants than are the beetles we have been considering. In the genus *Cucullia*, however, several species have larvae which in nature seem to feed exclusively on *Scrophularia* and/or *Verbascum*; the remaining British species of the genus all feed on Compositae. Another noctuid moth, the 'Frosted Orange', *Gortyna flavago*, has the Scrophulariaceae and the Compositae favoured by *Cucullia* as alternative food-plants. These Noctuidae indicate some particular similarity between Scrophulariaceae and Compositae. A tortricid moth, *Phalonia degeyrana*, has *Linaria* spp. and *Plantago* spp. as its alternative larval foods, thus supporting the evidence of Mecinini.

The herbivorous insects we have considered so far have not included leaf-miners or sap-suckers. Of the European leaf-mining insects reported to attack Scrophulariaceae, the species of Diptera and Lepidoptera are either restricted to the order (like the agromyzid *Dizygomyza verbasci*) or so widely polyphagous as to be of little indicative value. However, the two British species of the beetle genus *Apteropeda* (Chrysomelidae) appear to be restricted, as far as larval mines are concerned, to Scrophulariaceae, Labiatae and *Plantago*.

Among the sap-sucking Hemiptera, we may note the aphids, *Aphis verbasci*, whose recorded host-plants are *Verbascum* and *Buddleia* spp., and *Sappaphis gallica* which feeds on *Linaria* and *Antirrhinum* and is said to be more or less indistinguishable from *S. plantaginea* from *Plantago* spp. The reported food-plants of our four British species of shield-bugs of the genus *Sehirus* (Cydnidae) are as follows: *S. bicolor*, dead nettles (*Lamium*) and other Labiatae; *S. dubius*, *Thesium humifusum* (Santalaceae); *S. biguttatus*, *Melampyrum* spp. (Scrophulariaceae); *S. luctuosus*, *Myosotis* spp. (Boraginaceae).

Of the affinities which our insect evidence suggests for Scrophulariaceae, two are generally recognised by botanical systematists—those to Globulariaceae and Bignoniaceae. The Cionini, the only insect group providing evidence for these non-controversial affinities, testify even more strongly to affinities to *Buddleia* and Oleaceae which have been far less generally recognised. Many recent systematists have expressed doubts about the naturalness of the family Loganiaceae in which *Buddleia* is normally placed; Hutchinson [101] associated it with Oleaceae, while most others have placed it near Rubiaceae. We may note here that in India several hawk-moths of the genus *Macroglossum* have the loganiaceous *Strychnos* as an alternative food-plant to various Rubiaceae. In the phylogenetic system of Hallier [82], Oleaceae are considered to be directly derived from Scrophulariaceous ancestors, and both *Plantago* and *Globularia* are actually included in Scrophulariaceae; Loganiaceae and Labiatae are included in the same group Tubiflorae as these, and Bignoniaceae are considered to share the same Scrophulariaceous ancestors as Oleaceae.

Thus far, the evidence of the insects agrees remarkably well with Hallier's theories, but in suggesting remoter links to Boraginaceae, Campanulaceae and Compositae the insects support older conceptions of the unity of the 'Gamopetalae' which Hallier rejected.

A relationship of Scrophulariaceae which most systematists (including Hallier) have accepted, but which receives hardly any support from the insects is that to Solanaceae. One possible pointer in this direction is the food-plant selection of certain large hawk moths; in India the caterpillars of several species of *Acherontia* are recorded [12] to feed on various Bignoniaceae, Pedaliaceae, Verbenaceae and Labiatae as alternatives to their more normal solanaceous and convolvulaceous food-plants, and comparable phenomena are displayed by South American species of another sphingid genus *Protoparce*. The insects do not, of course, disprove a direct connection between Scrophulariaceae and Solanaceae; they indicate that if such a phylogenetic relation exists, the two families must have been separated by some change which was of decisive importance from the point of view of most herbivorous insects. It is conceivable that the development of poisonous alkaloids in the ancestors of modern Solanaceae might have been such a change. Note that none of the ordinary insects of Scrophulariaceae eat *Digitalis*.

The host-plant selection of the insects may be of some interest also in relation to the internal hierarchy of Scrophulariaceae. As far as most of the insects are concerned, the important dividing line seems to come between *Verbascum*, *Celsia*, *Scrophularia* and *Phygelius* on the one hand, and the main body of the family (including *Linaria*, *Antirrhinum*, *Digitalis*, *Pedicularis*, *Melampyrum*, *Veronica*, etc.) on the other; only in the genus *Gymnetron* do the host-plants embrace both these sections. Most botanical systematists, on the other hand, segregate *Verbascum* and *Celsia* from all the rest on account of their supposedly primitive and Solanaceae-like floral structure. It may be that the relationship to *Buddleia*, *Globularia*, Bignoniaceae and Oleaceae which the Cionini indicate in the *Verbascum–Scrophularia* group depends on the persistence in these genera of certain primitive features, not necessarily indicative of particularly close affinity, whereas the other group of Scrophulariaceae all descend from a common ancestor which had lost these primitive features and was ancestral also to *Plantago*.

The evidence of the host-plants may also be worth considering by entomologists studying the relationships of the weevil group Cionini itself. Most systematists have placed the tribes Mecinini and Cionini in more or less close proximity, and some have even united the two. The host-plant data we have considered are of definite value in supporting arguments for a relationship between the two groups. Within the Cionini, the evidence of the host-plants is in agreement with most systematists in setting *Cionus* and *Cleopus* rather apart from *Cionellus* and *Stereonychus*. Likewise, within the Mecinini the greater isolation of *Miarus* is indicated by the larvae [56] as well as by adult structure and the food-plants.

The host-plant selection of Cionini and other insects may prove to be relevant to classification within the Oleaceae; several other groups of insects provide evidence of similarities between *Fraxinus* and genera like *Ligustrum*, *Syringa* and *Jasminum*, and thus oppose Hutchinson's suggestion that the family is an unnatural one.

The food-plants of the leaf-beetles (Chrysomelidae) of the genus *Lema* (*sensu lato*) may be of interest from the point of view of botanists as well as entomologists. Species of the monocotyledonous family Commelinaceae serve as food-plants for a large and apparently 'central' mass of the *Lema* species, both in the Old World and in America. A rather distinct section among the Old World species (genus *Hapsidolema* Heinze, *Oulema* auctt.) is attached to Graminae, and there is a parallel but probably independently derived New World group of Graminae-eating species which partially repeats the characters of the Old World *Hapsidolema*. There is an Old World group (*Bradylema*, *Stethopachys*) living on various Orchidaceae, and one or two groups in both the Old World and the New have gone over to feeding on various Dicotyledonae, for example, the Neotropical *Plectonycha* on Basellaceae, a fairly extensive section of New World *Lema* on Solanaceae, and at least our familiar European *L. puncticollis* Curt. (*cyanella* auctt.) on thistles (Compositae). Both the Orchidaceae and the Graminae have been considered by some recent botanists to be more or less related to Commelinaceae. *Lema* and its allies belong to the sub-family Criocerinae, which also includes the genera *Ovamela, Pseudo-crioceris* (*Brachydactyla* Lac.), *Lilioceris, Crioceris* Geoff. and a few other doubtfully distinct ones. As far as is known, practically all of these feed on monocotyledonous plants belonging to the Liliaceae and allied groups (e.g. *Smilax, Asparagus, Dracaena, Dioscorea,* etc.; the most primitive of the forms with a recorded food-plant is perhaps *Pseudocrioceris,* which lives on *Dracaena,* no host-plant seems to have been recorded for the even more primitive-looking Madagascan *Ovamela.* Entomologically, *Lema* and its allies would seem to be more derivative than *Lilioceris,* etc., and presumably derivable from some extinct type of the latter group; if so, it suggests that the ancestors of *Lema* changed their food-plants from some early Liliaceous type to Commelinaceae, probably near the beginning of the Tertiary era. Unfortunately, there seems to be no agreement among botanists about the relationship between Liliaceae (*sensu lato*) and Commelinaceae, though there have been some suggestions of a relation between Commelinaceae and Graminae (cf. Eckardt [53]). It has recently been pointed out that some Commelinaceae have silica deposits in their epidermal cells like Graminae [193*b*].

The work of Ehrlich and Raven [55] is one of the most notable yet published on the evolutionary relationships between insects and their food-plants. One of their conclusions is of particular importance: 'We cannot accept the theoretical picture of a generalised group of poly-phagous insects from which specialised oligophagous forms were gradually derived. Just as there is no truly "panphagous" insect, so there is no universally acceptable food-plant; and this has doubtless always been true'. To put it in another way, the general evolutionary pattern has not been one of progressive restriction in the choice of food-plants, but one of evolutionary divergence with the food-plants, punctuated by switches—occasional in some lines, relatively frequent in others—to new types of food-plant. When such switches occur, the new food-plant will presumably be selected for some particular physiological or chemical similarity to the original one; such similarities may or may not be indicators of phylogenetic relationship between the plants showing them. Changes of range, either in the insect or in the food-plants, may often play a part in such food-plant switches.

11 Geographical Distribution and Classification

There lies the land where rolled the sea
O earth, what changes hast thou seen
                    TENNYSON ('In Memoriam')

The use of geographical distribution as a classificatory character is an old-established practice among museum and herbarium systematists. The specimens with which they deal normally bear some sort of locality label, and in large collections species of the more extensive genera are commonly grouped by geographical regions. More frequently than many of them would be willing to admit, systematists are apt to be influenced in their placing of new species by the geographical origin of the specimens. For example, where a Nearctic genus is very closely allied to a Palaearctic one, a systematist describing a new species from Nepal will probably attribute it to the Palaearctic genus without having made any critical comparisons between it and any of the Nearctic species.

The natural areas of distribution of plants and animals provided one of the strongest lines of evidence for organic evolution, and against the biblical creation story, in the controversies of the nineteenth century; the evidences for phylogeny which they offer are none the less important today, though frequently ignored or misinterpreted. The phylogenetic systematist should be profoundly interested in the natural distribution of his organisms, as it offers one of the most useful non-morphological lines of phylogenetic evidence available to us, particularly in determining the relative ages of groups at the lower classificatory levels.

In a celebrated book [205], the botanist J. C. Willis expounded a theory which, while it undoubtedly contains elements of truth, has only too often been the basis for conclusions which are almost certainly fallacious. Willis postulated that a new species will originate in a single small area, from which its descendants will spread progressively with the lapse of time. In the process they are likely to undergo evolutionary divergence, giving rise to a number of species, to a new genus, and eventually to successively higher groups. If this pattern is general, then the natural area of distribution occupied by a group should be an indicator of its age, and hence of its proper category in phylogenetic classification. In drawing this conclusion, Willis was thinking particularly of seed-plants, and specially of those without such specific long-range methods of seed-dispersal as the pappus of Compositae or the animal-carried seeds of many other groups, but he himself implied that it might be applicable to the less individually mobile types of animals and others have tried to apply it more widely in zoology.

The most obvious objection to the 'Age and Area' theory is that there are many groups, both of animals and plants, for which we have clear evidence that the area of distribution has shrunk rather than expanded with the passage of time. As examples I need only quote the Ginkgoaceae, the genus *Metasequoia*, and various arctic-alpine species in post-glacial Europe, or from the animal kingdom coelacanth fishes, rhinoceroses, and reindeer. A systematic group with a small geographical range may be a young one which has not existed long enough to have had the opportunity of extending its range, or it may be an old one which has become extinct over large parts of its former range; the higher the category of the group, the more likely it is that the latter explanation is the true one. In other words, instead of looking to the geographical distribution to determine the category of a group, we may need to consider its category in order to explain its geographical distribution.

Another defect of the 'Age and Area' theory is that, even where it is applicable, no general rate can be applied to the process of geographical spread, either between different groups in the same period and area, or between different times and areas for the same group. There are few if any species for which the surface of the earth can be considered as a homogeneous continuum; for all species, the world will appear as variegated, with favourable areas across which spread may be relatively rapid, interspersed with areas across which the species can spread only slowly if at all. The pattern of this mosaic has certainly changed, and is still changing, with time. With the merely possible exception (on one geophysical theory) of the major ocean basins, all types of barriers to plant and animal distribution have changed their position in the course of geological time. Furthermore, an effective barrier for one kind of plant or animal may offer little or no obstacle to the spread of another. Thus we are not in general justified in using the comparison of areas of distribution as a criterion for the relative ages of different groups. Only in so far as the ecological requirements, methods of distribution, and places of origin of the two groups are similar are the extents of their natural distribution likely to reflect their relative ages —assuming that in neither case has any significant contraction of range occurred.

Attempts have also been made to establish rules for determining the areas of origin of systematic groups from the patterns of distribution of their present-day members—like the 'Age and Area' principle, such rules have only too often proved misleading. It has been suggested that any group should be considered to have originated in that area where (a) it is represented by the greatest number of existing species, (b) its modern genera are most numerous, (c) it shows the greatest diversity at the higher classificatory levels, or (d) its most primitive living forms occur. These four rules frequently give conflicting results, and in those cases where they can be checked against fossil evidence, none of them appears to be reliable. Once again, we see that the attempt to treat natural history as if it were natural philosophy, to establish general principles from which to proceed deductively, often leads to error. This is not to say that no generally useful principles can be found for making phylogenetic deductions from patterns of distribution of organisms; useful rules of this kind are of a rather different, generally less abstract, kind than the ones we have so far considered. Two of the most general are that (1) any group whose existing representatives are widely separated geographically and not closely related to each other is certainly old, (2) a group which, as a whole and in its successive subdivisions, shows essentially continuous patterns of distribution, is a relatively young one and near the peak of its evolutionary success.

Some of the more isolated land masses of the world (the phenomenon is shown to a less degree by some enclosed water-masses, e.g. the Caspian Sea, Lake Baikal and Lake Tanganyika) offer particularly interesting evidence for the phylogenist, from which some very useful general principles can be developed—particular examples are Australia, Madagascar, New Zealand and South America. The endemic faunas and floras of all these include 'relict types'—last survivors of groups which were much more widespread in former times. Any higher category (e.g. of family or higher level) group which is now confined to one of these areas is almost certainly of a relict nature, as is any group whose living representatives are confined to two or more of these regions. Any group which is represented in Australia by endemic types which do not have their nearest modern relatives in the Indo-Malayan region can with high probability be deduced to go back at least as far as the Eocene period. Similarly, any group whose Madagascan representatives include endemic types whose nearest relatives do not occur in Africa is probably at least of Oligocene age, and probably older than this. For South America we have fairly solid geological (chiefly palaeontological; see Simpson [168]) evidence of a land connection with North America at the very beginning of the Tertiary era—a connection which seems to have been broken by the Lower Eocene period and was not fully reconstituted until the Pliocene. Some animals and plants, however, seem to have been able to reach South America during the intervening period, perhaps by 'island hopping' along a chain preceding the present-day Antilles.

In the case of New Zealand, the absence of any endemic land mammals (other than bats) is evidence against an uninterrupted land connection between it and any of the continents (except perhaps Antarctica) during the Tertiary era; what is perhaps even more surprising is that no trace of terrestrial Dinosauria has been found in the extensive plant- and coal-bearing beds of the New Zealand Jurassic series. The modern (and Tertiary) faunas and floras, however, provide evidence of an incomplete link, presumably in part an island chain, between New Zealand and the southern part of South America during the earlier Tertiary era. The fringes of the Antarctic continent must have formed a major part of this link, and there are indications that in the Eocene and Oligocene periods they were in lower latitudes than today and bore temperate forests, no doubt akin to the *Nothofagus* forests of New Zealand and southern Chile. This link permitted the passage of many plants, terrestrial insects, and other invertebrates, but not of any of the characteristic South American mammals of the lower Tertiary era. Since its effective severance, perhaps in late Oligocene times, the related species in Chile and New Zealand have diverged to such an extent that systematists often place them in separate, though closely related genera. New Zealand plants and animals which have their closest living relatives in Australia, like Madagascar ones whose nearest relatives live in Africa, may be assumed to have descended from ancestors which reached those lands during the Tertiary era, across wide seas.

If we are to deduce phylogeny from present-day distribution, clearly we may need information about the ways in which the geographies of different periods of the geological past would have differed from that of today. Naturally, for this information we look in the first instance to the geologists. They themselves, when trying to construct maps showing the distribution of land and sea, climatic belts, etc., for past periods usually rely heavily on information from palaeontologists, i.e. on data about the former distribution of plants and animals. This information suffers from three main sources of error (see Chapter 6). The first is that the fossil record for any particular type of organism at a particular time in the past is almost bound to be incomplete, the group is almost sure to have occurred in places where it has left no fossils. The second is that accurate time-correlation of terrestrial or fresh-water deposits in different parts of the world is very difficult. The third is, of course, that fossils rarely show all the characters which would be needed for the reliable classification of modern forms. An element of conjecture is usually present in the precise classificatory placing of fossil organisms —conjecture which may be dignified by the phrase 'considered judgement of an eminent authority in palaeontology' but does not thereby escape being affected by that gentleman's preconceptions. If like most authorities in the U.S.A. (at least, until very recently), he accepts the doctrine of the permanence of the continents and oceans, he is likely to use his judgement in such a way as to minimise either the closeness of the relations between, or the indications of warm climate in, the Lower Tertiary faunas and floras of Europe and North America.

Similarities which one authority accepts as evidence of genetic relationship may be interpreted by another as examples of 'convergence' between organisms occupying similar ecological niches in different continents. The accepted classifications of fossils, instead of providing independent evidence by which phylogenetic-distributional theories may be judged, may itself have been strongly influenced by such theories. In any case, the reader would be well advised to treat with some scepticism all hitherto published maps purporting to show the earth in the Eocene or earlier geological periods. The evidence on which they rest needs very careful scrutiny.

In order to illustrate the part which distributional evidence is likely to play in the deduction of phylogeny, we will consider two particular cases, one from the animal and one from the vegetable kingdom, which may offer some instructive analogies to each other. The zoological example is chosen from the Insecta rather than from vertebrate animals not solely on grounds of my personal knowledge, but also because it seems to illustrate some principles unusually well and will not for most readers have the staleness of familiarity.

Paussid beetles (using the term in its old sense, i.e. excluding the Ozaenini) will hardly be familiar to any but specialist entomologists. All the species are believed to live in obligatory dependence on ants, at least in their larval stages. The group is undoubtedly related to the large family of ground-beetles (Carabidae), with which the reader may be familiar, but both adult Paussidae and those few larvae of the group which have been described show marked specialisations enabling them to live in ants' nests. In the adult beetles, the antennae, by which the ants are reported to carry the beetles about, show varying degrees of thickening and consolidation, and a corresponding reduction in the beetles' own running powers is indicated by the thickening and short-ening of the legs, especially of the tarsi. Hairy pits, marking the open-ings of glands licked by the ants, are frequently present on the thorax, and the mouth-parts are more or less modified and reduced. Nearly all the species have well-developed wings, and many of them are attracted to lights at night. Paussid beetles have been found in all the main continental areas of the world (except Antarctica, of course) but no species of the group is known from any oceanic island.

Darlington [42] divided the Paussidae s.str. into two tribes, Proto-paussini (with only one genus *Protopaussus* with a very few species in the Indo-Malayan region) and Paussini. He made five subtribes of Paus-sini, Cerapterina, Pentaplatarthrina, Platyrhopalina, Ceratoderina and Paussina. Recent Cerapterina are known from tropical America, Africa, the Indo-Malayan region and Australia, extending to about latitude 35°S in Australia and well north of the tropics in Asia. The Pentaplatarthrina comprise only the genus *Pentaplatarthrus*, with several African species; the Platyrhopalina include several genera in the Indo-Malayan region, one of which extends into New Guinea. Ceratoderina have one endemic genus in Abyssinia and several in the Indo-Malayan region, while Paussina occur throughout the Indo-Malayan and Ethiopian regions, extending also to Madagascar and the Mediterranean fringes of Europe.

On Willis's principle, we might conclude that the Cerapterina are the oldest group of Paussini, and in this case the conclusion is supported by other facts, e.g. that the Australian Cerapterina are all of endemic genera and not closely related to Indo-Malayan ones, also that there are a number of fossil Cerapterina (but no other Paussini) recorded from the Baltic Amber of Europe (early Oligocene). This last fact both provides evidence of the antiquity of the group and indicates its former presence in the Holarctic region, which we should need to postulate in any case to account for its reaching South America. Theorists about centres of origin would probably stress that the Indo-Malayan region not only has representatives of all but one of the subtribes of existing Paussidae, but has also produced what is believed to be most primitive living paussid (*Protopaussus*); on the other hand, the Ethiopian region has the greatest number of species and genera of the group. Darlington's conclusion, that the ancestry of Paussidae is to be sought in the Indo-Malayan region, is not, in fact, one we can yet reject with any certainty, though I would suspect that the Palae-arctic is a rather more likely centre of origin of the group.

The Paussina contain the majority of the recent species, and the genera of the group show essentially continuous distribution patterns, which would suggest this as the most modern subtribe; the absence of Paussina from the Baltic Amber gains added significance from the fact that this subtribe is the only one represented in the modern European fauna. The Pentaplatarthrina, Platyrhopalina and Cera-toderina are in some degree intermediate between Cerapterina and Paussina, and no doubt represent earlier offshoots from the line leading to Paussina. This line appears to have originated too late to have been able to reach the New World by means of the warm-climate connection which seems to have been available earlier (? in the Eocene period) to a cerapterine type.

The taxonomically isolated *Protopaussus* appears, from its structural characters, to be even more primitive than any Cerapterina, and more or less annectant between Paussini and the Ozaenini (usually included in Carabidae but transferred to Paussidae by the present writer, 1955); presumably it should be regarded as a relict, representing a nearly extinct group. The most paussid-like of existing Ozaenini is probably the neotropical *Physea*, apparently living in obligatory dependence on ants as do true Paussidae; though Darlington suggests that Paus-sidae come from Ozaenine ancestors, he does not suggest that these ancestors were close to *Physea*, though failing to cite any very clear structural reasons why they should not be. I suspect that his judgement in this has been influenced, perhaps subconsciously, by zoogeographical considerations; in the same paper he expressly suggests that Paussidae originated somewhere in the Old World tropics, and implies that their similarities to *Physea* are attributable to 'convergence'. His thinking seems to have been influenced by the doctrine of the permanence of the continents and oceans, which few authors dared to challenge in America in 1950. Recent evidence from 'palaeomagnetism' and 'palaeoclimatology' has led to the widespread questioning of this particular dogma, even in America. If there has been significant movement of the continents during the Tertiary era, Darlington's conclusions are liable to be invalidated.

It seems quite likely to me that the original development of myrmecophilous habits in the Ozaenini, which gave rise to the Paussidae s.str., may have occurred somewhere in the then tropical or subtropical parts of the Holarctic region in the Paleocene—soon after the first appearance of ants. The oldest authenticated fossil ant appears to be the late Cretaceous *Sphecomyrma* [207a]; it is notable that no fossil ants have yet been recorded among the numerous insect inclusions in the supposedly Cretaceous Canadian amber from Cedar Lake. To postulate, as Darlington does, a Cretaceous origin for the true Paussidae, seems hardly logical when we consider that the ants themselves had only just come into being in the Upper Cretaceous.

The evolutionary line leading to Pentaplatarthrina, Platyrhopalina, Ceratoderina and Paussina no doubt arose from some cerapterine ancestor in the Old World, perhaps in the Palaearctic region at about the beginning of the Oligocene period (a supposed primitive representative of this line, *Eopaussus* Wasm., has been described from the Baltic Amber). The most primitive surviving offshoots of that line, the Pentaplatarthrina, are now confined to Africa but may well have occurred in Europe in the Miocene period (a number of the ants described by Emery from the Miocene amber of Sicily are related to modern African forms). Darlington expresses some surprise that the Paussina of Africa have succeeded in colonising Madagascar on at least two separate occasions, whereas the more primitive paussid types appear never to have done so. Mammalian evidence suggests that in the Oligocene period Madagascar was more easily colonisable than it has been for most of subsequent time—but there may well have been no Paussidae in Africa in the Oligocene period.

It is not easy to find a group of plants whose distributional pattern is closely comparable with that of Paussidae. I have chosen for study the dicotyledonous family Proteaceae which, in spite of some evident differences, really does offer some parallels to our previous example. A more extended discussion of its geographical distribution is provided by Kausik [113], who reaches conclusions somewhat different from mine. Like Paussidae, the Proteaceae occur mainly in tropical and subtropical continental areas, barely enter the Holarctic region, and are not recorded from any oceanic islands. The two families are also alike in being more prevalent in the drier types of habitat rather than in rain forests. Recent Proteaceae are all more or less woody perennials, mainly shrubs and trees; their flowers are reported to be insect-pollinated as a rule, but some are said to be visited by birds. Few of the species have any striking adaptations for long-range seed-dispersal; in many species the seeds are somewhat winged, which may assist short-range dispersal but would hardly enable them to be blown across seas. Succulent fruits, which might lead to wide distribution of the seeds by birds, are rather rare in the family, and only one or two genera have other adaptation (e.g. adhesive hooks on the fruits, etc.) which might facilitate dispersal by animals. The seed-capsules, however, are often hard and woody, and in some species might enable the seeds to survive a considerable period floating in sea-water.

Unfortunately, there is no very recent revision of the family on a world-wide scale, the most complete being that of Engler [58]. He divided the Proteaceae into two subfamilies and seven tribes; thus:

1. *Subfamily PERSOONIOIDEAE*
   (a) Persooniae—many Australian genera, one endemic genus each in New Caledonia, Madagascar and South Africa, one species of Persoonia in New Zealand.
   (b) Franklandiae—one genus in Australia.
   (c) Proteae—many genera, Australia and South Africa, one species in Madagascar.
   (d) Conospermae—two genera, Australian.

2. *Subfamily GREVILLOIDEAE*
   (e) Grevilleae—many genera, all endemic, in Australia, one of them with a species in New Caledonia, one genus Indo-Malayan extending as far as Japan, many genera, all endemic, in South America.
   (f) Embothriae—six genera, two in east and south Australia, one in north and east Australia and New Caledonia, one in New Zealand and New Caledonia, two occurring in both Australia and temperate South America.
   (g) Banksiae—two Australian genera.

It will be noted that Engler's subfamily Persoonioideae (which he considered to include the primitive forms of the family) is not represented in the New World, but that two of his three tribes of Grevilloideae are represented in South America; all tribes of the family are represented in Australia, and the Grevilleae are the only tribe represented in more than two major geographical regions. At least three tribes are represented in New Caledonia (where no Paussidae are known to occur) and two of them even reach the warmer part of New Zealand. The Grevilloideae have no African or Madagascan species, but both of the persoonioid tribes which occur in Africa also have species in Madagascar.

A considerable number of fossil plants—mostly foliage but also some isolated fruits—from the Tertiary deposits of Europe have been attributed to the Proteaceae; none of them, however, can be compared in reliability of determination with the Paussidae of the Baltic Amber. It is noteworthy that all these fossils which have been attributed to Recent genera have been referred to the Grevilloideae.

Except for the lack of African representatives, the tribe Grevilleae seems to be analogous in distribution to the Paussidae-Cerapterina. One very remarkable circumstance, hardly to be paralleled in the animal kingdom, is that species from New Caledonia, Queensland and tropical America have been attributed to one genus *Roupala*. The simplest explanation might be to postulate that *Roupala* must itself be practically as old as the Cerapterina, and would have occurred in the Eocene of the Holarctic region. If this is the true explanation, there is some hope that fossils of the genus may be found in Europe or North America. The only conceivable alternatives would seem to be some quite exceptional form of long-range trans-Pacific seed carriage, or that *Roupala* is an unnatural genus; it has been split in recent works. The Embothriae include several genera which extend into the temperate zone, and it may be that link between the New and Old World forms of the tribe may be an 'antarctic' one without analogues in the Paussidae.

In the case of the Paussidae, we reached the seemingly paradoxical conclusion that the group may have originated in the one major geographical region—the Holarctic—in which it is practically unrepresented today. If the Proteaceae exemplify a similar paradox, we might consider the Holarctic and Indo-Malayan regions as possible ancestral homes of the Grevilloideae. In general, the Australian region seems to occupy for Proteaceae a position rather analogous to that of the Indo-

Malayan region for Paussidae; only the Indo-Malayan region has species of as many as four of Darlington's five tribes of Paussini, and only in the Australian region are all seven tribes of Proteaceae represented.

At the beginning of the Tertiary era, it seems that the land-mass which is now peninsular India lay well south of the equator, and probably had some sort of links (not, however, complete land bridges) with Madagascar and Australia—much of it probably had the type of climate found in the present-day strongholds of Proteaceae, Australia and South Africa. Such a region might have been the habitat of ancestral Persoonioideae; subsequently the Indian mass seems to have moved northwards, across the equator and through the tropical rain-forest belt, a circumstance which might lead to the extinction of the Persoonioideae in the land-mass concerned. The Grevilloideae may have originated from early colonists on the northern side of the equatorial rain-forest belt, deriving from the ancestral persoonioid stock. Derivatives of this grevillioid stock may have spread through the Indo-Malayan region to reach Australia, and through the Holarctic region to America; presumably the group disappeared from the western Palaearctic region too early for it to be able to colonise Africa with the big northern intrusion of the Pliocene period.

This theoretical interpretation of the major distributional features of Proteaceae will obviously be invalidated if Engler's classification is shown to be seriously unnatural, or if fossil Grevilleae are found in early Eocene deposits in Australia or South Africa—it possesses what Popper regards as the essential property of a scientific theory, falsifiability. Proteaceae would be a very suitable subject for 'sero-diagnostics', and there is a reasonable hope that further fossils of the group will be recognised in future.

It is evident that Proteaceae have been more successful than Paussidae in colonising across considerable sea-barriers; the three tribes represented in New Caledonia, the two in New Zealand and the two genera of Embothriae which occur in both Old and New Worlds are sufficient evidence of this. It is probably true that most types of plants are better adapted to long-range passive dispersal than are most animal groups— the light air-borne spores of the lower plants are obviously advantageous from this point of view, but in the geological long run it seems that the resistant seeds of higher plants can achieve the same effect. It also appears to me that phytogeographers have hitherto been unduly reluctant to postulate the extinction of particular groups in whole geographical regions—such extinctions have undoubtedly occurred in many groups of animals, and I can see no reason why the phenomena should not be equally common in the vegetable kingdom.

Though a considerable number of fossils of foliage and fruits from Lower Tertiary deposits in Europe have been attributed to Proteaceae, mostly to Grevilloidiae, recent authors (e.g. Takhtajan [186a] and Kirchheimer [115a]) have nearly all adopted towards these attributions a highly sceptical attitude which they do not manifest in relation to many other fossil placements; this is probably an example of the phenomenon mentioned in Chapter 6, the classification of fossils being strongly influenced by pre-conceived ideas of the palaeontologist—in this case by ideas on phytogeography which neither Takhtajan nor Kirchheimer state explicitly. Kirchheimer cites as one reason for rejecting the attribution of European fossils to Proteaceae the absence in the deposits concerned of proteaceous types of pollen—but Samoilovitch (in Takhtajan, loc. cit.) refers to proteaceous pollen types from many deposits in Europe and the U.S.S.R. of ages from Late Cretaceous to Miocene, and figures some of them.

I append here a list of biogeographically important events from the middle of the Mesozoic era up to the present; many of the changes listed are postulated on the basis of scanty evidence, and it is almost certain that within the period covered there have been some important changes which are not mentioned in the list. The supposed events are listed in reverse order, from the present time backwards.

1. *PLEISTOCENE period*
Intermittent glaciations extending to sea-level in temperate latitudes in both hemispheres, with correlated lowerings of sea-level and changes in altitudinal-climatic zones of tropics and subtropics, and in sea-surface temperatures—attributable probably to fluctuations in solar radiation.

2. *PLIOCENE period*
Formation of present Central American Isthmus; establishment of a solid link between Ethiopian Africa and Palaearctic region; East Indian island chain, linking Australia with mainland of Asia, reaches its maximum development, northward movement of Australia and New Guinea to impinge on it.

3. *MIOCENE period*
Active mountain building in main 'alpine' ranges, probably associated with active horizontal movement of continental blocks; final rupture of any trans-Atlantic land links between Europe and North America; considerable northward movement of Eurasian land mass, less so of North America; India moves far to north and loses any links it may have had with Seychelles or Madagascar blocks; severance, through glaciation of Antarctic fringes, or sinking of islands, or both, of effective faunal and floral interchange between Australia–New Zealand and South America; Madagascar reaches its present degree of isolation; loss of effective island-chain link between North and South America.

4. *OLIGOCENE period*
North Atlantic rift rapidly opening up, but possibly a temperate link between Europe and North America remaining at first; Holarctic land-masses begin to move into higher latitudes; an island chain link (? via Antilles) between North and South America; Madagascar still partly linked by island chains with India (then lying south of equator) and perhaps with Africa; Australia lying further south than now, no effective links with Asia, but probably linked, like New Zealand, by an island chain to Antarctica and thus through temperature forests on its fringes to South America.

5. *EOCENE period (including Paleocene of American geologists)*
North Atlantic rift initiated, but a warm climate link persisting between Europe and North America for much of period; a land link between North and South America at the beginning, severed fairly early in the period; Holarctic land mass in much lower latitudes than now, latitudes in Europe probably at least 20° less than present ones; Australia already isolated but with an island chain connection, over which ancestral marsupials arrived from some other continent; Madagascar already separated from Africa, possibly connected with India at first; South Atlantic rift already too wide for any effective floral or faunal interchange between Africa and South America.

6. *CRETACEOUS period*
Final severance of any land links between Australia and other parts of the 'Gondwana' continent; probable severance of Madagascar from Africa (though not from India); South Atlantic rift rapidly widening,

severance of land links between Africa and South America; first beginnings of North Atlantic rifting.

7. *JURASSIC period*
Break-up of southern 'Gondwana' continent begins, rifts appearing between Africa and Madagascar–India block, and between the latter and Australia; connections at certain times between Northern 'Laurasia' continent and Gondwana continent, permitting transit of some terrestrial dinosaurs, seed-plants etc.

There is a reasonable likelihood that palaeontological, palaeoclimatological and palaeomagnetic evidence now being collected will make possible a much more detailed and accurate account of the biogeographically important changes in the geological past than is possible today.

On the other hand, the ever greater rate of change imposed on the earth's surface by human activity is having disastrous effects on the natural faunas and floras of all parts of the globe. The possibility of observing a 'natural' area of distribution of any kind of plant or animal becomes less every year. Fortunately, the general ranges of the larger vertebrate animals, and of the main types of trees and more conspicuous herbaceous Angiospermae were fairly well established by the explorers of the last century, at a time when human influence, outside the 'old' civilised regions of Europe and northern Asia, was vastly smaller than today. Comparison of the present ranges of such organisms with those recorded a hundred years or so ago may give us some idea of what is likely to have happened to other groups of whose past distribution we have far less knowledge. Naturally, the cryptogams and invertebrates received far less attention from the nineteenth-century bio-geographers, and for many groups of these the world faunas are still very imperfectly known.

The realisation that so much evidence bearing on the whole history of the earth, which could still have been collected in the last century, but was not, is now being progressively and irretrievably lost, is a truly tragic one for the zoogeographer and phytogeographer. More than most scientists, these should have a keen appreciation of the 'angst' which informs so much of the art of our day.

12 Heredity and Chromosomes in Relation to Evolution and Classification

It is a third corollary of integration that genetic systems can change while external form remains the same. This is shown by the evolution of the sex chromosomes but it can also be shown in other ways. Chromosomes can determine evolutionary discontinuities which are not morphologically visible. Two geographically separated varieties of *Hordeum sativum* give a vast array of segregation in their progeny which is not seen when parents with similar differences come from the same region. The same kinds of gene difference in two species of *Gossypium* have different properties of dominance. Certain cryptic species of *Drosophila*, although scarcely distinguishable in form, have chromosomes differently arranged and are intersterile. All these properties go to show that the genetic basis of form may change although the form itself does not. We need not suppose that the external stability of a *Lingula* depends on an unchanging complement of genes. Forms and their determinants are not necessarily related in the same way in different species at the same time, or in the same species at different times.

C. D. DARLINGTON [41a]

In the definition of the species which we considered in Chapter 3, genetical criteria played an important part, and when considering phylogenetic classification in Chapter 9, we saw that such a system should be based on genotypes rather than phenotypes. A phylogenetic systematist should thus be particularly anxious to obtain any possible information about genotypic similarities and differences among his plants or animals. At the species level or below it, such information can often be obtained through hybridisation and other breeding experiments, but for higher classificatory levels we usually have to rely on less direct evidence of genotypic differences. No doubt the most valuable for this purpose of currently available techniques is that of comparing the nucleic acids themselves, explained in Chapter 13; in the same chapter we shall also consider various techniques for the comparison of structure of proteins, which are believed directly to reflect genotypic differences. The technique of constructing 'chromosome maps' from linkage studies may also provide information about genotypic differences between species, but is open to two serious objections—first, that the amount of research involved in constructing such maps is enormous, and second, that it is often very difficult to establish which factors are truly homologous between different species. The study of 'giant chromosomes', considered later in this chapter, can to some extent replace the laborious procedures of chromosome mapping, but such chromosomes are hardly known outside the insect order Diptera.

Since the discovery of the importance of chromosomes as bearers of heredity, there has been a widespread expectation that the study of the visible structure of these organelles would reveal evidence of outstanding importance for phylogeny and natural classification. After fifty years during which numerous cytologists have studied chromosomes intensively in the most diverse groups of plants and animals, it must be confessed that the great expectations with which such investigations began have rarely been realised. This does not, of course, mean that chromosomal structure has been found to have no importance in relation to evolution and classification. Cytological evidence is being used to an ever-increasing degree as an aid in solving problems of phylogeny and classification at the species level and below it, in both botany and zoology. When we are concerned with classificatory problems at higher levels, however, the gross chromosomal patterns prove only too often to be distributed in such a way that the experienced systematist will tend to regard them as characters of rather low diagnostic value, comparable perhaps to patterns of coloration in animals, or leaf-form in plants.

In order to understand this state of affairs, we shall need to consider briefly the role of the chromosomes in evolution. At least in higher, multicellular, organisms, the chromosomes differ from most other classificatory characters (anatomical, physiological, behavioural, etc.) in showing no direct connections with particular habits or modes of life of the individual organism. Chromosomal characteristics seem to be related rather to properties of populations and lineages than of individuals—to things like degrees of in- or out-breeding, polymorphism within populations, rates of response to selection, etc. The effects of such differences will probably be manifested in rates rather than directions of long-term phenotypic evolution.

Linkages between factors carried on the same chromosome seem to be important chiefly in relation to variation within the species. In a given species, closely linked factors sometimes influence the same character, sometimes quite different ones; they are liable to be factors for which individuals in wild populations are often heterozygous, and to influence characters in which these individuals show polymorphism. Such polymorphism in natural populations is not usually an affair of random variations, but involves correlated variation in more than one character; chromosomal linkages are likely to be important in maintaining this correlation. The small inversions and reduplications which are sometimes found in wild populations may also be related to polymorphisms of this type. On the other hand, differences between species and higher groups are probably mediated by factors for which members of a given species are usually homozygous, mutants affecting them being kept at a low level by selection. Thus there is no need for linkage in order to maintain the stability of species-differences.

As examples of the use of chromosomal evidence in classification at the species and lower levels, one may cite the work of Manton [137] and others on ferns, and that of Smith [171] on coccinellid beetles of the genus *Chilocorus*. The 'giant chromosomes' of the salivary glands of Diptera offer quite exceptional possibilities for this sort of study—opportunities which have been exploited by students of *Drosophila* fruit-flies, of mosquitoes, of fungus-gnats of the genus *Sciara*, and of certain Chironomidae. When giant chromosomes of flies of different subspecies of the same species are compared, the banding sequences can generally be homologised in detail, permitting the recognition of inversion, translocations, reduplications and deletions, without reference to meiotic behaviour (though the giant chromosomes are commonly associated in pairs more or less as in meiosis, so that hybrid features can be seen in them). The recognisability of particular elements of the banding sequence has been used as a basis for phylogenetic deductions, by a method which can best be explained by the use of letters to denote identifiable elements of banding. Suppose that, among various subspecies, etc., of a given species we can distinguish the following banded sequences on a particular chromosome arm: (1) abcdefghijklmno, (2) abcdfeghijklmno, (3) dcbaefghijklmno, (4) cbadefghijklmno,

(5) abcdhgefijklmno, (6) dcbaefghijklmon, (7) klmnoabdfeghij, (8) dcbeafghijklmon. If these arrangements are compared, it will be found that, starting from number 1, three of the other arrangements can be reached by a single-step change, three more by two steps, and only one (number 8) requires three steps; from no other starting point could more than two of the arrangements be reached in a single step, and no other could give rise to as many as six of them by one or two steps. Arrangement number 1 clearly occupies a special position, and could reasonably be postulated as the ancestral one for the species. Infallibility could hardly be claimed for this sort of reasoning, but I think considerable force could reasonably be attributed to it. An additional point is that small reduplications may occur without marked effect on viability, whereas deletions are very often lethal—where an 'extra piece' can be detected in a chromosome of one form by comparison with the homologous one of another form, the difference is more likely to have arisen from a reduplication in the first form than from a deletion in the second.

In the salivary gland chromosomes of Diptera, homologous portions in the banding sequence may be traced between different (but closely related) species as well as between subspecies of one species; this is usually found only when the species concerned are sufficiently close to be capable of producing F_1 hybrids, and in *Drosophila* they will be members of the same 'species-subgroup' (see Chapter 4).

It has recently been discovered that giant chromosomes occur in the embryos of some Angiospermae; Nagl [145b] reports that though these normally do not exhibit the banding patterns which are so notable in many dipterous salivary gland chromosomes, at least in *Phaseolus vulgaris* they can be made to do so by exposing the developing seeds to low temperatures at the crucial stages. We may look forward to a great deal of new information which will be valuable for species-systematics from the development of this technique.

As pointed out by White [200], if reduplications are likely to be viable whereas deletions are almost always lethal, there should be a long-term bias towards increase in the length, or number, of the chromosomes. This effect may, however, as White also suggests, be partly offset by a tendency for 'surplus' portions of eu-chromatin to be converted into hetero-chromatin; there is evidence that deletions of hetero-chromatin may have little or no deleterious effect. We have no evidence, however, that the effective number of genes remains the same in the long run of evolution; if there are substantial differences in numbers of genes between different organisms, this will weaken the theoretical basis of the 'numerical taxonomy' of Sokal and Sneath (see Chapter 15).

Hybridisation experiments have produced systematically interesting results in many groups, in both animals and plants. The general rule seems to be that first generation hybrids can often be reared from crosses between species of the same genus, but are usually either completely sterile or show only limited fertility in back-crosses to either of the parental species; Haldane's rule, that in bisexual species with an x–y sex determining system, the heterogametic sex shows less viability in hybrids, seems to be generally applicable. Most cases in which phenotypically 'good' species yield fertile F_1 hybrids relate to crosses between species whose natural habitats are widely separated geographically. An interesting study of hybridisation in butterflies of the genus *Papilio* by Ae (1967, in [215]) reports instances of successfully rearing F_1 adults from crosses between species attributed by systematists to different subgenera, but very few cases of fertile hybrids between what systematists regard as 'good' species. In contrast, Hubbs (1967, in [215]) reporting experiments on freshwater teleostean fishes, records numerous cases of fertile hybrids between species universally accepted as good by taxonomists, and even some allegedly fertile inter-generic hybrids— though he questions the validity of the generic distinctions in such cases. He states that, at least in freshwater Teleostii, it is usually possible to obtain hatchable eggs from a cross between two species of the same family, and that a few cases have been known of young being hatched from eggs produced by a cross between species of two different families, though in such cases he is inclined to question the distinctness of the families concerned.

The results reported by Hubbs and Ae provide further evidence of the discrepant applications of classificatory categories in Vertebrata and Insecta, discussed also in Chapter 19. Botanical hybridisation experiments, reported by Maheshwari (1967, in [215]) suggest that in most flowering plants intra- and inter-generic crossability relations are more similar to those of insects than of Vertebrata, though there are notable exceptions among orchids and grasses (cf. Chapter 5). The effects of polyploidy are, of course, a complicating factor in crossability relations in many plant groups.

As was pointed out by Turner (in [215]) substances used in 'micro-molecular systematics', unlike most structural characters used by systematists, are 'usually governed by genes expressing dominance or recessiveness, the appearance of a compound generally reflecting dominance. Thus, the chromatographic pattern of an F_1 between two species of *Baptisia* with differing flavonoid patterns is expressed additively, i.e. the flavonoids found in the hybrid are the same as those found in both the parental types'. This circumstance provides us with a useful means for detecting natural hybrids.

The evidences of gene duplication in the course of long-term evolution are reviewed by Nei [145c], who points out that the DNA content of an ordinary cell of a mammal (or higher plant) may be 1,000 times that of a bacterium, and cites three processes—unequal crossing over leading to reduplications, forms of aneuploidy (dysploidy) and polyploidy—which can increase the DNA content. He notes, from the results of Hoyer, Bolton and McCarthy and of Margoliash *et al.* (see Eck and Dayhoff [47a]), that 'The frequency of nucleotide substitution in DNA is so low that there seem to be many duplicate genes in the genomes of higher organisms.' He further concludes that 'If a considerable part of the increase in DNA content has occurred by unequal crossing over of short segments of DNA, or by single chromosome doubling, rather than by genome duplication, it is to be expected that some segments are repeated more than thousands of times.'

Perhaps the most notable achievement yet in the use of chromosomal evidence for classification at higher levels is that of M. J. D. White [199a], on the insect order Diptera. The characters he found most useful for this were derived, not from the 'giant chromosomes' of the salivary glands, but from meiotic figures. White distinguished six principal types, as follows:

(1) Small x and y chromosomes showing 'distance pairing'; normal crossing over in the somatic chromosomes of the male.
(2) x and y chromosomes with a 'pairing segment' and forming a bivalent at meiosis, but distinguishable; crossing over present in male.
(3) x and y chromosomes not distinguishable (? fused with a pair of autosomes); normal crossing over in the male.
(4) Small x and y showing distance pairing; no crossing over in somatic chromosomes of the male.
(5) x and y chromosomes with pairing segment, forming a bivalent at meiosis; no crossing over in the male.
(6) y chromosome lost; no crossing over in the male; anomalous chromosome cycle.

White assumed that type number 1 was the ancestral one of the order, as it most closely corresponds to the usual state of affairs in other Endopterygota, and particularly to the older orders Neuroptera and Coleoptera. He found that only one family of modern Diptera, the Tipulidae, fell into this group, and it is notable that among the oldest known fossils of Diptera (from the Lower Jurassic) tipulid-like wings predominate. The second group is represented by the Ptychopteridae (Liriopeidae auctt.), while the third comprises the Limoniidae (which in classifications based on adult structure have always been placed very close to the Tipulidae), plus Psychodidae, Simuliidae, Culicidae and Chironomidae—the last four families associated in a group Culiciformia by many modern systematists. White's type four included Rhyphidae (= Anisopodidae), Bibionidae, Scatopsidae, Thaumaleidae, Blepharoceridae and Mycetophilidae; the most recent classifications of Diptera place most of these in a group Bibiomorpha, but Thaumaleidae and Blepharoceridae are considered to be allied to the Culicomorpha. Type five corresponds to the vast suborder Brachycera, and six includes only Sciaridae and Cecidomyidae.

In another place White [201] makes the unjustifiable claim that 'a classification of the Diptera based on the type of meiotic mechanism in the male is surely more meaningful than one derived solely from studies of external morphology'. It may be pointed out in this connection that, from the researches of S. G. Smith [171, 171a] and others on another large insect order, the Coleoptera, it seems fairly clear that no rational classification of that group could be founded on the meiotic mechanisms of the male. The meiotic figures resemble other systematic characters in that their classificatory value cannot be determined *a priori*. The data published by White, however, are sufficient to establish that in the Diptera these characters are of considerable systematic importance. Nevertheless, the first thing that will strike a practised systematist considering White's results is the inadequacy of the basis for such far-reaching conclusions. The species studied by White were far from constituting an adequate and representative sample of all the main groups of the order. Numerous systematically interesting types, such as Tanyderidae, Trichoceridae, Cylindrotominae (intermediate in several respects between Tipulidae and Limoniidae), Dixidae, Rachiceridae, Rhagionidae, etc., were not considered.

White's phylogentic scheme, whereby groups two and three are derived directly from one, and five and six directly from four, is unduly simple. The most perplexing problems are posed by group four; this includes Thaumaleidae and Blepharoceridae (both unquestionably allied to families in group three), together with most of the Bibiomorpha except for the families in group six. It is difficult to imagine the thaumaleid-blepharocerid condition arising from the psychodid-chironomid one, or *vice versa*; the only obvious intermediary between the two types is the tipulid condition. Nevertheless, there are clear indications that Thaumaleidae are more nearly akin to Dixidae and Chironomidae, and Blepharoceridae nearer to Psychodidae, than either is to any of the Bibiomorpha. The rational conclusion is that chromosomal evolution in the Nematocera has been more complex than White supposed; the picture will probably become clearer when information is available for such forms as Dixidae, Trichoceridae and Tanyderidae.

Another of White's conclusions, that the meiotic features of the suborder Brachycera indicate a relationship to his group four, agrees well with other lines of evidence; the family Rhyphidae is generally considered to be the most brachyceran-like of the Nematocera, and most nematoceran-like of the Brachycera particularly resemble the Rhyphidae and allied families of Bibiomorpha. White's evidence suggests that the division into Nematocera and Brachycera is not the most fundamental phylogenetic one in the Diptera; this question, however, is a very complex one, needing more evidence than is yet available for a satisfactory answer.

The information, available at the time, on chromosome numbers in animals was summarised and considered in another work of White [202]; he presents the information particularly in the form of histograms for particular classificatory groups, plotting numbers of species recorded for each chromosome number. That this way of presenting the information is liable to be misleading is implicitly admitted by White when he says, concerning Hemiptera, 'if some of the other families of Heteroptera had been investigated as extensively as these four the histogram for the order as a whole might have had a very different shape'. Even if the numbers were known for all species of a group, the histogram would hardly be the most useful way of presenting the results. The use of histograms is a statistical technique based on ideas of 'randomness'. Natural classification deals with the results of a decidedly non-random process of evolution; it is very difficult to see how one could define a random sample of a classificatory group. To quote a writer on the philosophy of science [157a], 'Only on the last resort, when the heavenly powers fail, should resort be made to the demons of the underworld, chance and probability.' In the language of cybernetics, if techniques appropriate to random distribution are applied to data which are non-random in some unanalysed way, there is liable to be serious loss of 'information'. White himself says that 'the species which have been studied by cytologists usually do not represent a random sample, particular sections of the group having been investigated more extensively than others'. What is required is not random sampling, but systematically representative study.

From White's figures and histograms it is evident that in many animal taxa below the level of orders (in invertebrates) or classes (in vertebrates) particular chromosome numbers have been found to be prevalent—there are many groups for which one might reasonably talk of a 'modal' number. In many, if not all groups, however, such modal numbers are liable to occasional sharp differences between forms which seem in every other respect to be nearly related. From the classificatory point of view, gross chromosomal patterns seem to be analogous to secondary sexual characters in bisexual animals (see Chapter 3). Systematists working in many groups are familiar with the value of secondary sexual characters in providing sharp differences between closely related species, and a particular type of sexual character is often found in most, though rarely in all, of the species of a more or less extended group.

Just as in the case of chromosomes, it is argued that secondary sexual characters are not closely linked with specific modes of life in animals, or as it is sometimes put, they are not 'adaptive'—such characters, it is argued, may be expected to show evidences of affinities which have been obscured by adaptive changes. Undoubtedly, both chromosome structure and secondary sexual characters sometimes do this, but they have the corresponding disadvantage that they are liable at times to manifest striking differences between forms which in every other respect are closely allied. Given an adequately wide (and deep) systematic knowledge of a group, and a proper appreciation of other lines of evidence for phylogeny in it, the additional information provided by studies of chromosomes may be of great value, even though the chromosomes alone would provide a very poor basis for classification.

Polyploidy, which as we shall see appears to have occurred in the evolution of many groups of plants, is considered by most authorities (including White) to be practically restricted to parthenogenetic and/or hermaphrodite forms in the animal kingdom. There are, however, many cases among animals with normal bisexual reproduction where the chromosome numbers among related forms strongly suggest polyploidy; thus in the butterfly genus *Polyommatus* (Lycaenidae) most species have a haploid number of twenty-two or twenty-three, but *P. bellargus* has forty-five and *P. coridan* about ninety. White suggests that 'the caryotypes of *P. bellargus* and *P. coridan* have been derived by fragmentation from the normal lycaenid caryotype', but does not explain why such fragmentation should produce the numbers forty-five and ninety, rather than thirty-one, thirty-six, fifty-seven, etc. Rather similar phenomena are shown, e.g. among the hamsters (*Cricetus* and *Mesocricetus*) in the Mammalia.

Sachs [162], reviewing the chromosomal patterns in certain cricetine rodents, noted that the haploid number was eleven in both *Cricetus cricetus* L. and *Cricetulus griseus* Milne Edwards, whereas it is twenty-two in *Mesocricetus auratus* Waterh. From this information, together with the fact that in all cricetines except *Mesocricetus* the female has four pairs of mammae, while *Mesocricetus* has seven to eleven pairs, also that the existing *Cricetus* are principally European while *Cricetulus* are mainly Asiatic, and that *Mesocricetus* occur in or near the main zone of overlap between *Cricetus* and *Cricetulus*—also that both the last-mentioned genera have been recorded as Pliocene and Pleistocene fossils, whereas *Mesocricetus* is not known as a fossil—Sachs concluded that *Mesocricetus* arose recently from an allotetraploid hybrid of a *Cricetus* and a *Cricetulus*. As an additional point in favour of this theory, we may note that one of the main differences between these two genera lies in the much stronger ridges on the skull of *Cricetulus*, and that in this respect *Mesocricetus* is more or less intermediate between the other two. Sach's interpretation has been criticised by White [202], on the grounds that Moses and Yerganian [145] found that the DNA content of the *Mesocricetus* nucleus was much less than double that of *Cricetus*, that *Cricetulus* (*Tscherskia*) *triton* has been found to have a haploid number of fifteen, and that another species of *Mesocricetus*, *M. brandti*, shows a haploid number of twenty-one. As a case of allopolyploidy, I think the verdict must at present be the Scottish one of 'not proven'.

Cases of evident polyploidy among the hermaphroditic Oligochaeta (earthworms) are cited by White; as an example of their occurrence in parthenogenetic forms we may quote many Curculionidae among the Coleoptera. The basic diploid number in this group seems to be twenty-two, but many named forms (it is perhaps better not to call them species) consist solely of parthenogenetic females, among which such numbers as thirty-three, forty-four, fifty-five and sixty-six have been recorded. There seems to be a tendency for these parthogenetic forms to occur beyond the normal ranges of their diploid (and bisexual) relatives, similarly to some of the polyploid forms reported in the plant kingdom.

A recent zoological attempt to draw phylogenetic conclusions from studies of chromosomes is provided by Kiauta's [114a] study of the dragonflies (Insecta-Odonata). He presents chromosomal counts of 236 species, including representatives of all the generally recognised existing families. A haploid number of thirteen is recorded for nearly half the species examined of the suborder Zygoptera, for the remarkable 'living fossil' *Epiophlebia* (the sole surviving representative of the Anisozygoptera), and for more than 60 per cent of the anisopteran species studied. Kiauta, however, rejects the straightforward conclusion that this number is the ancestral one of the order on the grounds that in the two families of Anisoptera (Petaluridae, Gomphidae) which have been recorded in Mesozoic fossils, and which on morphological grounds are regarded as relatively primitive forms, the recorded chromosome numbers are always less than $n = 13$; from this circumstance, and the exclusive or predominant occurrence of $n = 14$ in the large and 'advanced' Zygopteran family Coenagrionidae as well as in one of the dominant anisopteran families, Aeschnidae, Kiauta concludes that the original dragonflies had a lower chromosome number and that an increase in the number has tended to accompany increases in general 'specialisation'. He presents the chromosome numbers in relation to a dendrogram derived from Fraser, according to which the existing system of suborders of Odonata is phylogenetically unnatural, in that some of the Zygoptera (e.g. Lestidae and Calopterygidae) are shown as more nearly related to the Anisoptera (and Anisozygoptera) than to the rest of the Zygoptera. If Kiauta is right, it appears that the 'typical' haploid number of thirteen has developed polyphyletically on several different lines, and one might pertinently ask, what is so special about the number thirteen that so many dragonflies should independently increase their chromosome numbers to it? It seems rather more likely to me that Kiauta is wrong. Kiauta notes that, in the anisopteran family Libellulidae, though the majority of species have $n = 13$, some have twelve, and there are single species recorded with eleven, nine, five, three to four and three. This circumstance, and the report of $n = 7$ for one species of Aeschnidae in which the 'modal' number is fourteen, he accepts as evidences of secondary reduction in the number, and he even implies that this might come about at times by a direct process of halving the chromosome number. I am not aware that such a process has been established to have occurred in any other groups of bisexual animals.

One remarkable chromosomal feature is shared by only two major groups in the animal kingdom, these two being phylogenetically very remote from each other. The character in question is a reversal of the normal sex-determining system, so that the female becomes the 'heterogametic' sex; in the animal kingdom, this condition appears to be universal in the Aves among Vertebrata and in the Lepidoptera (including Trichoptera) among Insecta, but is hardly known elsewhere. In both Vertebrata and Insecta, the normal x–y sex-determining system, with heterogametic males, is practically universal and undoubtedly primitive. As far as I am aware, no one has suggested a plausible series of adaptive changes by which this reversal can have been brought about. If we are entitled to reject *a priori* the possibility of polyploidy arising naturally in groups with the normal x–y system, it seems to me that an equally strong *a priori* case could be made for the impossibility of reversing that system. If, despite our complete failure to understand how selection could have brought this about, a reversed system has manifestly arisen in at least two quite distinct lines, I think we should logically allow that polyploidy may occasionally have arisen in normal bisexual animals in ways of which we have at present equally little understanding.

The diploid condition itself presumably arose at some stage as a phylogenetic advance from a prior haploid genome. Bacteria seem to be essentially haploid, and a definite diploid vegetative stage is apparently lacking also in blue-green algae and in fungi; it appears among red algae, is present in brown and green algae and occurs in at least part of the life cycle of all higher plants as well as animals. Diploidy, it is generally believed, confers greater evolutionary plasticity by enabling the individuals of a species to carry a 'reserve' of recessive factors. However, the persistence in Bryophyta and many Pteridophyta of an extended haploid vegetative stage, clearly points to some compensating advantages in the haploid condition, which in Bryophyta at least is vegetatively dominant. It is noteworthy that the difference between the leafy moss plant and the sporogonium seems not to be determined by the haploidy of the cells of the former as compared with the diploidy of the latter; several experimenters have succeeded in producing diploid plants with perfectly normal gametophytic form, and there have been claims to have observed haploid sporogonia. In

Bryophyta, the sporophyte is in almost all cases completely dependent on the gametophyte for its nutrition; the major radiation of the group has been manifested in the haploid stage, whereas in Pteridophyta the corresponding radiation has been that of the diploid sporophyte. We might thus expect to see the relative advantages of haploidy and diploidy displayed in a comparison of the evolution of Bryophyta and of Pteridophyta. In fact, the most notable difference has been the tendency throughout for the Pteridophyta to produce larger and more complex types. Within the algae, also, there seems to be a tendency for the larger and more complex vegetative forms in the Phaeophyceae and Chlorophyceae to be diploid; in the higher fungi, a rather comparable phenomenon is the tendency in Ascomycetes and Basidiomycetes to develop tissues of binucleate cells.

The great majority of green plants do not have the x–y sex-determining system. In many plant groups, multiplications of the normal chromosome number can be brought about quite readily under artificial conditions and 'even-numbered' polyploids may be capable of normal reproduction. Polyploidy has played a more evident part in the natural evolution of the plant kingdom than it has among animals. On one hand, polyploidy can establish in one jump a breeding barrier like that between species (see Chapter 3), on the other hand, it may at times have permitted the bridging of existing sterility barriers. A tetraploid, crossed with the corresponding diploid form, normally produces sterile triploid progeny; it may, however, be possible to cross it with another tetraploid form, deriving from a diploid form which itself may not be crossable with the first one, to produce a fertile 'allotetraploid' hybrid; however allotetraploidy seems more often to have arisen from hybrids of diploid species (see 204*b*), for example in the ancestry of *Galeopsis tetrahit* and *Spartina townsendi*. Such occurrences would be much rarer, if they occur at all, in animals. A review by Stebbins [180] gives a good account of chromosomal evolution in plants.

From the study of such works as Darlington and Wylie's *Chromosomal Atlas of the Flowering Plants* [41], it is evident that there are some groups of plants in which polyploidy has not been an important factor in chromosomal evolution, and there are cases where the chromosome numbers themselves seem to have considerable classificatory value. A notable example is that of the Coniferae. Chromosome numbers for more than a hundred species, representing all the main divisions of the group, are given by Darlington and Wylie. Except for four species of

Podocarpus with nineteen or twenty pairs, and for the evidently poly-ploid *Sequia sempervirens* with thirty-three pairs, the conifers all seem to have haploid numbers of eleven, twelve or thirteen. The only ones with thirteen pairs are the four species (in two genera) representing Araucariaceae; eleven pairs are recorded for the isolated *Torreya*, for Taxodiaceae and for Cupressaceae; *Taxus*, *Cephalotaxus* and the Pinaceae all have twelve pairs as has one of the recorded species of *Podocarpus*.

The chromosome numbers of Angiospermae, as recorded by Darlington and Wylie, are much more diverse than those of the gymnosperms, and their variation shows far less evident correlation with established classifications. Darlington and Wylie can be accused of a certain arbitrariness in their assignment of 'basic chromosome numbers' to species—for example, in the family Proteaceae (which we have already considered in connection with geographical distribution) some of the species are recorded as having fourteen chromosomes, corresponding to a haploid number of seven; when another species of the family is stated to possess twenty-eight chromosomes, Darlington and Wylie assign to it the 'basic number' of seven, considering it as a tetraploid—yet in another genus, systematically considered as close to the one with twenty-eight chromosomes, the recorded number is twenty-six, and the 'basic number' is set down as thirteen, not six and a half! The view of these two authors concerning the reliability of the numbers they quote—that not more than about 3 per cent are likely to be erroneous—must be considered as an optimistic one.

An important recent review by Ehrendorfer [54] considers the value of cytological evidence for the classification and phylogeny of Angio-spermae. The general problems and methods involved in this are discussed and expounded relatively fully, and the family Dipsacaceae is considered in considerable detail from this point of view, some attempt being made to integrate the cytological evidence with other available lines. Ehrendorfer suggests that the Dipsacaceae, in their general trends of cytological evolution, more or less closely parallel many other families of semi-shrubby or herbaceous Angiospermae.

A review of the available data on the chromosomes of primitive woody Angiospermae, by Ehrendorfer *et al.* [54a] leaves the critical reader in some doubt whether the ancestral haploid number of the group was 7 (or 8?), 13, or even 19 as in Magnoliaceae. The authors themselves postulate an original number of 7, and suggest that four main stages can be distinguished in the chromosomal evolution of the group, involving successive stages of polyploidy and dysploidy. They seem to believe that evolutionary changes in the chromosomes have nearly always involved increases in their number, with dysploidy responsible for only an occasional reduction by one (e.g. from 13 to 12).

It is noteworthy that polyploidy seems to have occurred very rarely in the evolution of the woody Gymnospermae, among which the ancestors of the Angiospermae must presumably have lain, whereas the family of Angiospermae which Ehrendorfer *et al.* regard as on the whole the most primitive, the Winteraceae, exhibit the highest recorded chromosome numbers of the group, with a haploid number of 43 in several species. The number 7 is recorded only in two relatively advanced families, Annonaceae and Chloranthaceae, while 12 and 13 are the predominant numbers in the Gymnospermae.

Ayyangar (in [215]), considering the chromosomes of Cucurbitaceae, report a series of numbers closely paralleling that reported by Ehrendorfer *et al.*, the haploid numbers ranging from 7 to 22. Although 11 and 12 seem to be the commonest numbers in Cucurbitaceae, and are frequently also in the supposedly related families Begoniaceae and Datiscaceae, Ayyanger inclines, like Ehrendorfer *et al.*, to accept the lowest number (7, occurring in some species only of the genus *Cucumis*) as the ancestral one. He points out that at least 5 of the chromosomes in the 7-chromosome species of *Cucumis* appear to be composite, with constrictions or 'satellites'.

Among the more extensive animal groups whose chromosomes have been studied, for example, the insect order Coleoptera, it seems that evolutionary changes involving substantial reductions, as well as increases, in the chromosome numbers have occurred; in Coleoptera, for which an ancestral complement of nine pairs of autosomes and an x–y sex-determining system is fairly well established [171, 171a], the number of pairs of autosomes may be reduced to six pairs in Coccinellidae, to five pairs in Elateridae, and to three pairs in Chrysomelidae. It seems to me that whatever selective forces have been responsible for such decreases in the basic number in animals, are likely to have had their analogues in relation to the evolution of plants, and that the evident bias of botanists in favour of accepting the lowest chromosome number as ancestral in any group is likely to lead them into error.

The work of Manton and others on the chromosomes of pteridophytes [137] has revealed phenomena which generally resemble those found in Angiospermae, except that multiplication of the chromosome number tends to be carried further in the older groups; none of the major divisions appears to offer a stability of the chromosome number which could be compared with that of the Gymnospermae. Manton points out that allotetraploid hybrids seem to be quite common in pteridophytes, whereas autotetraploidy, if we are to judge from the occurrence of multivalents in meiosis, seems to be comparatively rare. She also points out that some of the supposedly polyploid forms have smaller chromosomes than related forms with lower chromosome numbers—and suggests that this may represent a fairly general process of reduction in chromosome size following multiplication of their number. If this process is normal, then White's main argument against the allegedly tetraploid nature of the Golden Hamster (*Mesocricetus auratus*) may be invalidated. Manton considers that changes in the chromosome number in Pteridophyta have been almost entirely increases, through polyploidy, polysomy or fragmentation, and that higher chromosome numbers are associated with slower evolutionary rates. The information published by Darlington [41] on gymnosperms and angiosperms can hardly be said to establish Manton's conclusions as valid for these groups. On the whole, the conservative systematist is likely to conclude that in plants as in animals chromosomes do not offer a very reliable basis for the recognition of natural groupings, and are hence to be considered as a classificatory character of rather low value. As already noted, nothing like the 'giant chromosomes' of the dipterous salivary glands has yet been found in plants (see p. 148).

The natural occurrence of processes which would reduce the number of genes, or amount of chromatin, per nucleus is not well established. Fusions between chromosomes, leading to a reduction of the chromosome number, are, of course, well known in many groups of plants and animals; this process would not, however, involve any change in the number of genes or amount of chromatin. The chromosomes of many organisms are well known to contain greater or less proportions of differentially-staining 'heterochromatin', which appears to be genetically inert for the most part; there is some evidence that deletions in regions of heterochromatin may not be noticeably deleterious to organisms carrying them. It can thus be plausibly suggested that 'surplus' euchromatin can be progressively inactivated and converted into heterochromatin, and in this form ultimately lost. There are indications that the immediate increase in amount of chromatin (or DNA) per nucleus as a result of polyploidy may subsequently be compensated by a reduction in the size of the individual chromosomes. It has also been claimed recently that, at least in the apomictic plant genus *Rubus*, diploid forms may occasionally arise spontaneously from a tetraploid stock [51*a*, 195*a*, 204*b*].

A priori, it is natural to assume that the much more complex organisation of the higher plants and animals as compared with things like green Algae or Coelenterata must require a larger number of genes for its determination and appropriate variation. There is indeed evidence that cells of higher organisms contain much larger amounts of DNA than do things like bacteria or yeasts. Sneath [172] compares, for a whole range of organisms, the number of histologically-recognisable cell-types with the DNA content of the gametes. His figures indicate a very approximate proportionality between the two; rather anomalous positions in his graph are occupied by the Angiospermae and the Amphibia, both of which have considerably more DNA in their gametes than would be expected from their numbers of cell-types. The gametes of *Zea mays* contain about as much DNA as those of *Homo sapiens*, but few would claim that maize equals man in organisational complexity. It may, however, be significant that both maize and man belong to essentially 'modern' groups, which have evolved rapidly during the Tertiary era; the Graminae, like the Primates, are probably of post-Mesozoic origin. Sneath concluded: 'There is a suggestion that the relation between complexity and genetic information is not proportional, but that rather more than a hundred times as much information is needed for a hundred cell types than is needed for one cell type.'

Among plants, there is a well-known tendency for old and evolutionarily conservative groups, such as eusporangiate ferns, lycopods and Psilotineae, to have very high chromosome numbers, a circumstance which might be taken as supporting Darlington's assertion that increase in the chromosome number offers immediate advantage at the expense of ultimate survival—though the groups in question *have* survived for a very long time.

In the epigraph to this chapter, Darlington suggested that, where organisms have persisted for a very long time without evident phenotypic change, the phenotypic constancy will have been maintained by selection over a steadily changing genetic basis; unfortunately, in the absence of a Wellsian 'time-machine' it is difficult to see how we could test this hypothesis directly. There is, however, indirect supporting evidence—for example, many instances are known of 'sibling species' which are quite or almost indistinguishable phenotypically but differ sharply in their genotypes. Thus the potato and corn eel-worms of the genus *Heterodera* (Nematoda) are not distinguishable by structural characters, though Cunningham has shown that they are readily separable serologically, and they behave in other respects as separate species. No doubt the lack of any precise quantitative correlation between phenotypic and genotypic differences is one of the causes of discrepancies between phylogenetic classifications and traditional, Aristotelean ones.

An interesting genetical study by Sondhi [175] may have some bearing on systematic and evolutionary problems. This worker started with a stock of *Drosophila subobscura*, homozygous for the mutant 'ocelliless', the phenotypic manifestation of which is the absence from the top of the head of the usual three ocelli and three pairs of long bristles. In some individuals of the homozygous stock there were, however, visible vestiges of some of these structures; by breeding selectively from these, Sondhi was able to develop a stock in which most of the individuals showed the 'wild type' phenotype, though crossing experiments with other strains showed them still to be homozygous for the original mutant factor. At this point, with three ocelli and the normal three bristles present, he experienced a difficulty in obtaining any additional structures through selective breeding, though eventually he succeeded in producing a stock in which more than three ocelli, and reduplications of some of the bristles, were common. His most notable achievement was to produce, on more than one occasion, flies with an additional pair of bristles, making four pairs in all, though still with the normal three ocelli. The additional pair of bristles always appeared at the same place—between the 'orbital' and 'ocellar' bristles —and were always curved backwards. Bristles in this position have not been found in any species of Drosophilidae, but are a normal feature in a closely related family Aulacigasteridae.

Several interesting conclusions are suggested by Sondhi's work. In the first place, it seems that characters, lost as a result of single-factor mutations, may be recovered through the agency of selection—even complex characters like dorsal ocelli, which would normally be regarded as subject to Dollo's Law (Chapter 9). Secondly, the same phenotype may be produced with a different genetic basis. Thirdly, 'new' bristles at fixed positions may arise as a result of selection, so we are not obliged to suppose that evolution in bristles of Acalypterae is always by reduction; if Aulacigasteridae really are related to Drosophilidae, it does not follow that the latter group must be derived from the former rather than *vice versa*. It may, of course, be that the remote ancestors of Drosophilidae possessed the extra pair of bristles as in *Aulacigaster*, that some trace of the genetic machinery for producing this character persists in modern species of the family, and is capable of being re-activated by selection.

13 Phylogenetic Evidence from Nucleic Acids and Proteins

A great deal of recent biological research has been concerned with elucidating the structure and functions of the nucleic acids, both the DNA of the chromosomes and similar bodies and the RNA which is believed to function as an intermediary between the DNA and the extra-nuclear protein-synthetic activities of the cell. The DNA molecules are now believed to be the physical embodiment of at least the major part of the genotype of the organism. As is indicated in Chapter 9, phylogenetic classification is concerned to express genetical relationships, and there is reason to suppose that these relationships would be most clearly seen if we could compare the genotypes directly. Just this is what we may hope to be able to do if a practical technique can be devised for unravelling the structure of the DNA molecules of numerous species. This, however, would be a forbiddingly complex and difficult task, unlikely to be achieved in the near future.

Of the techniques currently practised, perhaps the most promising from this point of view is that of measuring the degree of 'pairing' between single-strand DNA fibres of different species; a method of doing this has been described by Bolton and McCarthy [96]; see also Spiegelmann [177]. In most organisms natural DNA fibres are duplex, consisting of two strands held together by a large number of weak chemical bonds; the bonding between the strands depends on the presence of appropriate radicals at corresponding points on the two strands. Duplex DNA may be split into single strands, without otherwise altering it notably, by carefully controlled heat treatment, and can be preserved in this condition by embedding it in suitable gels. It is also possible to 'label' DNA molecules of organisms or of tissue-culture cells by the incorporation of radio-phosphorus. The Bolton-McCarthy technique involves the incubation together of single-strand 'long' DNA fibres of one species with 'labelled' single-strand DNA, broken up into short lengths by physical treatment, of another species. The fragments tend to adhere to the complementary portions of the long DNA strands; by subsequently centrifuging out 'long' DNA fibres from the mixture, and determining the extent to which these have become radio-active, a measure can be obtained of the degree of pairing between the two species of DNA.

Figures obtained in this way by Hoyer, Bolton, McCarthy and others agree on the whole very well with estimates of degrees of phylogenetic relationship based on other evidence—and quite significant degrees of pairing have been found between the DNA's of organisms as different as a man and a bony fish. Some seeming anomalies in the published results may be due to imperfections of technique; the results may be greatly influenced by the precise temperature and duration of incubation, by concentrations and other experimental variables. When the technique of this way of measuring DNA pairing has been perfected and standardised, it may provide systematics with the best of all single methods of determining degrees of phylogenetic relationship.

A recent report of Britten and Kohne [20a] indicates that, in all species of eucaryote organisms (but in no procaryotes) so far examined, a considerable part of the genome is composed of DNA sequences repeated many times over; their evidence suggests that such repeated sequences are liable to comprise 200 or more nucleotides, and to be repeated up to 100,000 or more times. Thus in DNA fibre-pairing experiments like those of Hoyer, Bolton and McCarthy, sections of the 'short' DNA are likely to pair at times with non-homologous parts of the 'long' DNA fibres. The functional significance of such repetitions in the DNA sequence, and their implications for the phylogenetic interpretation of 'hybrid DNA' experiments, have yet to be clarified. It seems, however, that it is not now reasonable to expect DNA fibre-pairing results to give as precise a measure of phylogenetic relationships as Hoyer, Bolton and McCarthy originally claimed.

The experienced systematist's presumption, based on the consideration of evidence from many different fields, that there are no techniques which provide infallible measures of affinity, is likely to be vindicated once again. It would be regrettable, however, if the DNA hybridisation technique were abandoned prematurely on this account. Natural history in general knows of no techniques which provide infallible answers to the questions it asks, and DNA fibre pairing may well be the best single method available to us for measuring phylogenetic affinity (or genotypic resemblance, if you prefer the term) between fairly distantly related organisms.

Two other currently practised methods of studying nucleic acids may yield results of systematic and phylogenetic importance. One is the determination of the proportions of the four normal bases— adenine (A), thymine (T), cytosine (C) and guanine (G) in particular species of DNA. In normal, double-strand DNA, each base is present in the same molecular proportion as its complement—each adenine being matched by a thymine and each cytosine by a guanine. The results of this study can thus be expressed in a single figure for each species of DNA, usually expressed as the ratio of C + G to A + T. The method and some of its results are discussed by Sueoka [183b] and Subak-Sharpe [183a]. There is no evident reason why evolutionary change in the C + G/A + T ratio should be unidirectional, and it seems quite likely that, of two species showing the same ratio, one has developed from an ancestor with a greater ratio and the other from one with a lower ratio, a form of biochemical convergence. The systematist would thus expect this ratio to be a character of rather limited value from his point of view.

The second, systematically rather more promising, method is the determination of what is called 'base doublet frequency', alias 'nearest neighbour analysis'. The deoxyribonucleic acid under test is used as a 'template', together with the enzyme DNA polymerase and a supply of the four usual kinds of nucleotides, in the synthesis of new DNA. One of the kinds of nucleotides is supplied labelled with radioactive phosphorus. After the synthesis has been completed, the resulting DNA is broken down enzymatically into single nucleotides, and in the process, the phosphate links will have been transferred from the nucleotides to which they were originally attached to their next neighbours. By determining the extent to which each of the nucleotides has become radioactive, it can be determined how often each of them serve as 'next neighbour' to the original radioactive nucleotide. The procedure is repeated using each of the four nucleotides in turn as the radioactive one, and will yield sixteen figures. As pointed out by Subak-Sharpe, four of the nucleotide doubtlets so indicated—AT, TA, CG and GC—would be identical with their complements in the double-strand DNA, while the remaining twelve would form six complementary pairs—AA and TT, CC and GG, AC and TG, AG and TC, GA and TC. We may thus get ten distinct figures for each species.

Base doublet figures have now been determined (Subak-Sharpe, personal communication) for several species of Vertebrata, ranging from a man to a fish, for three species of Echinodermata, for a crab, an insect, a Protozoan, a green plant, a fungus and a number of bacteria. The Vertebrata clearly stand out by their very low proportion of the CG doublet (which, as Subak-Sharpe points out, always codes for arginine), this character being least pronounced in the fish and approached to a slight extent in the Echinodermata. Like the cytochrome-*c* studies discussed later in this chapter, the base-doublet frequencies emphasize the diversity of the bacteria, in comparison with the Eucaryota, and hint that some of the photosynthetic bacteria may be nearer to Eucaryota than they are to some other bacteria. It seems likely that the Procaryota-Eucaryota division adopted by Klein and Cronquist will have to be abandoned in any system which pretends to be phylogenetic. Available data suggest, though they are by no means sufficient to prove, that base doublet frequencies will provide valuable evidence for relationships between the major divisions of the plant and animal kingdoms, as well as for the classification of bacteria.

The sequence of the bases along the RNA molecule is believed, as we have seen, to determine the sequence of the amino-acids in protein molecules built under the 'direction' of this particular RNA. Conversely, the sequence of amino-acids in a protein molecule is thought to be directly determined by a part of the genotype of the organism. The study of the amino-acid sequences of the proteins of an organism can give information, only one step further removed than that of the DNA itself, about its genotype. Considerable progress has been made recently in determining the detailed structure of some proteins, and in the case of a few of the smaller types of protein molecules, the precise structure has been unravelled for a number of species of organisms. It is already evident that such comparisons of homologous proteins can provide phylogenetic evidence of high importance.

The 1968 edition of Dayhoff and Eck's *Atlas of Protein Sequence and Structure* presents, in conveniently comparable form, all the data available at the time on amino-acid sequences of proteins, and the information presented is sufficient to convince a systematist that at least some proteins, notably cytochrome-*c*, will provide evidence of the highest importance for phylogenetic relationships in the higher levels of the system. Sufficient data are also provided to show that the globins and insulins deserve great attention from systematists concerned with

Vertebrata. It is regrettable that no such comparative information seems yet to be available for the proteins of green plants, and very little for invertebrate animals.

The general systematic value of this evidence can best be illustrated by considering in some detail the data provided for cytochrome-*c*. This protein, which is believed to play an essential part in the energy-transferring reactions in mitochondria, seems to be present and homologous in all eucaryote organisms, and Dayhoff and Eck record also an incomplete sequence of a probably homologous protein from a bacterium, *Pseudomonas*. There is, it appears, a sequence of no less than 103 amino-acids occupying homologous positions in all eucaryote organisms, and out of these 103 no less than 35 are the same in all the species recorded (excluding the *Pseudomonas*). In effect, this molecule presents to the systematist a highly complex piece of structure which is present and homologisable at no less than 103 points in all organisms from a yeast to a man.

In discussing this evidence, it may be convenient to borrow the geneticists' term locus, to indicate a homologisable position of an amino acid in the cytochrome-*c* molecule. Dayhoff and Eck point out that, besides the 35 loci which are occupied by the same amino acid in all the recorded Eucaryota, there are at least 6 loci where changes seem to have occurred at least five or more times in the evolution of the group. The authors show that, as a systematist would expect, 'mutability' at the various loci in the molecule is not randomly distributed. They also recognise the likelihood that 'back mutation' will have occurred at times in phylogenetic history, and actually postulate it for one locus at 'node 1' of the dendrogram.

Dayhoff and Eck point out that 'It is obvious that there is a good deal of information in the detailed nature of both the non-identities and the identities. In scoring identities, it would be inefficient to consider one amino-acid as good as another.' A systematist would add that it is likely to be similarly inefficient to consider one locus as good as another. In regard to amino acids, it may be worthy of note that arginine, the largest, most complex and most unstable molecule of the twenty, appears at only four loci in the recorded cytochrome-*c* molecules of Eucaryota. At two of these it is constant in all of them, at one it appears once only, in a yeast, while at the third it is replaced by the related lysine in all the Vertebrata. It may not be altogether a coincidence that, as we saw in Chapter 8, in Vertebrata the functions of arginine as a phosphagen are taken over by creatine.

Considering the data from a classical systematic point of view, rather than from the quasi-numerical-taxonomic one of Dayhoff and Eck, some interesting points emerge from their tables called 'Alignment 1' and 'Alignment 2'. Thus the 'old world monkeys' (Catarrhina) appear to be distinguished by the presence of alanine at the locus numbered 50 in alignment 2 (or 58 in alignment 1), where the representative of Platyrrhina (*Erythrocebus patas*) has glutamic acid and other mammals have aspartic acid. The four representatives of Primates are distinct from the other tabulated mammals in having isoleucine at locus 11 instead of valine, methionine at 12 instead of glutamine, tyrosine instead of phenylalanine at 46, and valine instead of alanine at 83. Study of alignments 1 and 2 suggests that there may be some degree of correlation between variations at these four loci (numbered 19, 20, 54 and 91 in alignment 2), and this in turn suggests that they may be functionally linked in some way. If and when the actual configuration of the cytochrome-*c* molecule is worked out, light may be thrown on this matter.

The Mammalia as a group are also readily distinguished in the cytochrome-*c* sequences recorded in alignment 1. They have glutamic acid instead of lysine at 112 (hence perhaps the culinary popularity of monosodium glutamate) and asparagine instead of serine at 111, lysine rather than aspartic acid at 108, and valine or proline instead of glutamic acid at 52. Loci numbers 52 and 108 also seem to show some degree of correlation in their variation. The birds, of which four species are recorded, are distinguishable by the presence of serine instead of alanine at 23 and serine instead of glycine at 97. These two loci also seem to show somewhat correlated variation, and 97 in particular would appear to be a systematically 'interesting' one which might be particularly linked with key changes in evolution.

The old Ungulata (Artiodactyla + Perissodactyla) seem to be distinguished as a group by having glutamic acid instead of alanine at 92 (100 of alignment 1), and Perissodactyla are separable by having lysine instead of glycine at 60 (68), also threonine instead of glycine at 89 (97) though this feature recurs in the artiodactylan guanaco. The only marsupial recorded is a kangaroo, and its cytochrome-*c* sequence shows no evident features by which to distinguish Marsupialia from placental mammals. The two cetacean species recorded, representing two different families of Mysticeti, likewise show no distinctive common features in their cytochrome-*c* sequence.

It is unfortunate that the only animals other than Vertebrata for

which cytochrome-*c* sequences are recorded are two insects; the
Insecta, like the Vertebrata, represent a particular and highly specialised
division of a single phylum of Metazoa; before drawing any definite
conclusions about the interrelationships of these two groups and about
the cytochrome-*c* sequences of ancestral Metazoa, the systematist
would require to know the cytochrome-*c* sequences in a number of
additional species, e.g. at least an echinoderm, a mollusc, a flatworm
and a coelenterate. Dayhoff and Eck indicate a common ancestor of
insects and vertebrates at 'node 12' of their dendrogram, but their
'sequences of common ancestors' table for this point leaves ten loci
undetermined, whereas for the common ancestor of the vertebrates
(node 11) they leave only three loci undetermined.

Only one green plant, wheat, is included in their tables, and it
appears as significantly nearer to the animals than it is to the three fungi
tabulated. This might be taken as evidence supporting the idea that
metazoan animals are related to green plants through a *Volvox*-like
ancestor, but much more evidence will be needed to substantiate such
a conclusion. It seems likely that the study of cytochrome-*c* will throw
much light on the relationships of the lower organisms, considered also
by Klein and Cronquist [117].

Dayhoff and Eck do not specifically discuss the relations between
their sequences for Eucaryota and the partial one recorded for an
apparently homologous molecule from the bacterium *Pseudomonas*,
and this species is not compared with the others in their 'alignments'.
It appears, however, that the loci which they number 12, 15 and 16
in the *Pseudomonas* sequence correspond to numbers 22, 25 and 26 of
alignment 1, including the two cysteines concerned with attachment
to the haeme group. A glycine at number 39 in *Pseudomonas* may
correspond to a constant glycine at 49 in the Eucaryota, asparagine
at locus 50 of the bacterium to a constant one at 60 in Eucaryota, and
a glycine at 54 may correspond to one which appears at 64 in all the
'sequences of common ancestors' given by Dayhoff and Eck. Thus of
the eighty-two loci recorded for the bacterium, only five agree with
constant ones in the Eucaryota, and one more with a nearly constant
one in the latter group. Dayhoff and Eck also print the sequence for
a fragment, thirteen amino acids long, from cytochrome-*c* of another
bacterium, *Rhodospirillum*; loci 1, 4 and 5 of this fragment clearly
correspond to 12, 15 and 16 of *Pseudomonas* and 22, 25, 26 of the
Eucaryota. This fragment differs from the corresponding part of
Pseudomonas, and agrees with the postulated ancestral sequence of

Eucaryota, in having threonine at 6 (17 of *Pseudomonas*, 27 of Eucaryota) and glycine at 10 (21, 31).

While *Pseudomonas* represents a relatively 'normal' type of gram-negative bacteria, *Rhodospirillum* is a member of the peculiar group of photosynthetic bacteria, whose place in the bacterial scheme of things is usually regarded as isolated and enigmatical. While it would hardly be justifiable to base far-reaching phylogenetic conclusions on such a small piece of the sequence as we have for *Rhodospirillum*, it does offer us a glimpse of a fascinating possibility—that some of the heterogeneous forms lumped as Procaryota may prove to be genetically nearer to Eucaryota than they are to other Procaryota. It has been seriously suggested that Eucaryota may have come from photosynthetic Procaryota. We need, not merely more knowledge of the *Rhodospirillum* cytochrome-*c* sequence, but also comparable data for blue-green algae, red algae, and various other bacteria.

Most of the data presented by Dayhoff and Eck were reviewed previously by Margoliash [139], who based his conclusions on a simpler numerical treatment, considering only the numbers of similarities and differences at corresponding loci between each pair of cytochrome-*c* molecules. His dendrogram, constructed on this basis, i.e. incorporating the principle of equal weighting advocated by Sokal and Sneath (see Chapter 15), differs at some points from that of Dayhoff and Eck, and is less consonant with other lines of evidence for the relationship of the forms concerned. Margoliash tried to establish an approximate proportionality between the number of loci at which two cytochrome-*c* molecules differ and the geological age at which their common ancestors separated, much as Hoyer, Bolton and McCarthy tried to establish a similar proportionality in respect of degrees of pairing in hybrid-DNA experiments. It seems unlikely to me that any such proportionality will be strong enough to be of much use in deducing phylogeny.

Most of the data on the haemoglobins of the mammalian order Primates presented by Dayhoff and Eck had previously been discussed by Zuckerkandl [212] and Buettner-Janusch and Hill [23]; the last-mentioned authors concluded that 'the hemoglobin data, however, at this stage are not crucial for reorganising phylogeny'. This conclusion would probably be reached by any systematist after study of the tables in Alignment 7 of Dayhoff and Eck. It may, however, become possible eventually to correlate some of the haemoglobin data provided by the biochemists with results of immuno-diffusion and 'fingerprinting' studies of vertebrate blood discussed later in this chapter.

'Systematic serology' or 'immunodiagnostics' stands to the chemical study of protein structure in much the same relation as the Bolton–McCarthy DNA fibre-pairing technique stands to the structure of the DNA itself. The 'antibodies' produced in a vertebrate's blood in response to an injected alien protein are somewhat comparable to the 'complementary' fibres of DNA, and probably become attached to their complements in much the same way. As in the case of the Bolton–McCarthy technique, results are liable to be considerably altered by slight differences in experimental circumstances or technique. However, the methods of systematic serology are simpler, and much more widely understood and applied than those of Bolton and McCarthy. Results are already available from a very large body of investigation of this kind, and it is clear that comparative serology, with critically standardised methods applied to appropriate material, can provide phylogenetic evidences of the highest value, at least as far as the lower and middle levels of the classificatory hierarchy are concerned.

It has long been known that when a foreign body or substance (antigen) of a proteinaceous nature is introduced into the blood-stream of a mammal or bird, it usually induces the formation in the blood of another substance (or substances) known as an antibody; the characteristic property of an antibody being to react with the substance(s) which occasioned its formation, with the formation of an insoluble precipitate (precipitin reaction). The usual procedure for demonstrating this reaction is as follows—an experimental animal (e.g. rabbit, horse or fowl) is injected with a suitable quantity of an isotonic serum containing the antigen(s), and a period of some days is then allowed for antibody formation to take place. A sample of blood is then drawn from the animal, the serum is filtered off from it, and a suitable quantity of this 'anti-serum' is mixed with a sample of the serum which was originally injected. Very soon a more or less flocculose precipitate begins to form, and the process is usually completed within a few hours. For a given amount of anti-serum, the amount of precipitin formed increases with the amount of antigenic serum up to a point and then falls off. Both antigens and antibodies are normally proteinaceous and too complex to be determinable by chemical means; there is no general method of estimating the concentration of antigen in a given serum, or of antibody in an anti-serum, other than by titrating them against each other. The speed and amount of antibody formation in response to a given dose of antigen varies between individuals of one species of animal, and even between different times in the one animal.

The systematic interest of the precipitin reaction derives from the fact that antibodies are rarely if ever quite specific: they will as a rule react to varying extents with a range of antigens. The reaction is ordinarily most intense with the specific antigen(s) which occasioned the formation of the antibody in question, but occurs to a greater or less degree with comparable preparations from organisms more or less nearly related to the one from which the original antigen was obtained. The simplest procedure is to prepare an isotonic serum from blood or other parts of one or more organisms of species A, to inject this into a suitable animal, and after a suitable lapse of time to draw from the latter a large enough sample of blood to provide serum for a considerable number of precipitin tests. The original injected serum, and sera prepared in exactly the same way from a series of organisms more or less related to A, are then subjected, in varying proportions, to precipitin tests against the 'anti-A' serum; the maximum amount of precipitin formation, or the maximum dilution at which perceptible precipitin formation occurs, is then recorded in each case. The maximum amount of precipitin formation between standard samples of anti-A serum and serum from species B, or the maximum dilution at which serum of B will still react visibly with anti-A serum, are taken as indices of the serological similarity between species A and B. For example, in Mammalia, an anti-serum against blood serum of a given species will usually react quite strongly with sera from other species of the same genus, to a moderate extent with sera from other genera of the same family, rather weakly with sera from species of other families in the same order, and very weakly if at all with sera of species of other orders.

It is possible, of course, to make reciprocal tests—to test anti-A serum against serum of B, and anti-B serum against serum of A; the results of reciprocal tests are by no means always identical, though usually fairly similar. Measurements of 'serological similarity' are inevitably erratic and approximate in practice, in order to obtain reliable and consistent results, it is usually necessary to replicate each test several times and to average the figures obtained. Experiments using rabbits as antibody producers usually give the same results as those based on horses or fowls, though the actual antibodies produced by different species in response to the same antigen are certainly different

—as is shown, for example, when anti-sera are subjected to electro-phoretic analysis. The principles and techniques of 'simple' comparative serology of this type are well explained by Boyden [17], Chester [29] and Moritz [144a].

One of the earliest extensive studies in systematic serology in the plant kingdom was that of Göhlke [76], who worked in Königsberg with Mez. Göhlke extracted protein material from powdered seeds of his plants by digestion with suitable saline solutions. His anti-sera were each prepared from a single rabbit, though he admitted that it would have been better to replicate the production of each anti-serum several times using different animals. Anti-sera were prepared against seed-proteins of *Petroselinum sativum* (Umbelliferae), *Brassica napus* (Cruciferae), *Papaver somniferum* (Papaveraceae), *Helianthus annuus* (Compositae), *Cucurbita pepo* (Cucurbitaceae), *Pyrus prunifolia* (Ros-aceae), *Lens esculenta* (Leguminosae), *Salvia officinalis* (Labiatae), *Juglans regia* (Juglandaceae), *Cannabis sativa* (Cannabinaceae) and *Corylus avellana* (Betulaceae). Each test was replicated several times with different amounts of anti-serum, and results were recorded for periods from twenty minutes to two-and-a-half hours. In almost all cases, marked positive reactions were recorded when an anti-serum was tested against a serum prepared from seeds of another species of the same family, and in most cases more or less definite, though weak, reac-tions were recorded with species of at least some related families. Some of these results were interesting and suggestive. Thus the anti-serum against *Corylus avellana* reacted quite strongly with sera from seeds of *Alnus* and *Betula*, but considerably less so with a *Carpinus*-seed serum —in fact its reaction with the latter was at about the same level as with *Castanea*, *Quercus* and *Juglans* sera, or with those of *Morus* and *Cannabis*. No reactions were recorded between this anti-serum and sera of *Myrica*, *Salix*, *Celtis* or *Urtica*, and only a very slight one with *Casuarina*. The anti-*Cannabis* serum reacted about as strongly with *Corylus* serum as the anti-*Corylus* serum did with *Cannabis* serum. The strongest reactions of the anti-*Cannabis* serum, apart from those with *Humulus* spp., were with Moraceae and Juglandaceae, and after them with Ulmaceae and Betulaceae; with most other families they were weak or absent. The anti-*Juglans* serum reacted with *Cannabis* and *Corylus* sera to about the same extent as anti-*Cannabis* amd anti-*Corylus* sera did with *Juglans* serum respectively; this anti-serum behaved a little anomalously in that it reacted to an almost equal degree with sera of Urticaceae and Cannabinaceae and also showed a marked reaction with *Myrica* serum as neither of the other two had done. Still more anomalous were its positive reactions with sera from several species of Polygonaceae. The anti-serum against *Papaver* reacted positively with sera from *Fumaria* and *Corydalis* as well as several other Papaveraceae, but hardly at all with sera of Cruciferae, Capparidaceae, Resedaceae, Berberidaceae or other families. The anti-*Brassica* serum on the other hand showed perceptible though weak reactions with several Papaveraceous sera as well as with *Fumaria* and *Corydalis*; it reacted quite strongly with *Cleome* (Capparidaceae), rather less so with *Reseda* spp., *Podophyllum* (Berberidaceae), *Viola*, *Magnolia* and *Passiflora*. More or less strong positive reaction of the anti-*Helianthus* serum were recorded with several species of Campanulaceae, weaker but quite definite ones with several Cucurbitaceae, little or none with Dipsacaceae, Valerianaceae, Rubiaceae etc. Correspondingly, the anti-serum against *Cucurbita* showed marked positive reactions with Campanulaceous sera, and with Compositae, but also with Loasaceae and Violaceae.

In setting Papaveraceae rather apart from Cruciferae and allied families, Göhlke's results were in line with recent biochemical evidence (see for example [185] and [111a]); likewise his evidence for an affinity of Umbelliferae to Cornaceae as well as a close one to Araliaceae has been supported by most recent work. The traditional view, supported by Göhlke's experiments, of a direct affinity of Compositae to Campanulaceae, is no doubt correct even though it receives very little support from the host-plant selections of herbivorous insects (see Chapter 10); on the other hand, his evidence for a connection between Campanulaceae and Cucurbitaceae has received less support from other studies, and might be suspected as a case of 'serological convergence'.

During the 1920's, the German botanical comparative serologists split into two opposing schools; the Königsberg school of Mez and Ziegenspeck continuing to assert the prime importance of the method for revealing phylogenetic relations throughout the plant kingdom, whereas the Berlin school produced a considerable body of results which cast doubts on the consistency and reliability of the method, at least where inter-familial and inter-ordinal relations were concerned. As an example of the work of the Berlin school, we may cite the study of Huhn [99], concerning the relationships of the dicotyledonous group Gamopetalae. He prepared anti-sera against leaf-extracts of several species of Gamopetalae and tested them against leaf-extracts of representatives of most readily available families of Dicotyledonae, gamopetalous and polypetalous. His published results were notable for their extreme inconsistency, positive reactions with a given anti-serum were liable to be distributed almost randomly among the families of both Gamopetalae and Polypetalae. Huhn concluded that the method was practically valueless for the purpose for which he tested it.

Subsequently Krohn [121], a pupil of the Königsberg school, also embarked on a study of the Gamopetalae, studying rather fewer sera than Huhn had done but providing a considerably fuller documentation of his methods and results, which were decidedly more phylogenetically suggestive than those of Huhn. Krohn's anti-sera were prepared against *Vaccinium* (Ericaceae), *Armeria* (Plumbaginaceae), *Buddleia* (Loganiaceae), *Hydrophyllum* (Hydrophyllaceae), *Asperula* (Rubiaceae), *Nolana* (Nolanaceae), *Doronicum* (Compositae) and *Faveolaria* (Styracaceae). Each was tested against sera of most of the families of Gamopetalae—no instance of a reciprocal test was recorded by Krohn, whereas Huhn reported discrepant results in several cases of reciprocal testing. Krohn's results complemented and sometimes confirmed those of Gohlke; thus Krohn's anti-*Doronicum* serum reacted positively with serum of Campanulaceae and also Cucurbitaceae—also *Calycera*, *Stylidium* and *Goodenia* but not with others of his sera. His results also provided further evidence for the unity of the group Tubiflorae and for the affinity of Labiatae and Convolvulaceae to that group.

A recent example of an extensive systematic serological study within a single family is provided by Jensen [111b] on Ranunculaceae. In this study sera were prepared from seeds of twenty genera of the family, and anti-sera were prepared from rabbits against ten of them. The immune reactions were studied in various ways, including the turbidimetric methods of Boyden, the double diffusion technique of Ouchterlony, and the so-called 'pre-saturation' method; the results obtained in these ways are compared and discussed extensively. A 'serological' classification of Ranunculaceae is proposed which, although not identical with any previously published morphologically based system, could probably be reconciled with structural evidence and also agrees fairly well with evidence provided by alkaloid chemistry (see Hegnauer) and chromosomal studies. The genus *Hydrastis* is placed in the family but given a subfamily of its own, while *paeonia* is excluded. Prof. Jensen's work probably marks an important step towards a truly phylogenetic system of Ranunculaceae.

Systematic serological studies among the Gymnospermae have been conducted by Kojima [119] and Koketsu [120], following the methods used by Göhlke. Kojima was mainly concerned to investigate the relations between major groups of Gymnospermae and the Angiospermae. He prepared anti-sera against *Cycas*, *Ginkgo*, several Coniferae, and three Angiospermae, *Magnolia*, *Pasania* (Fagaceae) and *Daucus*; against these he tested extract of fifty-two species of Dicotyledonae and a number of Gymnospermae. None of the three anti-sera against Angiospermae reacted visibly with any of the extracts of gymnosperms, but several of the anti-sera against Gymnospermae showed weak but perceptible reactions with extracts of one or more species of Dicotyledonae. These positive reactions occurred in representatives of all the gymnosperm groups studied, and involved different species (and often different families) of Dicotyledonae; a systematist could hardly fail to draw the conclusion that these reactions were of an essentially 'asystematic' kind and should not be taken as valid evidence for phylogenetic relationships. This is in accordance with the fact that the ancestral divergences the groups concerned were certainly much older than those between families of Dicotyledonae; it is fairly clear that this type of immunotaxonomy is not of much value as a measure of inter-familial relations in Angiospermae.

Koketsu was more concerned with inter-relations among the Coniferae themselves, though he also tested anti-sera against extracts of *Cycas* and *Ginkgo* in his studies, together with ones against two species of *Podocarpus*, *Torreya*, *Sciadopitys*, *Abies*, *Larix*, *Pinus* and *Chamaecyparis*. Where *Cycas* and *Ginkgo* were concerned, reactions with the Coniferae were mainly negative; the anti-*Ginkgo* serum reacted weakly but distinctly with the *Cycas* extract but the anti-*Cycas* serum did not react with *Ginkgo* extract; an anti-*Podocarpus* serum reacted quite strongly with *Ginkgo* extract, but the anti-*Ginkgo* serum did not react with extract of the same *Podocarpus* species. Wide discrepancies in reciprocal tests seem often to occur where the reactions are 'anti-systematic'. Between the representatives of Pinaceae—*Abies*, *Larix* and *Pinus*, reactions were moderately strong and reciprocally consistent as a rule; at this and lower levels, it seems, comparative serology using seed extracts as antigens may be expected to yield useful results.

It is evident, from the results of a number of comparative serological investigations with plant-seed extracts, that the amount of precipitin formation between a given serum and anti-serum is not necessarily proportional to the phylogenetic relationship between the plants in question; clear instances of 'anti-systematic reactions' are cited by

Moritz [144a]. We have to allow for the occurrence of a certain amount of evolutionary 'convergence' in serological characters, just as we do in structural characters used by systematists. Moritz indicates methods by which some of the resulting anomalies may be eliminated by the use of 'pre-saturated' anti-sera, very much as was done by Manski, Halbert and Auerbach in their study of eye-lens proteins discussed later in this chapter. This method was, in fact, tried already by Göhlke. One of Moritz's conclusions was 'it appears to me that no quantitative biosystematics will arise, nor ought it, because systematics, which has always been a biological science (why then a special branch of "biosystematics"?), is by its very nature a qualitative science'. A recent critique of serodiagnostis in Angiospermae is provided by Vaughan [214].

As examples of systematic serology applied to the animal kingdom, we may consider briefly the results of Leone [125a] on crabs, of Leone and Wiens [126] on the mammalian order Carnivora, and of Williams and Wemyss [204] on the Primates, in each case using blood sera. In the case of the Crustacea, it has been shown that the haemocyanins are responsible for almost the whole of the antigenic activity of the blood sera. Leone [125a] prepared anti-sera for nine species of European crabs, representing nine genera and eight families; these were tested against the sera of the same nine species plus four more, including three species of one genus. Of the forms studied *Dromia vulgaris*, the sole representative of the Dromiaceae, proved to be the most isolated serologically, in good accordance with established views on the classification of the Brachyura. The spider crab, *Maia squinado* proved to be the second most isolated form, a circumstance less clearly shown in the accepted system. The species of the families Portunidae and Xanthidae all showed more or less strong reactions with each other, and their sera reacted very similarly against those of other families, with the notable exception that *Maia* serum reacted much more strongly with anti-serum to *Portunus puber* than with any other anti-sera from this group. As *Maia squinado* and *Portunus puber* usually occur together, were collected in the same area for Leone's study, and the *Portunus* is known as a particularly aggressive carnivore, we might suspect here a case of the predator being antigenically influenced by its prey. Other conclusions from Leone's work are that Cancridae and Ocypodidae are less close to Portunidae and Xanthidae than these two are to each other; Grapsidae appear to be still less close to these, and Calappidae almost as distant from them as are the Maiidae.

Leone and Wiens prepared anti-sera against a series of species of Carnivora, and tested these against sera of many species of the order. One notable feature of their results is that they offered no support for the primary division of the order into Fissipedia and Pinnipedia; a fur-seal, representing the latter group, proved to be serologically closer to some Fissipedia than these were to other Fissipedia. The anti-fur-seal serum showed a particularly strong reaction with polar bear serum, suggesting a parallel to the case of the crabs *Maia* and *Portunus*. In general, the bear sera and anti-sera showed the highest average level of reaction with other sera tested, a fact which Leone and Wiens took as indicating that the bears are particularly primitive serologically; on the other hand, the Palm Civet (Viverridae), representing what most systematists would take as the most primitive type of those studied, was the most serologically isolated of them, while the Hyaena, also considered to be a rather primitive type, occupied a serological position somewhere between the viverrid and the bears. The Giant Panda proved to be serologically closer to the bears than to the racoons. On the whole, these results agree fairly well with current views on phylogeny of Carnivora. The most serologically isolated form, the viverrid, is also the type which is believed to have the oldest separate origin, probably coming from a line distinct since the late Eocene period. Of the remaining forms, the bears are considered to occupy a rather central position, with direct affinities to the dog-like groups, to the cat-like forms, and through the pandas and racoons to Muste-lidae; the Pinnipedia are considered to be an aquatic derivative from the last mentioned line; all these views receive some measure of support from the results of Leone and Wiens.

Williams and Wemyss prepared anti-sera against sera of man and the Rhesus Monkey (*Macaca mulatta*), and tested sera of a number of species of the order Primates against these. They found the Chim-panzee and Gorilla to be much closer to man than any of the other species studied, that serum of the Mandrill (*Papio*) reacted more strongly with anti-human serum than did serum of the Rhesus Monkey, while the New World monkeys were serologically considerably less close to man than are any of the Old World ones; the reactions of *Cebus* serum to anti-human and anti-rhesus serum were almost identi-cal. These results were largely confirmed and extended by the more elaborate investigations of Paluška and Korinek [148] considered below.

As an example of the results achieved by simple comparative

serology within a single genus we may consider the work of West, Horwood, Bourne and Hudson [196] on the sawfly genus *Neodiprion*. Their investigations involved six species of the genus, and anti-sera were prepared against extracts of the larvae of four of them; similar larval extracts of all six were tested against these anti-sera, and the results expressed as percentages of the homologous reaction, in respect of degrees of dilution of the serum at which the minimum perceptible reaction occurred. The resulting table was as follows:

ANTIGENIC SPECIES	ANTISERA AGAINST			
	sertifer	*nanulus*	*lecontei*	*banksianae*
N. sertifer	100	41	40	51
N. virginianus	90	69	—	63
N. nanulus	54	100	57	—
N. lecontei	50	73	100	58
N. banksianae	46	47	39	100
N. swainei	33	28	9	—

From a study of this table it is evident that, serologically at least, *swainei* occupies a rather isolated position in relation to the other five species, and the generally similar reactions of *nanulus* and *lecontei* suggest fairly close relationship; there are anomalies, however, for example the close relationship suggested by direct tests between *sertifer* and *virginianus* compared with their markedly different reactions with *nanulus*, the notably weak reaction of *lecontei* with *swainei*, and the discrepancy in reciprocal tests between *lecontei* and *banksianae*. It would be an exaggeration to say that the relations between the species of *Neodiprion* have become clear as a result of this study.

In the last decade or so, the technique known as 'double diffusion', developed by Ouchterlony and others, has tended increasingly to replace simple precipitin testing for systematic and other purposes, and a still further development has taken place, with the introduction of the technique known as immuno-electrophoresis.

It has long been realised, or at least suspected, that sera used in serological tests are liable to contain several different antigens and that anti-sera are liable to be correspondingly complex. That some of these components might be separated, prior to the precipitin reaction, by diffusion (as in chromatography) or electrophoresis, has only recently been realised. Simple diffusion is employed for separation in the

Ouchterlony double-diffusion test. Serum and anti-serum are placed in two small 'wells' in a plate of agar jelly, and left to diffuse outward simultaneously. Different antigenic components, and their corresponding antibodies, are liable to diffuse at different rates, and we may observe the formation of several opaque lines of precipitin between the wells, each line representing the meeting of a particular antigen with its antibody. The distance between the two original wells on the plate is rather critical; the further apart these wells are, the more widely separated the precipitin lines are likely to be, thus favouring fine discrimination, but the more dilute will both antigen and antibody be at their meeting point, so that weaker reactions are likely not to occur at all. By using three or more wells instead of two, it is possible to test two or more sera simultaneously against a given anti-serum. In such tests, it may be possible to observe the junction of a specific precipitin line, from the interaction of one serum with the anti-serum, and the corresponding line in the reaction of the anti-serum with another serum. If both sera are from the same species, the 'homologous' lines form a smooth, rounded junction, a phenomenon known as the 'identity reaction'. Where the two sera are from distinct, not closely related species, few if any of the lines are likely to show this identity reaction: instead, the 'homologous' lines will probably cross one another angularly where they meet.

The immuno-electrophoretic method (see Williams [203]) involves a preliminary rapid electrophoretic separation of components of the antigenic serum along a line at right-angles to the subsequent diffusion, followed by diffusion towards a simultaneously diffusing anti-serum in an elongate trough parallel to the axis of electrophoretic separation. By this technique, more precipitin lines can usually be distinguished than in Ouchterlony tests, and there is more possibility of identifying particular antigenic components of a serum by the direction and distance of their electrophoretic movement and by the application of staining techniques, as in chromatography.

While simple precipitin testing gives a result which can be expressed as a simple number, thereby realising the ideal of the numerical taxonomists (Chapter 15)—a single figure representing the total 'similarity' or 'difference' between a pair of species—the results of Ouchterlony or immuno-electrophoretic tests are much more complex, and from a systematic point of view comparable with comparative anatomical data on fairly complex structures. For the phylogenetic systematist, such results should be worth a good deal more than those of simple precipitin testing. As an example of the results achieved so far by these more elaborate serological techniques, we may consider the work of Paluška and Koriňek [148] on Primates, which can be compared with the results obtained by Williams and Wemyss studying the same group by simple precipitin testing. Paluška and Koriňek prepared, from rabbits, anti-sera against sera of man and of the Asiatic monkeys *Macaca mulatta* and *M. cynomolgus*. Against these they tested sera of man, Chimpanzee, *Papio anubis*, *P. hamadryas*, *Theropithecus gelada*, *Macaca mulatta*, *M. cynomolgus*, of the New World monkey *Cebus capucinus*, of the African Galago (a Lemuroid), of *Myotis* (a bat), of cattle and of pigs. In Ouchterlony type double diffusion tests, each serum was used at a geometrically increasing series of dilutions; the degree of dilution of the serum and the number of observed precipitin lines was recorded for each test. For each serum tested against a given anti-serum two figures were recorded, the first being the maximum number of precipitin lines observed, the second the maximum dilution at which any detectable line was produced. With anti-human serum, human serum gave seven lines and fifteen, Chimpanzee seven to thirteen, *Papio hamadryas* seven to twelve, *P. anubis* seven to eleven, *Theropithecus* seven to twelve, both *Macaca* species five to twelve, *Cebus* four to nine, the Galago three to eight, *Myotis* two to seven, cattle two to six and pig two to four.

These authors also carried out a number of double diffusion tests combined with electrophoresis, according to the method of Grabar and Williams [203]. In this way more precipitin lines could be distinguished, and it was possible to attribute particular lines to more or less specific antigenic fractions. In these tests, both human and Chimpanzee sera gave fourteen lines with anti-human serum, *Macaca mulatta* gave ten, *M. cynomolgus* nine, *Cebus* six or seven, the Galago three or four, *Myotis* three; with *M. rhesus* anti-serum, human serum gave six lines, both *Macaca* species eight, *Cebus* four, the Galago three, and *Myotis* two; with *M. cynomolgus* anti-serum, human serum gave eight lines, Chimpanzee nine, *M. cynomolgus* twelve, *M. rhesus* eight, the Galago four, *Myotis* three.

In simultaneous double-diffusion tests in agar plates, pairs of sera were tested against a particular anti-serum. Testing Chimpanzee and human sera against an anti-human serum, four of the seven distinguishable lines in each interaction showed the 'identity reaction' when they met; these four lines were attributed to albumin, gamma globulin, beta-1-globulin, and beta lipoprotein. In the remaining Old World monkeys, only the line attributed to beta-1-globulin shows the identity reaction with the corresponding line from human serum.

On the whole, the serological evidence so far published confirms accepted ideas about the phylogeny of Primates, but there is at least one apparent anomaly: the African *Papio* and *Theropithecus* seem to be decidedly closer serologically to *Homo* than are the Asiatic *Macaca* species. Current ideas on phylogeny, based rather heavily on 'cheek-teeth systematics' (cf. Chapter 6) postulate a rather old separate origin for the anthropoid line, and would link the African and Asian monkeys much more closely with each other than with the anthropoids. It is quite conceivable that current ideas are wrong, and that the anthropoids may be a basically African line, having a common ancestor with the baboons more recently than the latter have with the macaques. However another type of explanation of the anomaly is possible. The Chimpanzee, like the *Papio* and *Theropithecus*, is an African animal, and recent discoveries strongly indicate that our own ancestry lay in the same continent. Even today, chimpanzees and baboons are liable to live in rather close contact with each other. It seems likely that for a considerable period the ancestors of all these animals were exposed to practically the same parasites and diseases—parallel adaptations to these may have conditioned some parallel developments in the antigenic nature of the hosts' sera. A recent paper by Damian [39] provides clear evidence that the antigenic properties of parasites may be influenced by those of their hosts; something like Newton's Third Law might well be applicable to such interactions.

A very interesting instance in which comparative immunological studies lead to phylogenetic conclusions is the work of Manski, Halbert and Auerbach [125] on lens-proteins of the vertebrate eye. The protein components of the vertebrate eye-lens seem to have been unusually stable in the long-term course of evolution so that, to quote Manski *et al.* 'an anti-serum prepared against lenses of any of the vertebrate species gives precipitin reactions with lenses of all other vertebrates'. They are thus able to produce serological evidence for relationships at much higher levels than are usually practicable. Their investigations were limited to the soluble (crystalline) proteins of the lens, excluding the insoluble (albuminoid) ones; most of their results are tabulated from simple serological tests, the results of which are classified as positive, slight, or negative. However, they also made a number of tests with the immuno-diffusion method which helped to elucidate and confirm the results of the simple precipitin tests. The

G

special feature of their work was the systematic application of the technique of treating an anti-serum with a particular type of antigen in order to inactivate particular types of antibody before testing it with other anigenic sera known as pre-saturation.

Their method is perhaps best explained by reference to the theoretical interpretation they developed from their results. Taking a lamprey as a representative of the most primitive surviving vertebrate type, they suggest that its lens preserves the protein constitution of the lens of the ancestors of all other living vertebrates; in support of this assertion they show that an anti-serum to soluble lens proteins of the lamprey, if it is first treated with lens-protein serum of any of the other vertebrates they studied, can be rendered unable to form precipitin with lens-protein sera of any vertebrates, even the lamprey itself. Similarly, they found that an anti-serum to lens protein of a shark, after treatment with lamprey lens-protein, retained the power to react with lens protein sera of all vertebrates other than the lamprey—hence, they conclude, a new component came into the constitution of the lens at the beginning of the Gnathostomata line and has persisted in all vertebrates originating from that line. Other new proteins are similarly shown to have come in at the amphibian, reptilian, and mammalian stages. There are also indications that at a few points in phylogeny specific proteins have been lost from the constitution of the eye.

Reviewing the results of these and similar investigations, I think the reasonable conclusion is that, if proper precautions are taken to ensure that only 'homologous' antigens are used, comparative serology provides one of the best available guides to phylogenetic relationships in the lower and middle ranges of the taxonomic hierarchy, coresponding to common ancestors within the Tertiary era, e.g. within orders of mammals or families of dicotyledons or insects, but its results tend to be inconsistent and unreliable at higher levels. There are howewer some types of antigens, such as cytochrome-c (see Chapter 13) and vertebrate eye-lens proteins which are unusually stable and can provide valuable serological evidence for relationships at much higher levels. The Mez-Ziegenspeck 'Stammbaum' (see Alston and Turner [3] p. 33), based on an eclectic (there is probably substance in the Berliners' claim that the Mez was adept at finding specious reasons for rejecting experimental results which did not fit into his preconceived picture) use of serological results at unduly high taxonomic levels, can probably be dismissed as of little evidential value.

It is an interesting paradox that comparative serology, in its original form of simple precipitin testing, provided at the outset what the numerical taxonomists see as their ultimate taxonomic goal—a single figure to represent the degree of difference between any two species. Progress in systematic serology has been in the direction of breaking this figure down more and more into its constituent elements, by the development of techniques of chromatography, immuno-diffusion, immuno-electrophoresis, etc. The apparent goal of this progress is a situation comparable to that of classical, anatomically-based systematics, in which groups are defined by specified combinations of characters.

14 The Systematic Value of Characters of Immature Stages

The phylogenetic (and, by implication, the classificatory) significance of the characters of developmental stages in animals and plants has been the subject of much recent controversy, in which the issues have been bedevilled by dogmas and counter-dogmas. Without doubt, Haeckel, in expounding his 'biogenetic law', greatly exaggerated the extent to which the ontogenetic development of structure in animals 'recapitulates' the main stages in the structural development of the ancestral series of adult forms. I believe, however, that de Beer's reversal of the Haeckelian doctrine has been carried to at least equally absurd excesses. It is clear enough that evolution by 'paedomorphosis' or 'neoteny' has occurred at times in both animal and vegetable kingdoms; the adult, reproductive stages of such animals as lice and perennibranchiate newts, and of such plants as duckweeds (Lemnaceae) or the celebrated *Welwitschia*, have structures resembling those of immature stages of their ancestors.

It appears to me that paedomorphosis may have played a part in the development, not merely of the perennibranchiate forms, but of the entire class Urodela. Scarcely anyone seems to have suggested that the tail, which is one of the main features distinguishing adults of this group from the Anura (frogs and toads), might be a persistent larval feature rather than a genuinely ancestral adult one. If this is the case for larval gill-pouches and external gills, why not for the tail? On this hypothesis, Urodela could be a relatively modern development from genuinely anuran ancestors. Such geological and distributional (see Chapter 11) evidence as we have agrees quite well with this theory; fossil Anura are known at least as far back as the Jurassic period, but no indubitable pre-Cretaceous fossils of Urodela have been recorded—and such fossils of the group as have been found are all from the Holarctic region, to which Urodela are now confined (except for evidently recent incursions into the Neotropical region and into the mountains of south-east Asia). Such a geographical limitation, in a group which is quite numerous and apparently successful in the modern world, is hardly to be paralleled at anything like the accepted classificatory level of the Urodela. It may well be that the neotenous tendencies, exceptional regenerative powers, large cells, and apparently recent origins of Urodela are more or less related phenomena.

Apart from this evidently exceptional case of the Urodela, neoteny in animals seems generally to be associated with small size, structural simplification, and generally regressive evolution. A typical example may be seen in the parasitic lice among Insecta; these are almost certainly a relatively recent offshoot of the winged free-living Psocoptera, and preserve in the adult stage many features found only in the nymphs of the winged forms. Other possibly (though by no means certainly) neotenously-derived groups in the animal kingdom include Larvacea among the Tunicata, some of the Acoela among Turbellaria, Mystacocarida among Crustacea, and Pselaphognatha among the Diplopoda.

It is also true that the immature stages of many animals show characters, not present in their adults, which we have good reason to believe to have occurred in the adults of more or less remote ancestral types. On any reasonable estimate, this phenomenon seems to be far more widespread than is neoteny; it can be traced to a greater or less extent in practically every major group of the animal kingdom, and is known also in many plants. In general, I believe that when the immature stages of one kind of animal show resemblances to adults of some other kind, this should be taken as *prima facie* evidence that the first kind of animal has descended from ancestors resembling the second. The reverse relationship, implying neotenic evolution, may also be possible but is in general much less probable. Like practically all other types of evidence for phylogenetic history, apparent 'recapitulation' in development does not warrant certainty, but may help a great deal in establishing high probabilities. An interesting entomological instance is discussed by Emerson [57].

I believe that there has been, in the major evolution of the animal kingdom, a tendency for new habitats and modes of life to be pioneered by the adult stages, and consequently for adaptations to these new conditions of life to appear first in the adults. This tendency could hardly be manifest in the same way in non-motile multicellular plants, and this, I think, may be the reason why Haeckelian recapitulation is a less evident phenomenon in the vegetable kingdom than it is in animals. However, I think it is established that a certain amount of ontogenetic recapitulation is traceable even in higher plants—one need quote only the relatively normal juvenile leaves in various types whose adult foliage is very highly modified or reduced, and perhaps also Bailey's ontogenetic sequence from protoxylem to metaxylem (see Chapter 9). A possible example at a higher level, more comparable to the tadpole larva of Amphibia, is that of the protonema stage in the development of the gametophyte of many foliose Bryophyta—this may be a genuine carry-over from a remote alga-like ancestor.

An important factor working against the manifestation of proper recapitulation in the animal kingdom has been the widespread tendency toward accelerated development, often correlated with the development of more yolky eggs. In some groups of the animal kingdom, there are suggestions of a repeated pattern of this sort—first, an ontogeny involving a more or less recapitulatory larval stage, with quite different structure and habits from the adult, then the development of very yolky eggs with the elimination of this larval stage in favour of a 'direct' development, then the adoption of a radically new mode of life in the adult stage, leading to the appearance of a marked metamorphosis with the old type of adult structure (and habits) now appearing as a larval stage, then perhaps an elimination of this larval stage leading to a new type of direct development. The most obvious example is furnished by the Chordata, which no doubt originally had bottom-living adults and a free-swimming larva of the tornaria type; from this the elimination of the ciliated larva may have produced a more or less direct development similar to that of *Amphioxus*; thence arose a macrophagous agnathan ancestor which retained the *Amphioxus* type of organisation in an ammocoetes-like larval stage; then a step to a fish-like form with direct development; then the adult becoming adapted to terrestrial life, as an amphibian, retaining the fish-like organisation in the larva—and finally the elimination of the aquatic larva from the postembryonic development of the Amniota.

A developmental feature which I believe may have a far more profound phylogenetic significance than has yet been generally realised is the asymmetry in the formation of the coelom in those Echinodermata and Chordata in which it is formed enterocoelically. I am not aware that anyone has suggested a functional reason for this inward asymmetry in larvae whose outward bilateral symmetry is normally almost perfect when it is manifest. It seems probable that we are here dealing with an element of Haeckelian recapitulation, pointing to a remote ancestral form which was asymmetrical in the adult stage. The Echinodermata indeed are believed to have developed their radial symmetry on the basis of an original asymmetrical posture following fixation, as is seen today in the ontogeny of some Asteroidea; very few authors, however, have considered the possibility that Chordata come from similarly asymmetrical ancestors. Yet there are many indications of a particular affinity between Echinodermata and Chordata, not merely in the echinoderm-like larvae of Enteropneusta, but also in the phosphagens (see Chapter 8), and in the presence in these two phyla alone of a mineralised mesodermal skeleton. Several palaeontologists have noted the remarkable similarity between isolated skeletal plates of the earliest known Vertebrata (Heterostraci of the group Astraspida) and those of certain supposed Echinodermata, the Heterostelea, Carpoidea or Homalozoa, of the lower Palaeozoic period. These Heterostelea were highly asymmetrical animals—why should not they, or something like them, have been ancestral to the main vertebrate line? This suggestion has, in fact, already been put forward by Gislen [73], and developed recently by Jeffries [1111].

Another developmental characteristic which may have phylogenetic significance is the spiral cleavage pattern seen in the early development of many groups of invertebrates. This normally gives rise to perfectly bilaterally symmetrical larvae, in contrast to the perfectly bilaterally symmetrical type of cleavage seen in primitive types of the echinoderm –chordate group, where the resulting larva always exhibits internal asymmetry. There are reasons for thinking that both echinoderms and chordates come from ultimate ancestors which were markedly asymmetrical in the adult stage. Is it possible that spiral cleavage is indicative of a remote ancestor with a more or less spiral organisation in the adult stage? Among the Pre-Cambrian fossils described by Glaessner [74] at least two (*Tribrachidium* and *Mawsonites*) show more or less distinct indications of spiral organisation. It is easy to imagine spiral organisation arising from the radial symmetry of Coelenterata or Ctenophora.

Like practically all generalisations in natural history, the statement that conditions in earlier developmental stages are closer to ancestral ones than are those of the adult organism is liable to exceptions and may be seriously misleading when this is not taken into account. A well-known instance concerns the origin of bone in vertebrate animals. In most Vertebrata, cartilage appears as the ontogenetic precursor of bone, and on this account it was long thought that the ancestral vertebrates had a cartilaginous skeleton in the adult stage, as do modern Cyclostomata and Elasmobranchii. Modern interpretations of the fossil record make it appear very unlikely that this is true; both Cyclostomata and Elasmobranchii are believed to have come from ancestors with extensively ossified skeletons. Another notable instance concerns the eyes of the higher insects (Endopterygota). In these, the dorsal ocelli (simple eyes) appear only in the adult stage, the larval eyes being precursors of the compound eyes of the adult. Yet dorsal ocelli are certainly an ancestral feature of the entire hexapod line of Arthropoda, and are probably homologous with the 'nauplius eyes' which in Crustacea are essentially larval organs, appearing in ontogeny before the compound eyes. A somewhat similar phenomenon is manifested in the excretory organs of the Crustacea themselves; in many Entomostraca, the glands of the second antennae appear only as larval organs, functionally replaced in the adult by those of the second maxillae—but the typical malacostracan condition is the converse one, maxillary glands in the larva and antennary glands in the adult. Similar examples could be drawn from the development of the vascular system in higher plants.

On the assumptions which I have outlined, the phylogenetic systematist considering information about the developmental stages of his organisms needs to answer two main questions. The first is, to what extent are the special features of the immature stages recapitulatory, reflecting former adult conditions? The second, to what extent are the non-recapitulatory, 'caenogenetic' features of the young stages phylogenetically and taxonomically significant in their own right? For example, the unsegmented tarsus of the larval legs of higher insects (Endopterygota) probably reflects the condition in the earliest terrestrial ancestors of the hexapod line; on the other hand, the total absence of thoracic legs in larvae of Diptera tells us nothing about ultimate ancestry but is an excellent classificatory character for the order. This condition may confidently be predicated for the ancestral species of all Diptera, and seems to be strictly subject to Dollo's Law (see Chapter 9), no exceptions to it being known, despite the vast number and diversity of the Diptera.

There is, of course, nothing novel in the use of characters of developmental stages in classification; as examples, it may be sufficient to quote the long-established division of ferns into Eusporangiatae and Leptosporangiatae, according to the method of development of the sporangia, the Dicotyledonae–Monocotyledonae division of the Angiospermae, the division of winged insects into Exopterygota and Endopterygota according to the mode of metamorphosis, and the Amniota as a division of Vertebrata.

In those insects (Endopterygota) which exhibit what is termed a complete metamorphosis, the larval and adult stages manifest a degree of phylogenetic and taxonomic independence of each other which is hardly equalled by the developmental stages of one and the same organism elsewhere; the hydroid and medusoid stages of Coelenterata, or the gametophytes and sporophytes of Bryophyta and Pteridophyta, which present analogous systematic problems, are usually considered as independent organisms. In the case of Endopterygota, a rather special form of Dollo's Law seems to be applicable—the loss of adult-like structures in early larval instars seems rarely to be reversible in later larval development. Apart from the 'imaginal discs', a full-grown larva of Endopterygota is no more similar in functional structures to the adult than is a newly-hatched one, and may often be less so. This is manifest, for example, in things like numbers of ocelli, segmentation of antennae and legs, mouth-parts, and sclerotisation of body segments; according to Hinton [94] there may be some exceptions to this generalisation in the development of the tracheal system. Reductions in any or all of these characters in the larva have as a rule no influence on the corresponding adult characters.

The regular systematic use of larval characters is a comparatively modern development in entomology, and undoubtedly its full potential has by no means been generally realised. The order Coleoptera, within which my own studies have largely been confined, can probably be taken as in this respect typical of Insecta-Endopterygota. My experience indicates that, as a general rule, the larvae are nearly equal to the adults in classificatory value; some natural groupings are much more clearly expressed in the larval stage, for example, the subfamilies of Elateridae, while others, such as the main divisions of Carabidae, are more evident in the adult structures. It may not be altogether accidental that, of these quoted examples, the Elateridae characteristically have long-lived larvae and short-lived adults, whereas in Carabidae larval development is usually fairly rapid and adult life long. The Insecta-Endopterygota provide the best example in either the plant or the animal kingdom of the systematic usefulness of characters of pre-adult and post-embryonic developmental stages. In most other organisms whose ontogenetic development involves changes of form, we do not find such abrupt metamorphoses as those of Endopterygota, and it is difficult or impossible to define homologous developmental stages really sharply. Hence it is difficult to base clear-cut systematic distinctions on the characters of such stages. This is true to a greater or less extent of most groups of animals which possess larval stages, such as Polyzoa, Mollusca, Annelida, Crustacea, Echinodermata and Amphibia. In none of these groups has a fully developed 'larval classification' been produced, and in none of them do systematists in practice place much reliance on larval characters in making classifications.

To some extent, of course, the failure to use larval characters in the classifications of these groups is attributable to the difficulty of obtaining the necessary information. In most of them it is far more difficult than in Endopterygota to rear larvae in the laboratory. We know, however, that the major groups of Echinodermata can as a rule be recognised immediately by their free-swimming larvae, and it is probable that a really thorough study of the trochophore larvae would make a very valuable contribution to an improved classification of the Annelida. The characters of stages such as the cercaria and the cysticercoids are recognised as having systematic value within the Platyhelminthes. The characteristic larval stages provided the first and almost the only evidence for placing Rhizocephala in the Crustacea–Cirripedia.

The torsion which characterises the molluscan class Gastropoda, and which is brought about relatively quickly during the metamorphosis of the larva, is a rather problematic phenomenon. Many malacologists have followed the verses of Garstang [69], in supposing this condition to have originated as a larval adaptation, rather than through selection operating in the adult stage. However, the fact that in what are universally agreed to be the more primitive forms of existing Gastropoda, ontogenetic torsion takes place in two well-separated stages, casts doubt on the Garstang hypothesis. It is difficult, in any case, to think of another instance where a major structural change affecting the adult organism can be attributed to selection operating purely on the larva. To me it appears far more probable that torsion originated, phylogenetically, as an adult adaptation which has been subject to later ontogenetic acceleration, in order that the process may be completed by the time the larva is ready to take up a snail-like mode of life on the bottom. This theory of the origin of torsion could also be expressed in verse, perhaps like this:

> The ancestral snail's large fore-twisted shell
> Was buoyed up o'er her back by air inside—

The air was lost, the heavy spiral fell
    And dragged, lop-sided, when to creep she tried.
Hence twisted was the shell, whose weighty coil
    Was pulled behind; the pallial space now turned
To lie in front, and, resting after toil
    The snail to tuck her head in quickly learned.

In the vegetable kingdom, we have seen that the primary division of the Angiospermae has long been based on the cotyledons of the embryo. Carr and Carr [27] have recently pointed out that, in some of the Dicotyledonae, the cotyledons may bear appendages which appear to be homologous with stipules; these structures may be present in some forms whose adult foliage is exstipulate, or absent in some others where the mature leaves possess stipules. Carr and Carr state that cotyledonary stipules are present in all species examined of the order Myrtales, though in the family Myrtaceae, many species of *Eucalyptus* and *Angophora* lack stipules in the adult foliage; in Salicaceae, the adult leaves possess stipules, but these are lacking in the cotyledons and the first true leaves of the seedling. Violaceae, Rosaceae and Fabaceae are said to lack cotyledonary stipules, but the true leaves in them are stipulate from the beginning. *Hedera* (Araliaceae) is reported to have stipules on the cotyledons but never on the true leaves, while those Onagraceae examined lacked stipules in all stages. The cotyledonary stipules of Myrtaceae, forming a transverse row across the base of the petiole, are said to resemble the adult ones of *Combretum*, belonging to the nearly related small family Combretaceae. Assuming that Carr and Carr are right in their controversial homologisation of some of these structures with stipules, it appears that their presence in early developmental stages may be of considerable systematic importance; the phenomena may even involve some degree of Haeckelian recapitulation. In another dicotyledonous family, Gesneriaceae, it has been shown by Burtt [24] and others that some degree of inequality of the cotyledons is a very widespread feature, and provides a valuable basic classificatory feature in the family; in some cases the inequality is carried over into the later true leaves, thus manifesting an element of paedomorphosis. Vasilchenko [193b] provides a good conspectus of the seedlings of woody plants.

Among the Gymnospermae, a sharply differentiated juvenile form of foliage is characteristic of many Cupressaceae, which have adult foliage of a specialised type; the juvenile foliage tends to resemble the adult form in species of *Juniperus*, and may well be in some degree recapitulatory of an ancestral type. The presence or absence of a protonema stage, and its form when present, have for some time been accepted as characters of some value for the higher classification of the Bryophyta; here again, developmental stage may be to some extent recapitulatory, though few botanists have seriously considered the possibility. Recent biochemical work on fungi has strongly indicated that flagellation of the spores, when present, provides characters of the highest classificatory importance.

The available data on the embryonic development of the Angiospermae are reviewed in an important work by Davis [46] following particularly on the studies of Maheshwari. This work, of course, deals only with the stages involved in the seed itself, and provides no data on the seedlings or any elements of Haeckelian 'recapitulation' in ontogeny. The data provided by Davis on embryonic development in Scrophulariaceae in general support conclusions of the relationships of the family based on structure of the adult plants and on chemical and serological characters. In having a unitegmic and tenuinucellar ovule, Scrophulariaceae agree with most other families of the Gamopetalae and differ from the large majority of Polypetalae; the combination of these features with a polygonum-type embryo sac and embryogeny of the onagrad-type is common to Scrophulariaceae, Orobanchaceae, Acanthaceae, Buddleiaceae, Globulariaceae, Oleaceae, Plantaginaceae, Labiatae, Verbenaceae, Myoporaceae, Selaginaceae, Callitrichaceae, etc., but separates these families from Solanaceae, Boraginaceae, Polemoniaceae, Primulaceae, Ericaceae, Rubiaceae, Valerianaceae, Compositae, etc. Embryological characters are often useful in indicating the true relationships of types in which adult structure is highly reduced or modified, e.g. Lemnaceae (embryologically close to Araceae), Callitrichaceae (embryologically near Scrophulariaceae, etc.) and Adoxaceae (embryologically near Caprifoliaceae), etc. It is noteworthy that, by the embryological characters recorded by Davis, Cucurbitaceae do not belong in the Gamopetalae, and have more in common with Begoniaceae than with Campanulaceae.

Developmental features in the fields of biochemistry and physiology (see Chapter 8) may possess phylogenetic and systematic importance quite comparable to that of the structural features we have so far considered. The appearance of ammonia as the final excretory product in early developmental stages of Amniota has a similar significance to the presence of gill-slits and cardinal veins in the same embryos. The early developmental stages of freshwater eels manifest, in their physiological adaptations to marine life, an element of ontogenetic recapitulation; ichthyologists are agreed that eels were originally marine fishes, and no species of the group is known to breed in fresh water. The haemoglobin of human embryos differs from that of adults in many respects; in at least one of these, the presence of only two polypeptide chains in the molecule, it resembles haemoglobin of the adults of some of the lower Primates, such as the lemuroids. Among Passeriform birds, even in the most specialised seed-eating and herbivorous groups, the nestlings are fed almost exclusively on insects, etc., the presumed food of the adults in ancestors of this group.

15 Numerical Taxonomy

Ninety-six identical twins operating ninety-six identical machines!
The voice was almost tremulous with enthusiasm.

ALDOUS HUXLEY [102]

We have already (Chapter 2) defined a natural classification as one in which members of a given group may be expected to resemble each other in characters other than those needed in order to refer them to that group; the phylogenetic definition of a natural classification is considered in Chapter 9. We need now to consider a third, and quite distinct, possible definition. Many recent writers have suggested that a natural classification should be one in which organisms are grouped in accordance with their total degrees of difference and similarity, when *all* characters in which they may differ are taken into account. If this, which we may call the statistical definition, is to be applied in practice, several conditions need to be fulfilled. First, we need an objective definition of a unit character: without this, no objective quantification of similarity and difference is possible. The nearest approach to such a definition which I have been able to find in the writings of the numerical taxonomists is the following, from Sneath and Sokal [173]: 'a taxonomic character of two or more states, which within the study at hand cannot be further subdivided logically (except for the subdivision brought about by changes in the method of coding the states)'. In the paper from which this definition is quoted, two or three examples of unit classificatory character are given, e.g. the hairy vestiture of mammals, their non-nucleated erythrocytes, and the presence or absence of particular setae on the bodies of flies. The logical indivisibility of these characters is not immediately apparent: for example, the transformation of reptilian scales into hairs can hardly have been an all-or-nothing process; one can easily imagine stages in the degeneration of the nuclei of erythrocytes; and particular setae on the body of *Drosophila* may vary in size, be present on one side only, or be present or absent in different individuals of one species.

A second requirement for the statistical definition of natural classification is that all the variation within a group should be resolvable into a finite number of such unit characters. The third requirement for the satisfactory applicability of the methods of numerical taxonomy is that the same number of characters should be present in all the forms to be treated. Such an assumption may seem not unreasonable when the forms to be compared are different genera and species of bees, as in the work of Sokal's colleague Michener—though even then there is the problem of the parasitic bees, in which pollen-collecting and wax-manipulating organs are liable to be lost altogether. If, however, the Sneath–Sokal methods were applied in an attempt to establish the proper place of the Lemnaceae among Monocotyledonae, the third assumption would be hard to establish, and serious difficulties would probably be encountered.

The theory of 'statistical' classification is that, supposing the three pre-conditions are fulfilled, we could base our taxonomic hierarchy on the number of characters common to the members of particular groups. Sokal and Sneath are at pains to point out that the groupings they are advocating would be 'polythetic', in that a group would be based, not on a specified list of characters present in all its members, but on a given *number* of characters in common when any species of the group is compared with any other, regardless of the nature of the characters. Species A, B and C would all be included in one group if species A had at least fifty characters in common with B, species B at least fifty in common with C, and species C at least fifty in common with A, even though the characters common to A and B, to B and C, and to C and A, were largely different in each case. The establishment of groups on this basis would be a very laborious procedure for the human systematist, but can be greatly facilitated by the use of suitable computers. One paradoxical result of such a procedure is that it might happen that species A might have more characters in common with a species D in another group than it has with species C in its own group—D being excluded from the first group because it has less than fifty characters in common with B and C. If our classificatory hierarchy were established on this basis, the incorporation into it of a newly discovered species would be a complex problem probably requiring the use of a specially programmed computer. And no species could be given any definite place in the system until all the requisite characters were 'recorded' for it.

The advantages claimed for numerical taxonomy are objectivity and mechanisability. To quote Sneath and Sokal again: 'It is the aim of numerical taxonomy to develop methods by means of which different scientists, working quite independently, will and must arrive at identical estimates of affinity between two organisms, given the same characters on which to base their judgements.' In our day it is becoming almost an established principle that any human function which cannot be performed by a machine is 'subjective' and has no place in science; Sneath and Sokal probably think of themselves as engaged in a gallant attempt to 'save' systematics by giving it a properly scientific basis. A further quotation from these authors will exemplify their attitude to phylogeny: 'The proponents of numerical taxonomy consequently insist on the separation of the taxonomic process from phylogenetic speculations' —here their essentially technological approach to systematics is implied in the word 'process', and the word 'speculations' may well reflect the traditional American 'history is bunk' attitude.

The first of the two Sneath–Sokal quotations in the last paragraph is worthy of closer study. Its last phrase is ambiguous—one might take it to mean, 'given that it is the same pair of organisms that are being compared'; but I think they really mean 'if the same characters are *taken* into consideration by both scientists'. Referring back to their previously quoted definition of a unit character, we note the epithet 'taxonomic'; the only meaning I can suggest for this word in this context is 'already used by systematists'. The character, the basic unit of numerical taxonomy, is apparently to be taken over uncritically from the practice of their unscientific, non-numerical, predecessors. The Sneath–Sokal declaration of aims might be paraphrased as follows: 'if all systematists dealing with a given group adopt the same conventions about which characters in it are to be regarded as indivisible units, if they all confine their attention to the same set of characters, and if they all process their data in the same ways (i.e. those prescribed in the Sneath–Sokal works), they will all arrive (not very surprisingly) at the same conclusions'— a result which these authors treat as self-evidently desirable.

If Sneath and Sokal require that characters should be 'logically indivisible' if they are to be usable for numerical taxonomy, they will generally find themselves restricted to the use of very few characters. Elsewhere in the same paper they speak of the need to utilise 'large numbers of characters chosen at random rather than a few selected ones'. Whatever 'chosen at random' could mean in this context (could it be that these mathematically conscious authors are using the word random in a loose, undefined way?), it is not likely to be the same as 'chosen for their logical indivisibility'.

In spite of their apparent concern for and dependence on the established conventions and usages of systematists some passages in the paper of Sneath and Sokal appear to be decidedly subversive of established practices, e.g. 'We hold that a "natural" or orthodox taxonomy is a general arrangement intended for general use by all kinds of scientists; it cannot therefore give greater weight to features of one sort, or it ceases to be a general arrangement.' From this and other passages it would seem that established classifications in most groups, being based almost exclusively on those characters which are easily observable in preserved museum specimens, would fall under the stigma of 'giving greater weight to features of one sort'. If I understand them correctly Sneath and Sokal are arguing that a 'general' classification should be based as much on characters of physiology, behaviour, biochemistry, etc., as on traditional anatomy—a point of view which is not likely to gain much support from the existing taxonomic 'establishment', though phylogenetic systematists will sympathise with it.

Our original definition of the statistical 'natural classification' would be open to the grave objection that no organism could be properly placed in such a system until *all* of its characters had been studied. In face of this obvious impracticality, Sneath and Sokal suggest, in effect, that a sufficiently large 'random sample' can deputise for the totality of characters; this idea is implicit in their doctrine of the 'factor asymptote'. They claim '(a) that the more characters we study the more information we will accumulate; (b) that a random sample of the characters should represent a random sample of the genes of the organism; (c) that as we include more and more characters the rate of gain of new information for classificatory purposes will decrease. After a certain number of characters has been recorded the inclusion of further characters will become unprofitable from the point of view of classification.' Clause (a) of this may seem tautologous at first, but presumably the word 'information' is being used here in the currently fashionable technical sense, and the clause could be re-phrased as 'the more characters we study, the more similarities and differences we shall find within the group'. In clause (b) we meet again the loose use of the word 'random'. The essential content of the hypothesis is embodied in clause (c), which might better be worded in some such way as this: 'as we include more and more characters in our study of a given group, the classificatory importance of the new information (in the same sense of this term as in the first clause) will progressively decrease.'

Suppose that we have studied ten 'unit characters' in a given group and (using Sneath–Sokal methods or such others as we may prefer) have constructed a classification of the group based on this 'information'. If we then proceed to study ten more characters in the same group, we shall probably find more or less considerable modifications of our original 'ten-character' system necessary in order to make one which uses all twenty characters. Proceeding further to study ten more previously unutilised characters, we shall then be in a position to construct a 'thirty-character' system, which will probably involve rather less modification of the twenty-character system than that did of the ten-character one. If we went even further, to the construction of a forty-character system, this would probably involve still less modification of the thirty-character one, and so on. At some stage in this process, Sneath and Sokal suggest, the study of further characters will cease to be worthwhile. Without actually saying so, they seem to imply that this stage has already been reached in most groups of animals (plants are not considered by them)—at least, their examples always utilise already established classificatory characters and they nowhere suggest that it would be desirable, for the purposes of numerical taxonomy, to investigate new characters.

Given the basically statistical approach to classification which Sneath and Sokal assume, this idea of the 'factor asymptote' seems quite plausible however we interpret clause (b). If we look at the results of actual systematic study, however, it is not easy to find good supporting examples. In the animal kingdom, we may consider the analysis which Sturtevant [183] made of the distribution of thirty-three 'unit characters' in *Drosophila*. Most of the characters used in this analysis were derived from the external structures of the adult flies, and had long been used by systematists dealing with the genus, but a number of additional characters, drawn from the eggs, the larvae, and the internal characters of the adults, were studied. Sturtevant found a wide range in degrees of correlation between individual pairs of characters, from high positive values through zero (i.e. random association) to high negative ones. Some of the characters showed significant correlations, positive or negative, with most of the other ones, while some showed significant correlations with only two or three others. A noteworthy result was that, of the three characters showing the strongest correlations with the other ones, not one was a 'traditional' taxonomic character in the genus. This suggests that, if a 'factor asymptote' exists in *Drosophila*, it would be necessary to study many more characters than Sturtevant's thirty-three in order to approach it. Sturtevant's results in fact support the principle of Aristotelean classification, and are opposed to the 'randomness' which is presupposed by the Sneath–Sokal techniques. Similar results have been obtained by Stebbins [179] and Sporne [178] in analyses of the distribution of characters among families of dicotyledons. As a general point of scientific technique, it may be pointed out that to treat data which are inherently non-random by techniques appropriate to random distributions is liable to involve serious loss of 'information'.

A very interesting test case for numerical taxonomy has been the recent study by Rohlf [159] of larval and adult classificatory characters in mosquitoes (Diptera-Culicidae) of the genus *Aedes*. This investigator studied forty-eight species of the genus, forty-five of them American, and considered both adults and larvae (these being insects with a 'complete metamorphosis') of each species. Seventy-one different 'unit characters' were numerically coded for the larvae of the forty-eight species, and their distribution was mathematically analysed on the lines prescribed by Sneath and Sokal; the resulting measurement of the degrees of relationship among the forty-eight species of larvae was expressed in a 'dendrogram'. A similar procedure was adopted in relation to seventy-seven different 'unit characters' of the same forty-eight species of adult mosquitoes, and the results expressed in another dendrogram. Rohlf's own conclusion from the comparison of the two dendrograms was that 'highly significant (though not large) correlations were found between larval and adult relationships. Despite the correlations, there was also significant heterogeneity between larval and adult relationships'. As an illustration of the degree of discrepancy between the two dendrograms, we may note that the primary division in the larval one set off species numbered 40, 41, 42, 43, 46, 47 and 48 from all the rest; the corresponding division in the 'adult' dendrogram segregated species 20, 22, 28, 33, 39, 40, 42, 43, 46, 47, 48, 44 and 45 from the remaining ones, and its first subdivision split off 20, 22, 28, 33, 39, 40 and 42 from 43, 46, 47, 48, 44 and 45. In the larval dendrogram species 20 is grouped with 2, 22 with 33, 28 with 27, 39 with 8, and 44 with 45; in the adult one 41 is placed with 10 and 38. Rohlf presents a second pair of dendrograms, in which the same data for the larvae and the adults are analysed in respect of 'distance coefficients' instead of 'correlation coefficients'. These two dendrograms differ from each other to much the same degree as the first pair, but differ much more strikingly from either of the first pair. The prime division in the larval 'distance coefficient' dendrogram, for example, separates species 15 from all the rest, and the second one places species 47 and 48 apart from the rest; the first division in the distance-coefficient dendrogram for the adults sets species 43, 46, 47 and 48 apart from the rest; the next division of this residue proceeds to segregate species 15. In the larval distance-coefficient diagram, species 43 and 46 are placed with 40, 41 and 42 in a division of the sixth grade.

From Rohlf's work, it is evident that if a 'factor asymptote' exists for the species of *Aedes*, neither the seventy-one larval characters nor the seventy-seven adult ones are enough to bring the analysis near to it. In this connection it must be pointed out that the mosquitoes are an exceptionally well-studied group, and that the number of characters which Rohlf was able to use in his analyses was remarkably large. His work, to say the least, provides no support for the idea that the general adoption of Sneath–Sokal methods by systematists would make for greater stability and objectivity in classificatory systems. The practical alternative to phylogentic classification would still appear to be the Aristotelean rather than the 'numerical' variety.

Results somewhat comparable with those of Rohlf were obtained by Heiser, Soria and Burton [90] in a numerical taxonomic study of a series of species, subspecies, varieties and hybrids of the genus *Solanum*. Seventy-five 'operational taxonomic units' and fifty-eight 'unit characters' were used in their analyses. They compared the numerically-derived dendrogram with previous 'classical' systematic conclusions of their own and with the results of a large series of hybridisation experiments among the taxa included in the study. One conclusion was: 'it can be seen that the data on hybrid fertility do not completely support our previous concept of relationships nor that shown by numerical analysis, but that on the whole it more nearly agrees with our earlier concepts.' Their final conclusion was that 'although numerical taxonomy does not live up to the optimistic claims made by its proponents, it can provide useful information for the taxonomist'.

In addition to criticisms based on fundamental principle, the Sneath–Sokal conception of a 'unit character' will encounter numerous practical difficulties. How, for example, should one treat metamerically repeated structures, such as legs of myriapods, parapodia of Polychaeta, stamens of angiosperms, etc.? In a group of insects, the Coleoptera, which I have studied, there are three pairs of legs, at the end of each leg is a tarsus, which in most groups is composed of five segments. This number is, however, liable to reduction: you may find species with four segments in the tarsi of the front legs and five-segmented tarsi on the other two pairs of legs (tarsal formula 4-5-5); there are others with a 4-4-5 tarsal formula, some with all the tarsi four-segmented, some with the formula 5-5-4, others with 5-4-4, and so on. And the formula may differ between the sexes of a given species. Considering the data from an evolutionary point of view (i.e. indulging in what Sneath and Sokal would call 'phylogenetic speculation'), it appears that in some groups the reduction of the tarsal formula from 5-5-5 to 4-4-4 takes place in a single step (as it does, e.g. in certain *Drosophila* mutants), but in others 5-5-5, 4-5-5 and 4-4-5, or 5-5-5, 5-5-4, 5-4-4, are successive steps on the way to the 4-4-4 formula. Is the tarsal formula 4-4-4 to be considered as a 'unit character' or not? Similar considerations would apply, e.g. to modifications of one or more of the original five stamens in Scrophulariaceae.

How, on the Sneath–Sokal system, should we treat characters which are liable to secondary loss? In the classification of the orders of insects, for example, a fundamental distinction is based on the presence or absence of wings in ancestral forms—insects which have wings, or are derivable from ancestors which had them, are separated from those whose ancestors have always been wingless. There are, however, whole groups, commonly given the status of 'orders', which are entirely wingless yet are invariably classified among the basically winged section (Pterygota) of the insects. Wings, it may be added, when present offer numerous 'characters' in their venation, etc., which are used by systematists—what happens to these characters when a wingless group like the fleas (Siphonaptera) is being compared with related winged ones? Sneath and Sokal would presumably take the absence of wings in fleas and lice simply as a 'unit character' common to them and the primitively wingless orders (Apterygota).

For a comparable case among the vertebrates, we might take the absence of limbs in snakes, the amphibian Gymnophiona, and the lampreys (Cyclostomata). In the last mentioned group this condition is believed to be primitive, like the winglessness of apterygotan insects, but all authorities are agreed that snakes and Gymnophiona are derivable from ancestors which possessed true pentadactyl limbs. Such considerations would presumably, as 'phylogenetic speculations', be excluded from consideration in the Sneath–Sokal system, and limblessness would be counted by them as a unit character relating snakes and Gymnophiona to lampreys. In botany, a similar procedure would be to treat the perianth-less condition of the flowers in certain dicotyledons (e.g. some 'Amentiferae') as a unit character in common with Coniferae; unisexuality of the flowers would presumably be treated in the same way.

If the unit character on which the mathematical systematics of Sneath and Sokal depends is shown to be an arbitrary and conventional unit, incapable of objective definition, this does not mean that the sort of analysis they advocate is worthless. The attempt to apply it may have a salutary effect in forcing systematists to consider critically the bases of their systems, may well make for clearer thinking, and more careful and logical definitions of groups; it may even, by suggesting that there is a factor asymptote which might be approached if enough further characters were studied, stimulate serious taxonomic research. Also, by giving systematics a suitably mechanised and 'contemporary' look, numerical taxonomy may help systematics to survive through a period which is particularly unfavourable to it. The attempt to apply Sneath–Sokal techniques widely and consistently in actual classification will inevitably reveal the difficulties and inconsistencies of this approach, and should certainly lead to the eventual rediscovery of the advantages of Aristotelean and even of phylogenetic classification for purely practical purposes.

In some respects the outlook of Sneath and Sokal is nearer to that of phylogenetic systematists than to those of most 'formal' systematists. For example, they say: 'We view monotypic taxa or very numerous ones with equanimity. Their occurrence does not lead us, respectively, into lumping or splitting.' This may be contrasted with the principle, advocated and practised by many systematists up to the present time, that the isolation of a taxon should be inversely proportional to the number of species it contains. Sneath and Sokal also envisage a situation in which groups, established by 'numerical taxonomy', are not easily definable formally by reference only to one or two characters; they discuss various methods of treating such situations for the purposes of practical discrimination, e.g. in keys. Furthermore, they envisage more or less radical changes in our current rules of nomenclature and look forward to a time 'when name changes were no longer permitted for reasons of priority, and author citations would become unnecessary.'

In many respects the outlook of Sneath and Sokal strikes the present writer as decidedly unrealistic. Their pessimism about the knowability of phylogenetic history is matched or even exceeded by their optimistic faith in the future possibilities of computers, looking forward to machines 'which do not merely retrieve information, identify specimens, or give us their synonymies but which will also tell us if we have found a new organism ('Congratulations! You have discovered a new phylum'), suggest a suitable name, question the taxonomist's veracity, and learn by experience as new organisms are discovered and their particulars are given to it'. Their estimate that 'one problem of today is the flood of new characters now available which must be handled in some fashion' is quite misleading. The main barrier to the classificatory use of so many of these 'new characters' is not their multiplicity but the extreme fragmentariness of the available information on them, together with difficulty or impracticability of the average systematist investigating them for himself. We certainly do not possess a large mass of new systematic information about organisms just waiting to be incorporated in classification; the vast bulk of the new information relates to a small number of species of 'standard laboratory organisms'.

In their later book, [174], Sokal and Sneath to some extent retreat from the rigour of the attitude expressed in their 1962 paper. In the earlier work they considered only the analysis of the numbers of characters common to pairs of taxa, which they refer to as the Q technique. In considering the ways of treating matrices (checker-board diagrams in which taxa are listed along one dimension and classificatory characters along the other—see Chapter 18), they say 'matrices of this sort can be examined from two points of view. The association of pairs of characters can be examined over all OTU's (taxa). This is called the R technique. The converse practice, the association of pairs of OTU's (taxa) over all characters, has been called the Q technique'. Later they write: 'If there is redundancy in character information it might be possible to isolate the factors from a R correlation matrix and to calculate factor scores for each OTU (taxon) on the factors obtained. We could then employ only those characters which provide independent information on the taxa concerned. This would reduce the number of character scores on the basis of which OTU's could be classified. It would not, however, reduce computational effort, owing to the tedium of factor analysis. It might be argued that after such an analysis factors would, in fact, be weighted in terms of the independent amount of information they contain. This might open the door to a numerical taxonomy based on weighted characters.' This should be compared with the following from an earlier chapter of the same book: 'Equal weighting can therefore be defended on several independent grounds; it is the only practical solution, it and only it can give the sort of natural taxonomy we want, and it will appear automatically during the mathematical manipulations.' The numerical taxonomic study by Rohlf of the mosquito genus *Aedes*, already discussed on pp. 201–2, was based entirely, according to the Sneath–Sokal prescription, on the use of Q-type correlations. If, as seems likely to me, this work proves to be of little or no value as a contribution to the serious classification of *Aedes*, the fact is probably attributable to the particular methods used. The same data, analysed for R-type correlations, might have yielded highly significant results, of great value for entomological systematics and at least suggestive in relation to questions of phylogeny and function.

16 The Non-congruence Principle

Few systematists have been content to distinguish a particular classificatory group from related ones by means of a single character, even though they may prefer to proceed on this basis when constructing analytical keys for determination purposes. There is a widespread feeling that classifications in which each division is based on a single character are liable to be artificial and unsatisfactory. Almost all reputable systematists, when defining a classificatory group, will try to bring in as many characters as possible; this is an old-established tradition which goes naturally with the Aristotelean approach to systematics. The attitude is not, however, really consistent with the aims of phylogenetic classification, as we shall see.

The attempt to base the formal definition of a group on more than one character frequently leads to difficulties. Let us consider, as a botanical example from the higher categories, the definition of the Angiospermae. This is usually based on a series of characters—the enclosure of the ovules in closed carpels, the presence of true vessels in the xylem, the bisexual flowers, the perianth, the presence of a pollen-tube instead of free-swimming spermatozoids, the leaves of 'megaphyllous' type without circinate vernation, etc. All these characters coexist in a sufficiently large percentage of the Angiospermae to constitute a satisfactory 'Aristotelean essence' of the group; nevertheless, no two of them occur together in all cases. The closure of the carpels is usually accepted as the systematically 'best' character among them; if we ignore a somewhat similar enclosure of the megasporangia in the otherwise very different group Hydropterideae among the ferns, the character is peculiar to the Angiospermae and occurs in all of them. The nearest thing to an exception is constituted by those few forms (e.g. the recently discovered *Degeneria*) in which the carpels are open at first and close comparatively late in development.

If the dividing line between Angiospermae and Gymnospermae is drawn on the basis of the carpels, exceptions to all the other distinguishing features of the Angiospermae occur on one or another (sometimes on both) sides of this line. True vessels, for example, are present in Gnetales (usually placed in Gymnospermae), as well as in systematically remote groups such as Equisetaceae and Selaginellaceae, while some undoubted Angiospermae lack them. Bisexual flowers were present in the extinct Bennetitales, and there is some vestige of the condition in the existing *Welwitschia* and *Gnetum* (Gnetales); on the other hand, unisexual flowers occur in many groups of modern Angiospermae. Something like a perianth is found in *Welwitschia* and *Gnetum*, and was present in the Bennetitales. Fertilisation by a pollen-tube is found in Gnetales and Coniferae among Gymnospermae, in which group both *Ginkgo* and *Gnetum* have megaphyllous leaves without circinate vernation.

The distribution of these exceptions is worthy of further consideration. On the gymnospermatous side of the line, they are concentrated particularly in the Gnetales and the extinct Bennetitales. This constitutes a fairly strong indication that, of all surviving groups, the Gnetales are the closest relatives of Angiospermae; there is indeed only one really fundamental character—the lack of closed carpels—to separate *Gnetum* from the latter group.

Within the Angiospermae, true vessels are lacking in *Tetracentron*, *Trochodendron*, and Winteraceae, also in certain highly modified aquatic or parasitic types; the catkin-bearing trees include many types with unisexual flowers (found also in diverse other families), and some with little or no trace of a perianth; leaves may be entirely lacking in some parasitic or xerophytic types, and may lose their megaphyllous character, e.g. in Casuarinaceae. In so far as these exceptions tend to be concentrated in any one group, that group is the old 'Amentiferae', the catkin-bearing trees. Until recently there was a strong school of thought which regarded these 'Apetalae' as the most generally primitive group of modern Angiospermae, but most modern authorities are agreed that the peculiar features of this group are derivative rather than primitive, the result of early adaptation to pollination by the wind rather than by insects. The most difficult question concerns the 'homoxylous' wood of *Tetracentron*, *Trochodendron* and Winteraceae. Almost all recent authorities are agreed in considering these as among the more primitive of existing Dicotyledonae. If the vessel-less wood in them is a primitive character, then presumably the original Angiospermae did not have vessels, which must then have been developed independently in Gnetales and Angiospermae. This conclusion would weaken, though it would not destroy, the case for a close relationship between Gnetales and Angiospermae. On this hypothesis, one would expect the most fundamental division in Angiospermae to come between the primitively vessel-less forms and the rest—but apart from the 'homoxyly', there seem to be no characters which indicate a particularly profound division at this point.

On the whole, it seems more probable to me that homoxyly is a secondary rather than a primitive condition in *Trochodendron* etc.; in the 'lower' Angiospermae, the vessels would be a less 'stabilised' character and one more easily lost. If so, then it seems that all exceptions within the Angiospermae to our basic characterisation of the group are attributable to secondary loss of characters; the 'real' exceptions all lie on the gymnospermatous side of the line. Using the closed carpels as the basic criterion thus gives the group a smaller extent than any of the other characters we have considered would; if we used the presence of vessels in the wood, the Gnetales would be included in the group, if we relied on bisexual flowers with a perianth, the Bennetitales too would be added, while the criterion of a pollen-tube would add the Coniferae to these.

This exemplifies what I believe to be a general principle in the classification of organisms—that the extent of a classificatory group, defined by the possession of some one character A, is not the same as it would be if a different character B were made the criterion of membership. In other words, different classificatory characters (unless they are extremely closely linked functionally) are rarely coincident in their distribution. Further instances of this principle can be drawn from the data used in Chapter 18, for matrix analysis. The principle is one I have elsewhere [35] called the 'Non-congruence Principle'. In most cases where the distribution of independent characters appear to be 'congruent', we are dealing with what systematists call 'isolated' groups, intermediate forms having become extinct.

A good example of a group which, when only the modern forms are considered, is 'isolated' in this sense is the Mammalia. These may be distinguished from other living Tetrapoda by a whole series of characters, e.g. the lower jaw suspension, the hairy skin, the mammary glands of the female, the single (left) aortic arch, the heterodont dentition, warm blood, and the possession of a secondary palate. As far as modern animals are concerned, practically the only exceptions to congruence in the distribution of these characters are the warm-bloodedness of birds, and the hairlessness and apparently homodont dentition of whales. If the fossil evidence is considered, however, the position becomes very different. The heterodont type of dentition appears to have arisen in early Triassic therapsid reptiles which do not seem at first to have possessed any other of our basic mammalian features; the secondary palate seems to have developed rather later in the same stock, which still later gave rise to small forms (Ictidosauria) which seem to have been the first to possess a hairy skin; last of all came the mammalian jaw suspension. Another general mammalian character, viviparity, evidently came still later, since it is not present in the most primitive living forms (Monotremata) which possess the mammalian jaw-suspension. When the fossil evidence is taken into consideration, it seems that the case of Mammalia may be quite analogous to that of Angiospermae.

Examples of this type could be multiplied among the Insecta; as a typical one we may consider the great group Curculionoidea (weevils) among the Coleoptera. Typical modern members of this group have a whole series of characteristic features—in the adults, for example, a rostrate head, no free labrum, fused gular sutures, geniculate antennae, closed middle coxal cavities, two basal ventrites connate, only two ovarioles to each ovary, characteristic form of the aedeagus, etc. However, no two of these characters are completely congruent. The rostrate head, for example, clearly arose before the labrum was lost, and the loss of the labrum preceded the fusion of the gular sutures, the last-mentioned change preceded the development of geniculate antennae, and this was followed by the typical curculionid aedeagus.

The probable explanation of the non-congruence principle is that characters of this sort usually evolve successively rather than concurrently. If this is the case, then an explanation should probably be sought in terms of the nature of genetic variation and selection. Human plant and animal breeders have found empirically that the most effective selection operates on one character at a time [cf. 204b]. Results may be expected if a line of plants is selected for a particular flower-colour, or for an increased number of petals, but the attempt to select simultaneously for both would not be expected to succeed. Similar limitations probably affect natural selection. As we have seen in Chapter 3, single 'unit character' differences between species of plants and animals appear normally to be very 'polygenic' in their genetic basis. In order to accomplish quite a simple phenotypic change, selection has to bring about quite a complex reorganisation of the genotype—and there is probably a limit to the complexity of genetic differences which can be effectively selected. It is of course possible, even likely, that after selection has operated for some time in changing one phenotypic character, it will switch to affecting some other one; a prolonged bout of selection in adaptation to a changed mode of life may influence a whole series of characters in relatively quick succession. Doubtless this has occurred in the ancestry of some 'isolated' groups, e.g. in the origin of tetrapod vertebrates and of Archegoniatae (true land plants). In such cases, intermediate forms are unlikely to survive to the present day, and they might be very scarce even as fossils.

An apparent non-congruence in the distribution of two classificatory characters may result from the secondary loss of one of them in some forms, as well as from their non-contemporaneous origins; this may be called 'secondary non-congruence', to distinguish it from the primary kind. In practice, as we have seen in the case of the homoxylous Angiospermae, it is not always easy to distinguish secondary from primary non-congruence. Among the mammals, the loss of the tooth-differentiation in the Cetacea is a good example of secondary non-congruence with the mammalian features of the circulatory and reproductive systems, etc.; the egg-laying Monotremata exemplify primary non-congruence with these same features. Though examples of secondary non-congruence are well known in Angiospermae, it is far less easy to find clear instances of primary non-congruence among them—this I believe indicates a generally lower level of phylogenetic understanding in flowering plants as compared with vertebrate animals. Possible examples include the polypetaly of Clethraceae in the generally gamopetalous Ericales, the presence of genuine leaves in *Pereskia* alone of the Cactaceae, and perhaps the absence of betacyanins in Caryophyllaceae among Centrospermae (see Chapter 8).

The general criterion is that, if the absence of a normal character of a group in some member of it is attributable to primary non-congruence, then that member should show other indications of being unusually distant from the rest of the group and should possess other exceptional and primitive features. Thus the Monotremata show a less perfect homothermy, and a more reptile-like pectoral girdle, than other mammals, which could be taken as evidence in support of the primary nature of their non-congruence over viviparity; on the other hand, the secondary nature of the homodont dentition of Cetacea is supported by the absence in that group of any other seemingly primitive and reptile-like features. As applied to the homoxylous Trochodendraceae, etc., this criterion is unfortunately indecisive. A further criterion which can often be used to distinguish primary from secondary non-congruence is Dollo's Law (see Chapter 9).

The effects of the non-congruence principle are frequently manifest when different systematists propose classifications of the same group of organisms based on different characters, and they have underlain many protracted classificatory controversies in the past. An interesting zoological example may be found in the Coelenterata–Hydrozoa. These organisms exist in two sharply distinct forms—a vegetatively-multiplying sessile polyp (hydroid) stage, and free-swimming sexual individuals known as medusae. Among the polyps of Hydrozoa, two well-marked types have been distinguished—the Gymnoblastea in which the individual polyps are not enclosed in horny cups, and the Calyptoblastea in which each polyp has its protective surrounding hydrotheca. The main division which is drawn in hydrozoan medusae is that between Anthomedusae and Leptomedusae, distinguished by general form, position of gonads, and type of sense-organs present. To a very large extent, Anthomedusae develop from gymnoblastean polyps, and Leptomedusae from Calyptoblastea, but a few known Hydrozoa, such as the genus *Melicertus* Agassiz, have Medusae of the leptomedusan type arising from gymnoblastean polyps.

There is more than one possible phylogenetic explanation of this state of affairs. Perhaps the simplest of them is to postulate the primitive-ness of the naked gymnoblastean polyp and the anthomedusan type of medusoid; on this hypothesis, *Melicertus*, etc., would be considered as primitive offshoots of the line leading to Calyptoblastea, its medusoid characters having developed before the hydroid ones. It is also con-ceivable that *Melicertus* and its allies are the most primitive of living Hydrozoa, from which the Gymnoblastea have diverged mainly in the medusoid stage and the Calyptoblastea in the hydroid; this hypothesis would favour three rather than two primary groups of the class. We are hardly in a position to exclude a third possible hypothesis, the con-verse of the first-mentioned one—that calyptoblastean polyps and lep-tomedusan medusoids are primitive, and that *Melicertus*, etc., represent an intermediate stage in the development of the Gymnoblastea. The best hope for a decision between these hypotheses lies probably in serological or other such protein studies. On the first hypothesis, such forms as *Melicertus* should be more closely allied to Calyptoblastea than to Gymnoblastea, on the second hypothesis they should be about equidistant from both, on the third hypothesis they should be closer to Gymnoblastea.

In my own studies of the classification of the Coleoptera, a group of insects with a complete metamorphosis and a larval stage which is as taxonomically independent of the adult as the hydroid is from the medusoid in Hydrozoa, similar problems have frequently been encountered. As was pointed out by van Emden [56a], classifications of Coleoptera based on larval characters are liable to coincide only partially with those based on adult characters. For example, among the Curculionoidea already mentioned, the family Apionidae is readily separable from Curculionidae in the larval stage, but there seems to be no single character which will provide a satisfactory distinction between the two families in adult beetles; conversely, the Scolytidae, long recognised as a distinct family in the classifications of the adults, can hardly be separated from Curculionidae in the larval stage.

A botanical example perhaps comparable to that of the Hydrozoa concerns the major divisions of Bryophyta. The alternation of sporophyte and gametophyte generations in these is comparable with that of hydroids and medusoids in Coelenterata. The major divisions of Bryophyta accepted in most modern works are Anthocerotae, Hepaticae and Musci. In the dominant gametophyte generation, the most striking distinction is that between the thalloid forms (Anthocerotae, many Hepaticae) and the leafy ones (Musci, some Hepaticae). In the sporophytes, there are three main types corresponding to the three classificatory divisions, which, in fact, are based mainly on sporophyte characters. In the sporophyte generation, Anthocerotae are nearer to some Musci than to any known Hepaticae, while the *Anthoceros* gametophyte resembles those of some Hepaticae and is quite unlike anything known in Musci. The phylogenetic interpretation of this situation is just as problematic as that in the hydrozoan case we have just considered; as in that instance, useful fossil evidence can hardly be expected, and protein chemistry or serology are probably the most promising sources for evidences of the true relationships. Some modern authors, for example, Khanna [115] consider that the thallose gametophyte in Hepaticae is a derivative feature, wheras in Anthocerotae it may be primitive; others consider that the leafy gametophyte of *Jungermannia*, etc., has been developed independently of that of the Musci. However, it may be interpreted, there is a clear non-congruence between the sporophytic and gametophytic classifications of the Bryophyta.

At least in the cases of the Hydrozoa and the Curculionoidea we have considered, the classificatory non-congruence between different phases of the life-cycle seems to result from the fact that important changes in the modes of life of the different stages do not normally take place concurrently—if selection is bringing about active change in the medusoid stage, the hydroid is liable to remain practically stable, and vice versa. In such instances, there is no 'rule of thumb' method for deciding which stage provides the better basis for classification. For example, it is not necessarily true that sexually reproducing stages (medusoids, adult beetles, gametophytes) are better indicators of relationships than juvenile or vegetatively-reproducing phases.

Examples of non-congruence in the incidence of individual characters could be multiplied indefinitely from either the plant or the animal kingdom. Purely as illustrative instances we may cite (1) the Cycadaceae, in which the genus *Cycas* differs from the rest in not having the megasporophylls aggregated into cones, (2) the Scrophulariaceae, in which the genera *Verbascum* and *Celsia* differ from the norm in having five perfect stamens, (3) the Cruciferae, in which a longitudinally-partitioned 'siliqua' type of fruit occurs in almost all except for the small group Euclidiae, (4) the Australian marsupials, among which the bandicoots (Peramelidae) have the dentition of the Polyprotodontia and the syndactyly of the Diprotodontia, (5) the Pecora among Ungulata [Haltenorth—83], in which horns or antlers are almost universal except for the small group of musk deer (*Moschus*), (6) the presence of an apionid-type aedeagus in one small group (Erirrhinini) of the true Curculionidae among Coleoptera. Any experienced systematist will be able to add examples from his own group.

Such examples of non-congruence have long been a source of perplexity and controversy among systematists with the Aristotelean approach; for the phylogeneticist, they provide welcome evolutionary evidences and, once he can distinguish primary from secondary non-congruence, they pose no particularly difficult classificatory problems. Considering only the zoological examples in the last paragraph, and assuming in each case that in them the non-congruence is primary, the phylogenetic systematist would conclude that the main dividing line in Australian marsupials should be shifted so that the Peramelidae went into the same division as Diprotodontia, that the musk deer should be set apart as a primary division of Pecora, and that Erirrhinini should be similarly separated from the rest of Curculionidae.

One particular case in which non-congruence might be expected has been treated by the methods of 'numerical taxonomy' (see Chapter 15). Rohlf [159] tried to develop numerical classifications of adults and of larvae of the mosquito genus *Aedes*. When similar problems have been treated by the methods of traditional systematics, the result has usually shown a fair degree of conformity between the larval and the adult systems, but with a persistent tendency for the main dividing lines to be drawn in slightly different places in the two systems. As between Rohlf's numerical systems of adult and larval *Aedes*, no such simple pattern of non-congruence is apparent; the very complex differences between the systems of the two stages are summarised in Chapter 15.

The obvious conclusions to be drawn from the non-congruence principle is that the definition of a phylogenetically natural group should in essence be based on a single character—but with the proviso that such single, basic characters are very liable to secondary loss within extensive groups—i.e. that secondary non-congruence is likely to occur. The definition thus requires to be extended to include clauses which will make possible the recognition, as members of the group, of those forms in which the 'basic' character has been lost. Thus for example, the major group of modern Insecta is the Pterygota, defined essentially by the original possession of wings, but the practical definition of the Pterygota has to allow for the inclusion of all those forms, such as lice and fleas, in which the wings have been lost without trace. Similarly, the Pteropsida are basically distinguished from other vascular plants by the possession of the 'megaphyllous' type of leaf, but the group will include many modern forms in which leaves are totally lost or so modified that their megaphyllous character is obscured. Complications of this sort, annoying to the tidy-minded bureaucrat, are part of the price we shall have to pay for a phylogenetic classification; they will usually make it necessary to use a number of characters, not just one, in the practical definitions of groups. These characters will not, however, be used in any simple, routine additive way such as is the rule among the numerical taxonomists; they will be organised in a series of alternatives, embodying knowledge of the unique evolutionary history of the group concerned.

H

17 Some Special Classificatory Problems

In classifying composite organisms like lichens, in the practically necessary task of constructing some sort of system of 'natural communities' of plants and/or animals, in dealing with things like pollen grains (especially fossil ones) and various types of micro-fossils (e.g. conodonts), also in the practically important 'fungi imperfecti', special difficulties are liable to be encountered. In this chapter some of them are briefly considered, beginning with the fungi imperfecti.

The major division of the fungi are defined in relation to the characters of the spore-bearing ('perfect') stage, whose production normally involves some form of sexual fusion. In many types of fungi, especially among Ascomycetes, the 'perfect' stages are rarely encountered, and there are numerous species whose spore-bearing stage has not yet been identified. Such forms, which cannot be placed properly in the established system of the fungi, are known as 'fungi imperfecti'— within this division genera and species are defined purely on vegetative and conidial stages. Some species may be known under two generic names, one for its spore-bearing form and the other for its conidial stages; the well-known genus of moulds, *Aspergillus*, is defined on characters of the 'imperfect' stages—where the 'perfect' form of species of *Aspergillus* is known, it is attributed to the genus *Eurotium*. Theoretically, it should be possible to develop a natural classification of fungi, using characters of all stages, as well as serological and other evidences, into which the fungi imperfecti could be incorporated; in practice, it is likely to be some time before this ideal is realised, and mycologists are likely to retain the category of 'fungi imperfecti', and a system of names within it, for some time yet.

The problem in this group is no doubt to some degree analogous to that encountered by zoologists in the Coelenterata. By no means all the known hydroid (asexual) and medusoid (sexual) stages have been correlated with each other, and there are doubtless instances where the two stages of a given species are still known by different generic names (and perhaps even placed in different families). In the case of the Hydrozoa, however, the difficulty is mainly, if not entirely, one of simple ignorance; there is no general tendency for one of the stages (hydroid or medusoid) to be eliminated from the life-cycle. Many of the fungi imperfecti have effective alternatives to the normal sexual process involved in the production of the spore-bearing stage, and the latter may really be nearly or quite eliminated from their normal life-cycle.

Lichens (see Lamb [123]), as is now well known, are composite organisms, the essential components of which are some kind of fungus —usually an ascomycete—and some species of green (or blue-green) alga. The fungus provides the organisational framework and perhaps bears the main responsibility for maintaining mineral supplies (including water); the alga contributes the essential products of photosynthesis. For all lichens which have been critically investigated, the fungal component has proved to be more or less readily cultivable in isolation, and in some cases it has been alleged to exist in nature as an independent organism; probably all the species of algae which have been identified as components of lichens are also known as free-living organisms. If both the components of a lichen may be given their separate classificatory places, in the systems of fungi and of algae, is there any need for a separate classification of lichens? Practically everyone who has had much practical experience with lichens has answered this question in the affirmative, although Donk [51] expresses some reservations on the matter.

Nearly all lichens have special arrangements for their indefinite propagation as such, and there are strong indications that a genuine evolution of lichens has taken place. Species and genera of lichens, it seems, may have been produced by gradual evolutionary divergence much as has happened in other organisms, even though the group as a whole is doubtless polyphyletic and thus unnatural. Genotypic evolution within the lichens has probably been almost entirely in the fungal component, and has produced types of fungi which apparently do not occur in nature except as components of lichens. Changes in the algal components seem to have been a matter of occasional switches from one species of algal component to a quite different one.

It seems probable that Hennig's conception of phylogenetic classification is not strictly applicable to lichens; the group, however, is undoubtedly susceptible to 'natural' classification along Aristotelean lines. Existing classifications of lichens are of this type, and with the increase of knowledge of the group they should be capable of further improvement. We may be able to reconstruct to a considerable extent the phylogenetic history of their fungal components, and incorporate these into a phylogenetic classification of fungi in general; this may be preferable to the attempt to make a 'phylogenetic' classification of lichens themselves.

It has been pointed out that some of the chemical substances produced by lichens, e.g. pentacyclic triterpenoids, are not known to be produced by either algae or fungi by themselves, and chemosystematists could well claim that such substances would provide a basis for an independent classification of lichens as such. Another problem which arises in the classification of lichens (see Neelakantan and Rao in [215]) is that of the treatment of chemically different but morphologically indistinguishable forms. The reproduction of lichens is a complicated matter, which makes it difficult to apply the standard genetic criteria for species to them. Lichenologists tend to treat as species only those forms which are morphologically distinguishable; chemically distinguishable types within a 'species' are usually termed 'forms' or 'races' rather than subspecies.

There are certain animals which are similar to lichens in that they can only live by the incorporation and exploitation of algae in their tissues—well known examples are the familiar laboratory *Hydra viridis*, many reef corals, and some Turbellaria among the flatworms (Platyhelminthes). None of these have, however, been considered to pose any special classificatory problems; they are incorporated in the general system together with related non-symbiotic forms. Consideration of these symbiotic associations leads naturally to the problems of classifying plant and animal communities, considered in the last section of this chapter.

The classificatory value of pollen grains is well reviewed in a recent article by Erdtmann [61]. His assertion that pollen grains 'appear as an alter ego of the parent plants, reflecting the "macro-individuals", the trees, shrubs, herbs, etc., that produced them' may convey a somewhat misleading impression, but he counteracts it by citing a number of instances where very similar pollen grains come from very different plants, e.g. the sapindaceous *Diplopeltis* and various Umbelliferae, *Linum* and Plumbagineae, *Artemisia* and Salicaceae, etc. The general picture presented by Erdtmann is, however, systematically (and, no doubt, phylogenetically) significant. Bryophytes and pteridophytes have spores with single openings, directed towards the centre of the 'spore-tetrad'; in Gymnospermae, most Monocotyledonae, and some primitive types of Dicotyledonae, the opening is still single, but is directed *away* from the centre of the tetrad; in some Monocotyledonae and the main mass of Dicotyledonae, the opening(s) are somewhere in the middle rather than at either end, and generally more than one. An interesting point about multiple openings is that the number of them tends to be increased in polyploid forms—for instance, where a 'diploid' species has grains with three openings, a tetraploid form of it will probably have four openings, and a decidedly larger grain-size.

It may be permissible to consider a pollen grain—which represents a highly reduced gametophyte—as an independent organism rather than as a mere part or developmental stage of a plant. The classification of pollen grains by their structure may offer a genuine analogy with similarly based systems of unicellular plants or animals (e.g. Diatomaceae, Foraminifera). In the case of pollen grains (at least, of modern ones) we are, however, in a position to check this classification against that of the higher plants (sporophytes) from which they derive. Even the most enthusiastic palynologist would hardly claim that a pollen grain was equal in value as an indicator of relationship to an entire seed plant. The comparison may provide us with a valuable means of estimating the probable degree of naturalness of morphological classifications of unicellular organisms as compared with those of higher plants or animals.

The palynologists have not yet found any convenient and consistent structural features by which to distinguish the pollen grains of such well-characterised groups as Coniferae, Gnetales, dicotyledons and monocotyledons; if confronted with a wholly unknown type of modern pollen grain, they would probably claim to be able to attribute it with some probability to one or other of these groups—but would have to admit that all the characters they relied on for this purpose are liable to more or less frequent exceptions. A special problem met with in some flowering plants is that of 'pollen polymorphism'—the production by a given species of two or more quite distinct types of pollen grain, usually associated with some degree of 'heterostyly' in the flower. For practical purposes, the palynologists have devised an elaborate terminology and artificial classifications of pollen grains (see, for example, Erdtmann [60]). By use of such methods (often codified in a punched-card system) competent palynologists claim to be able to identify almost any pollen-grain from a region known to them, as far as the genus and often to the species. It is, in fact, at about the generic level, or a little above it, that artificial classifications of pollen grains begin to coincide largely with whole-plant systematics, though some higher groups, up to the level of families, may have more or less distinctive pollen types.

The ability to determine the pollen grains of modern species of plants is useful not merely in connection with studies of such flower-seeking insects as bees but also in connection with all kinds of studies (archaeological, climatological, ecological, etc.) of geologically recent (Pleistocene and post-Pleistocene) times. Examples of such uses of pollen analysis are given in such works as Zeuner's *The Pleistocene Period* and Godwin's *History of the British Flora* [210, 75].

Fossil pollen grains also occur in pre-Pleistocene deposits, and are often so well preserved as to be practically comparable with modern ones. When dealing with pollen grains from these older deposits it can, however, no longer be assumed that they will be attributable to still living species, nor even always to modern genera. The artificial systems devised for identification of modern pollen grains, become unreliable, and may be seriously misleading, when the attempt is made to apply them to fossil pollens from the earlier Tertiary (and *a fortiori*, from the Mesozoic); in this a general weakness of artificial systems is manifest. A classic instance concerned certain pollen grains from the Jurassic coal-bearing beds of Brora, Scotland. These were at first, with unjustifiable dogmatism, attributed to the dicotyledonous family Nymphaeaceae, despite the fact that there are no well-authenticated macroscopic fossils of Angiospermae from deposits older than Cretaceous. Later study (see Hughes [98]) has led to the conclusion that this Brora pollen was gymnospermous, probably of some sort of conifer, with the grains somewhat distorted by pressure. It is unfortunately rare for macroscopic plant fossils to include pollen-bearing organs with recognisable pollen in them; it is to be hoped that in future more will be added to the present small number of fossil pollens which are more or less certainly associated with macroscopic fossil sporophytes.

Distinctive types of fossil pollen grains (or spores) can be (and actually are) used as 'zone fossils' without any need for their proper botanical attribution, and for this purpose an artificial classification of fossil pollens, comparable to that of modern types, may be justified. Especially in wind-pollinated trees, pollen grains are liable to be carried long distances from the parent plants down the prevailing winds, which is an advantage from the point of view of their use in stratigraphic correlation, even though it complicated the problem of associating them with the appropriate types of sporophytes.

The case of the Brora pollen grains, quoted above, is by no means unique—there have been many instances where fossil pollens have been claimed to establish the presence of particular types of seed-plants in geological eras older than any known macroscopic fossils of the group concerned. As a matter of scientific principle, I think that, at least for pre-Tertiary deposits, fossil pollens should not be accepted as establishing the presence of a particular group of plants unless macroscopic fossils of it have been found in deposits of at least equal age.

There have been some attempts recently to develop the study of fossil fungal spores (or supposed fungal spores) as a special branch of 'micro-palaeontology', as shown in an interesting paper by A. Graham [79]. This author suggests that fungal spores may prove to be of value for purposes of correlation, ecological analysis, etc., in the more recent (e.g. Pleistocene) deposits. Fungal spores are generally simpler structures than pollen grains of higher plants, and to found far-reaching conclusions about phylogeny of fungi on the fossil spores would be open *a fortiori* to the sort of objections that can reasonably be brought against theories of seed-plant phylogeny based on study of fossil pollen. However, used with due caution, fossil fungal spores might offer at least some addition to our deplorably scanty information about the past history of this important group of organisms.

The classification of bacteria is a very difficult and important problem which cannot be adequately considered in this book, not least because I lack the necessary specialist knowledge. Two general statements can, however, be made—that it is rarely if ever possible to identify a bacterium purely on structural characters, and that the unit for determinative work is not the individual but the culture. Furthermore, many types of bacteria show marked plasticity in outward appearance, in relation to differing cultural conditions, etc., and evidence is increasing to show that many also show a surprising amount of plasticity in their physiological-biochemical characters. In the absence of the normal type of sexual process which occurs in plants and animals, and in the presence of such peculiar phenomena as 'transduction', the criteria for species laid down in Chapter 3 are hardly applicable to bacteria. In view of all these difficulties, it is not surprising that current understanding of the classification of bacteria lags a long way behind that of plants or animals; in fact, the amount of systematic organisation achieved in such a work as Bergey's *Handbook* [20] should rank as a notable human achievement. The main hope for the future in this field no doubt lies with the application of serological and protein-chemical methods; really extensive application of these may make possible something like a phylogenetic system of bacteria.

The classification of Protozoa undoubtedly presents special difficulties, resulting from three main causes—(a) the relatively small number of conveniently observable characters in the individual organisms, (b) the difficulty of preserving reference specimens in which even the comparatively few generally used classificatory characters can be observed, (c) the plasticity which many of the species display in relation to environmental influences. Jahn and Bovee (in [215]) in expressing dissatisfaction with the existing systems of the group, blame their deficiencies on the lack of a 'logical' systematic method hitherto, and prescribe the procedures of 'numerical taxonomy' (see Chapter 15) as a cure.

I see no reason to believe that previous systematists working with Protozoa have been any less logical than the rest of us in their methods; if their achievements to date seem less impressive than those of say the entomologists, this is because they have had greater difficulties to contend with. If the protozoologists had full comparative data on a sufficiently large number of different characters to be in a position to apply Sokal–Sneath methods, they would be in a position to make a much better classification using traditional methods. DNA base-doublet frequency determinations, and the analysis of cytochrome-*c* sequences (see Chapter 13) offer the best hopes for new data on the higher classification of Protozoa, and comparative serology (cf. Loefer in [215]) should be useful for the lower levels; the main difficulty in applying such techniques will be that of establishing sufficiently pure cultures of the species concerned.

Several different views on microbial classification were expressed by contributors to a symposium on the subject published in 1962 [216]. Corliss provides an excellent review of the problems of Aristotelian classification in the Protozoa, but does not consider the possibility or desirability of a phylogenetic system. The even more formidable problems of making a rational classification of viruses are considered by Wildy and by Pirie, who both conclude that it would be premature to make the forms of Linnaean classification obligatory in dealing with these problematic entities (it would be begging questions to call them organisms). The use of chemical characters in the classification of bacteria is interestingly considered by J. De Ley, who advocates the procedures of numerical taxonomy, expounded in the same volume by

Sneath. Bisset, on the other hand, considers that the classification of bacteria, as of other organisms, should aim at the Darwinian phylogenetic ideal, and is concerned mainly to make phylogenetic deductions from available comparative data on these organisms.

An outstanding desideratum of modern biochemistry is a satisfactory classification of enzymes. The techniques of 'protein systematics' discussed in Chapter 13 open up the possibility of constructing a genuine phylogenetic system of enzymes, by way of making dendrograms on the lines of the one for cytochrome-*c* in Dayhoff and Eck [47*b*]. The adoption of such a system will almost certainly lead to a change in the methods of denotation of enzymes, whereby 'non-homologous' (in the sense of Florkin [65]) ones catalysing the same reaction will come to bear different names.

The practical demands of the oil companies have stimulated a recent boom in 'micro-palaeontology', with the resultant demand for the naming and classification of various types of 'micro-fossils'. These include tests of such Protozoa as Foraminifera and Radiolaria, diatoms, coccoliths, various spores and pollen-grains (which we have already considered), hystricospheres and conodonts. The last-mentioned type of fossil poses particularly interesting and perplexing problems for the systematist, and deserves special consideration here.

Conodonts [154] are small tooth-like Palaeozoic fossils; they resemble vertebrate bones and teeth in their phosphatic composition, but differ from them in their minute size and in microscopic structure. At least one fossil has been reported in which a number of conodonts appear to be attached to a jaw-like structure—but the true nature of this specimen is still problematic. It is quite usual for a number of conodonts, of several different forms, to occur together, constituting a 'conodont assemblage'. It is generally believed that such an assemblage represents a single fossil organism which possessed no hard fossilisable structures other than the conodonts. Isolated conodonts are, however, much commoner than assemblages; as a result of their generally good preservation and characteristic shapes, many recent 'micro-palaeontologists' have used conodonts for stratigraphical purposes. For this purpose, it is desirable to have some system of designation and classification of conodonts, just as palynologists have found it useful to make one for pollen grains. The important difference here is, however, that a single 'biological' species is probably represented by a number of different types of conodonts, while most (but not all) species of seed-plants have only one kind of pollen grain.

Ideally, it might be best to concentrate attention first on the 'assemblages' and to ignore isolated conodonts—to study the ranges of forms within individual animals, with a view to discovering some rationality in the variation of conodonts within a biological species. If this could be done, it might eventually become possible to recognise genuine genera and species among conodonts. Micro-palaeontologists, however share the general characteristics of human beings, including the usual quota of original sin. Theirs is a fashionable, vigorous 'growing point' in geology, and they are not very willing to refrain from using such attractive looking material as conodonts out of concern for the long-term interests of pure science.

Even if we accept the practical case for the designation and classification of conodonts at the present time, this does not necessitate the application to them of Linnaean binomial nomenclature. It would be perfectly possible to develop a system of designation of conodonts by formulae, thereby avoiding any difficulties with the International Code of Zoological Nomenclature. Unfortunately, the practice of naming and defining 'genera' and 'species' of conodonts is already well established. Proposals have been made to exclude conodont nomenclature from the jurisdiction of the International Code of Zoological Nomenclature altogether, and to obtain official recognition of them as 'parataxa'. It is difficult to see any logical justification for this proposal and it seems unlikely to win scientific acceptance. If there is a practical need for an artificial classification of conodonts (and other microfossils), why must this be expressed in the forms of Linnaean nomenclature? Why not use something analogous to the Dewey system for classifying library books?

Field naturalists and biogeographers have long maintained that, in nature, the various species of plants and animals are not as a rule randomly distributed in relation to each other; they usually occur in more or less definable 'communities'. If we are to make scientific use of this concept, it will be desirable to establish definitions and classifications of such communities. Recent ecologists, in their approach to this problem, have manifested two radically different attitudes. On the one hand there are those who, like Dice [48], accept the reality of communities as a datum, and are mainly concerned to devise methods for their delimitation and description; on the other hand, many, like Goodall [77], reject any such prior assumptions and, basing their studies on standard-sized 'random' sample areas, use mathematical methods in an attempt to determine the degrees of correlation in the incidence of particular pairs of species. Studies adopting this latter approach have never demonstrated the occurrence of significant discontinuities, and thus have provided no basis for the delimitation of definite communities. The methods of this school are well criticised by Daubenmire [45], who remarked 'It appears to me that if one selects any of several appropriate methods, one can demonstrate a continuum anywhere. The crux of the problem, as I see it, lies in the validity of the methods of gathering data, or in their subsequent manipulation, if not in both.' Daubenmire quotes a number of North American instances of natural vegetation falling into quite sharply defined communities, but notes that human action almost everywhere tends to change natural communities in such a way as to blur distinctions between them.

It is hardly imaginable that the enormously complex adaptive differentiations of recent (and past) plants and animals could have arisen other than through prolonged interactions in biotic environments which were both highly complex and relatively stable; a stable and richly diversified pattern of natural communities would provide appropriate media for evolutionary diversification, which in turn would produce further differentiation of the communities. It seems highly probable to me that most species of animals and plants have developed their peculiar characteristics as members of such communities, the understanding of persistent relations within which will thus be essential for the understanding of the species themselves. Throughout the world today, the remains of natural communities are being modified at an ever-increasing rate; their study should be the highest of priority tasks for ecologists.

Experience has already shown that the adoption of the 'Goodall' approach is not likely to produce rapid advances in our understanding of relationships within natural communities; in the urgent circumstances of today, surely Dice's approach is the appropriate one. He himself asserted that 'statistical methods for classifying vegetation are not practical. On the contrary, the classification of communities must be based largely on the good sense of the experienced field ecologist. Quantitative methods for describing communities are, nevertheless, extremely useful and should be employed wherever possible.' If the ecologists of today spend most of their time devising theories and techniques in the computer laboratory, while the last remnants of earth's natural communities vanish unstudied and unexplained, posterity will surely find our ecologists guilty of their own form of 'trahison des clercs'.

The usual procedure in defining a natural community (cf. Braun–Blanquet [19]) is to name one or two 'dominant' species and a number, perhaps of the order of ten, of other relatively constant component species, all of which might be expected to occur in any reasonable-sized sample area of the community. How large a sample would be 'reasonable' will vary with the type of community; if a tropical rain-forest is to be defined by its trees, samples of at least thousands of square meters will be required, whereas one or two square meters might provide an adequate sample of a mountain-top *Rhacomitrium* heath. Practical difficulties arise, of course, at the margins, which are rarely quite sharp; natural communities usually grade into adjacent ones through more or less narrow 'ecotones'. The problems which arise are comparable to those experienced by systematists dealing with geographical subspecies and clines (see Chapter 4).

Numerous natural communities have been recognised and named in many parts of the world, and there is a strongly felt need for a practically useful classification of them. The problem, however, presents difficulties of a peculiar sort, comparable perhaps to those of classifying social groups in *Homo sapiens*. Not only do natural communities tend to have ill-defined edges, they are apt also to show progressive changes over distances, as for example do salt marshes when traced from the Guadalquivir estuary to the north of Norway.

It is often suggested (e.g. by Lambert and Dale [124]) that the methods of Sokal and Sneath's 'Numerical Taxonomy' (see Chapter 15) should be particularly suited to the classification of natural communities. However, many of the objections to the use of this method for classifying organisms, which we have already considered, apply also to its use for classifying communities. For simplicity and practicality, numerical taxonomists prefer to operate on simple presence/absence distinctions for each species in communities, rather than on numerical measures of abundance, and thus would tend not to recognise 'dominant' species as such. Furthermore, they would treat each species alike, regardless of size and life-form; the dominant species of tree in a forest community would carry no more classificatory 'weight' than a soil mite or nematode species which occurred regularly in the same community. It might be possible to introduce 'weighted' numerical methods to overcome some of these drawbacks, but these would inevitably make the procedures more complex and difficult.

Up to the present, the most useful classifications of natural communities have been devised by the botanists, based on vegetation types and dominant species. It may seem paradoxical to some that a 'temperate salt-marsh' community in Norfolk may have not a single species in common with a similarly named community in Patagonia, whereas the Norfolk salt-marsh might have several species in common with a differently-named but physically adjacent community.

Zoologists have been less successful than botanists in naming and defining natural communities; the static and easily-visible character of most higher plants lends itself much more readily to the necessary recording than do most animals. Zoologists, wishing to define a community for synecological studies, usually do so on the basis of an already defined vegetational community; they study, for example, the fauna of a mangrove swamp, or a *Eucalyptus* savannah, rather than a *Periophthalmus* or an Emu-Kangaroo community. Undoubtedly, animals may influence plant communities as well as *vice versa*; the presence of large herds of grazing Ungulata may stabilise a grassland community which would otherwise develop into forest, and the relative abundance of specialised herbivores may strongly affect the outcome of competition within the plant community.

The classification of domesticated animals and plants is a practical necessity, but can hardly be based on the phylogenetic principles we have developed for the 'wild' forms. Artificial hybridisation has been an important factor in the establishment of very many, perhaps the majority, of domesticated species, and nearly all of them have been subjected to intensive artificial selection which has produced more or less striking phenotypic differences from their wild ancestors. Domesticated plants frequently show doublings of the chromosome number found in their wild ancestors, with the effect that hybrids between them and the wild forms are sterile triploids. In cases where the true ancestry of a domesticated form is more or less certainly known, it might be possible to devise some form of nomenclature which would indicate this, but this would inevitably be very cumbersome—and for many, probably most, domesticated varieties we do not have reliable or detailed knowledge of their ancestry. In such cases, an artificial, 'phenetic' classification is the only practicable solution; this is a field in which the methods of numerical taxonomy could well be applied, though it is doubtful whether practical plant and animal breeders would be content with a situation in which no cultivar could be put in its appropriate classificatory category until *all* its characters had been duly coded and put through a computer.

18 Taxonomic Research

For the Snark was a boojum, you see.

LEWIS CARROLL

If the classification of plants and animals were regarded as essentially analogous to that of books in libraries, there could hardly be any place for taxonomic research in the sense in which the term is used in this chapter. From time to time, no doubt, new systems for classifying books are proposed, but any research involved in their preparation will be on readers rather than on books. None, as far as I am aware, published doctoral theses on their efforts to establish the proper place of this or that group of books in the Dewey system. Systematic research in biology is normally aimed at establishing the proper places of particular groups of plants or animals in the natural system; it is an enquiry in search of either the Aristotelean essences of things (cf. Chapter 2) or their evolutionary ancestry (Chapter 9). Particular sorts of books, as we have seen, can hardly be said to possess Aristotelean essences, nor do they result from phylogeny in the biological sense, even though most books show traceable derivations from previous ones.

The Aristotelean essence, as defined in Chapter 2, might be considered as a statistical conception, and thus as implying resort to mathematical methods. In practice, the vast majority of systematists who have worked by Aristotelean principles have been content to say that this or that character is found in 'the vast majority' of species of a group, or in hardly any of them, without quoting any figures. The recent school of 'numerical taxonomists' (see Chapter 14), on the other hand, have not attempted to establish Aristotelean essences; their general criterion has been that, between any two members of a group, there must be a certain minimum of common characters, irrespective of what those characters are. The study of correlations in the incidences of different characters ('R type correlations' in the language of Sokal and Sneath), which might lead to Aristotelean classification, has not yet been taken up to any great extent among them.

A very common pattern in systematic research is the attempt to establish the true relationships of some taxon whose position is uncertain in the existing state of our knowledge. Examples which spring to mind are the Madagascan Aye-aye (*Chiromys*) among the Mammalia–Primates, or the genus *Tropaeolum* among Angiospermae–Dicotyledonae. The first stage in such an enquiry is to review the available comparative information (possibly by constructing a matrix as discussed later in this chapter), and to consider what relationships appear possible on the basis of the data. The phylogenetic systematist will translate each possible relationship into a hypothesis about ancestry, and will try to work out the implications—including functional, chronological and distributional ones—of each hypothesis in as much detail as possible. From each such hypothesis he will try to make deductions about the conditions which should obtain in characters which are as yet unrecorded for some or all of the forms concerned. The final stage of the research would be the attempt to verify these deductions observationally or experimentally.

In the case of *Chiromys*, for example, one possible theory is that it is a true lemur, of common origin with the other Malagasy ones, whose peculiarities result from adaptation to a mode of life unusual in the group; the other is that it is a surviving representative of a type which has been distinct almost from the beginning of the Primates stock, and possibly less closely related to the other Malagasy lemurs than they are to Primates of the rest of the world. In this case, the most obvious routes to a solution would be by way of protein-chemical (e.g. of haemoglobins or cytochrome-*c*) or serological studies. The difficulty would be that *Chiromys* is an exceedingly rare animal today, and that any attempt to obtain living specimens of it for such purposes would easily precipitate its extinction. It might be necessary to confine investigations to such characters as could be observed in material of *Chiromys* at present preserved in the world's museums. The skeleton, of course, furnishes large numbers of classificatory characters, but these have been fairly thoroughly studied already; the physico-chemical study of material from the skins might yield more significant new information.

In the case of *Tropaeolum*, the apparent possibilities are (1) a position in or near the Geraniales, with no real affinity to Cruciferales, (2) a position in or near the Cruciferales, with no real affinity to Geraniales,

(3) a real relationship to both Geraniales and Cruciferales, (4) an isolated position, without close relations to either of these orders. If we could show that *Tropaeolum* has some primitive characters which are not known in any Cruciferales but are found in some or all Geraniales, while it possesses no primitive characters which are not found in at least some Geraniales, this could be taken as very strong evidence for the first possibility, but might also be reconcilable with the third, or even the fourth. If, on the other hand, *Tropaeolum* proves to have one or more primitive characters which are unknown in Geraniales but found in some or all Cruciferales, and does not have any primitive features which are not found also in the latter group, then this supports the second possibility, but could likewise be reconciled with the third. If it comes to appear that either Cruciferales come from geranialean ancestors, or Geraniales from cruciferalean ones, by way of forms akin to *Tropaeolum*, then the existing system is clearly not a phylogenetic one. If *Tropaeolum* comes from a cruciferalean ancestor (akin perhaps to Resedaceae and Capparidaceae) and is itself directly allied to a geranialean ancestor, then presumably there has been a loss in the geranialean line of the ability to produce the characteristic sulphur-containing substances of the Cruciferales. Such a loss, if it occurred, might be attributed to the development of some new form of chemical defence, which proved more efficacious than mustard oils, etc. On the other hand, if *Tropaeolum* comes from a geranialean ancestor and is kindred to the ancestors of Cruciferales, we need not assume either a polyphyletic development or a secondary loss of these sulphur compounds. If *Tropaeolum* is geranialean and unrelated to Cruciferales, or if it is not directly related to either group, the development of these sulphur compounds must have been polyphyletic.

In the case of *Tropaeolum*, as in many similar botanical instances, the major difficulty in this type of reasoning is that of deciding to what extent and to what characters Dollo's Law is properly applicable. Because of this uncertainty, it is particularly desirable to have evidence from protein or nucleic acid chemistry or serology in such cases; it may help a great deal in deciding which characters are primitive, which are polyphyletic and which are liable to secondary loss.

Another familiar pattern in taxonomic research is the comparative study of some piece of structure, physiological function, element of behaviour, or chemical component, throughout a representative series drawn from a particular taxon. Such studies can usually be relied on to yield new information of classificatory (and phylogenetic) importance, though unless, as is rarely the case, those conducting them have a really thorough understanding of the existing system of the forms concerned, it is only too likely that they will lead to erroneous conclusions. The soundest general attitude in such studies is to accept the existing system of the group as a working hypothesis, to convert it into a dendrogram, and then to try to account for the different conditions found in the character you have studied in terms of this dendrogram. If, in order to do this, you find yourself obliged to postulate excessive amounts of polyphyly, serious breaches of Dollo's Law, or very implausible changes of function, then you may begin to question the existing system. The next step will be to see what minimum alterations in the existing system would remove the major anomalies in respect of your evidence, and then to consider whether such alterations would introduce as great or greater anomalies in respect of the evolution of other, already recorded characters. Only if the new anomalies seem less than the old, will you be justified in proposing in print a new classification.

In such studies, it is very important that the series chosen for study should be adequately representative of the taxon, with examples from each of its main subgroups and each of its main adaptive types. As such studies are commonly made by young botanists and zoologists at the beginning of their scientific careers, poorly situated in respect both of status and of personal connections for obtaining material of rare and out-of-the-way plants or animals, the actual studies made are only too often deficient in this matter of coverage. In practice, it is vitally important for young biologists undertaking such studies to obtain the patronage of established experts in the groups concerned. If the leading systematists are unable or unwilling to conduct such investigations themselves, they should recognise a duty to make material available to those who wish to do so.

The third standard pattern in systematic research is to attempt to improve the existing classification within a given taxon. The process is sometimes referred to as 'revising' the taxon. There are two main ways in which you may attempt to improve the classification of a group, one is to seek for a deeper phylogenetic understanding of the characters used in the existing system of the group, the other to seek for new characters in which the subdivisions of the taxon may differ. Of course, in order to discover new classificatory characters, you will be obliged to make comparative studies of species in the group. Such comparative studies need not be confined to visible structure, you might well gain equally useful (or even more useful) information from studies in comparative chemistry (e.g. of 'essential oils', etc., in plants) or in some form of systematic serology. Comparative serology, like nucleic acid fibre-pairing (see Chapter 13), has the advantage of providing something which may be considered as a measure of the classificatory 'distance' between two forms, but does not for this very reason lend itself well to consideration in conjunction with ordinary types of comparative information, e.g. in matrix analysis.

Matrix analysis is a method commonly used today, and is undoubtedly useful at times, even though it provides no infallible method of achieving a natural or phylogenetic system. As practised by systematists, the initial stages of the method involve the preparation of what Americans call a 'checker board diagram', from which, with the aid of a good deal of arithmetical computation, several different types of numerical matrix can be derived.

Let us suppose that you wish to review the subordinate classification within a given family. You first decide how far down the classificatory hierarchy you wish to take your analysis; the decision will probably depend on the numbers of subaltern groups at different levels. Matrix analysis is comparatively easy when the number of taxa and of characters involved is not very large; numbers up to about twenty of each are quite easily handled, but to treat fifty characters in fifty taxa would be exceedingly laborious. If the family you have chosen contains not more than thirty genera, you will probably base your analysis on the characters of the genera. You review all the characters in which these genera are known to differ among themselves, and select a manageable number of them, as far as possible those which can be expressed in a fairly clear-cut alternative 'either-or' form, and preferably which can be used to distinguish more than one genus from the rest. In the initial stage of matrix analysis it is possible to use a character which has three or four alternative conditions, but this will make for difficulties in later stages.

At this stage, you may discover that some of the characters which are known to differ between certain of the taxa (genera) in your group have not been recorded for some of them; this may provide the cue for studies of the organisms themselves, in the course of which you may well discover some additional characters which could be analysed in your matrix. When you have accumulated data on a sufficient number (say twenty) of suitable characters for all the taxa concerned, the first checker-board diagram can be made. Along one edge of the

matrix, you set out in order the names of the taxa (genera), and along
the edge at right angles, the characters you are using in the analysis.
Each square in the checker-board will then correspond to a particular
character in a particular taxon (genus). If the character is a simple
either-or one, it can be indicated by a white or a black square; it
is possible to indicate an intermediate condition by a grey or cross-
hatched square, and a variable condition within the given taxon
(genus) by a part-black square.

If the taxa and the characters involved in the analysis are taken as
'given', the operator remains free to vary the serial order of both taxa
and characters, and may also make his own decisions about which
condition of a given character is to be indicated by a black and which
by a white square. Some would argue for the 'objective' rule that the
condition in the majority of the taxa is to be indicated by a white
square and the minority one by a black square. A better rule, where it is
possible to apply it, is to indicate the primitive condition in the group
by a white square and a derivative one by a black square, but, especially
in the early stages of such studies, it may be very difficult to decide
which condition is primitive. In any case, the aim at this stage is to
produce as clear-cut as possible a pattern, with the black squares forming solid coherent blocks. When you have completed an initial diagram
of this sort, it may become apparent that some of the characters
included in your analysis have contributed little or nothing to the
emergent pattern; these characters may be omitted or replaced by
others in a revised version.

A matrix of this kind is illustrated in fig. 1. The taxa concerned are
the eighteen superfamilies (excluding Stylopoidea) of Coleoptera-
Polyphaga, for which fourteen selected characters are tabulated—both
numbers being small enough to make for relative ease in this type of
analysis. White squares indicate what I believe to be primitive conditions in all or at least more than 90 per cent of the species of the superfamily in question; a black square denotes a derivative condition in
more than 90 per cent of the species of the superfamily; a half-black
square indicates a derivative condition in at least 10 per cent but not
more than 90 per cent of the species. Cross-hatched squares represent
intermediate conditions in all or nearly all of the species of the superfamilies concerned.

It will be noticed that three groupings are fairly clearly indicated
in the pattern—superfamilies 1–3 (Staphyliniformia mihi), 7–11
(Elateriformia mihi) and 14–18 (Cucujiformia mihi). Each of these

CHARACTER / TAXON	1 HYDROPHILOIDEA	2 HISTEROIDEA	3 STAPHYLINOIDEA	4 SCARABAEOIDEA	5 EUCINETOIDEA	6 DASCILLOIDEA	7 BYRRHOIDEA
1 ADULT: NO FUNCTIONAL SPIRACLES ON ABDOMINAL SEGMENT 8		▨		◣	◣		
2 ADULT: AEDAEGUS OF COCUJOID FORM OR DERIVABLE FROM IT							
3 LARVA: MAXILLA WITH UNDIVIDED MALA			◣		◣		
4 ADULT: MALPIGHIAN TUBULES CRYPTONEPHRIDIC							
5 ADULT: HIND COXAE FLAT OR NEARLY SO	▨	▨	▨	▨		▦	
6 ADULT: ANTENNAE CLAVATE	▨	◣		▨	◣		▦
7 ADULT: TARSI PSEUDOTETRAMEROUS							
8 LARVA: MANDIBLES WITHOUT DISTINCT MOLAR PART	◣	◣	▨				▨
9 ADULT: WINGS WITH RADIAL CELL MARKEDLY ELONGATE							
10 ADULT: ONLY FOUR MALPIGHIAN TUBULES	▨		▨	▨			
11 LARVA: SPIRACLES WITHOUT A CLOSING APPARATUS				◣		▨	▨
12 LARVA: MAXILLA WITH ARTICULATED GALEA	◣	▨					▨
13 LARVA: LABRUM REPLACED BY A RIGID NASALE	◣		◣				
14 LARVA: ARTICULATED UROGOMPHI PRESENT	▨	▨	▨				
TOTAL POSITIVE CHARACTERS IN THE TAXON	$4\frac{1}{2}$ / 5	7	$4\frac{1}{2}$	4	$1\frac{1}{2}$	$1\frac{1}{2}$	$3\frac{1}{2}$

Fig. 1a

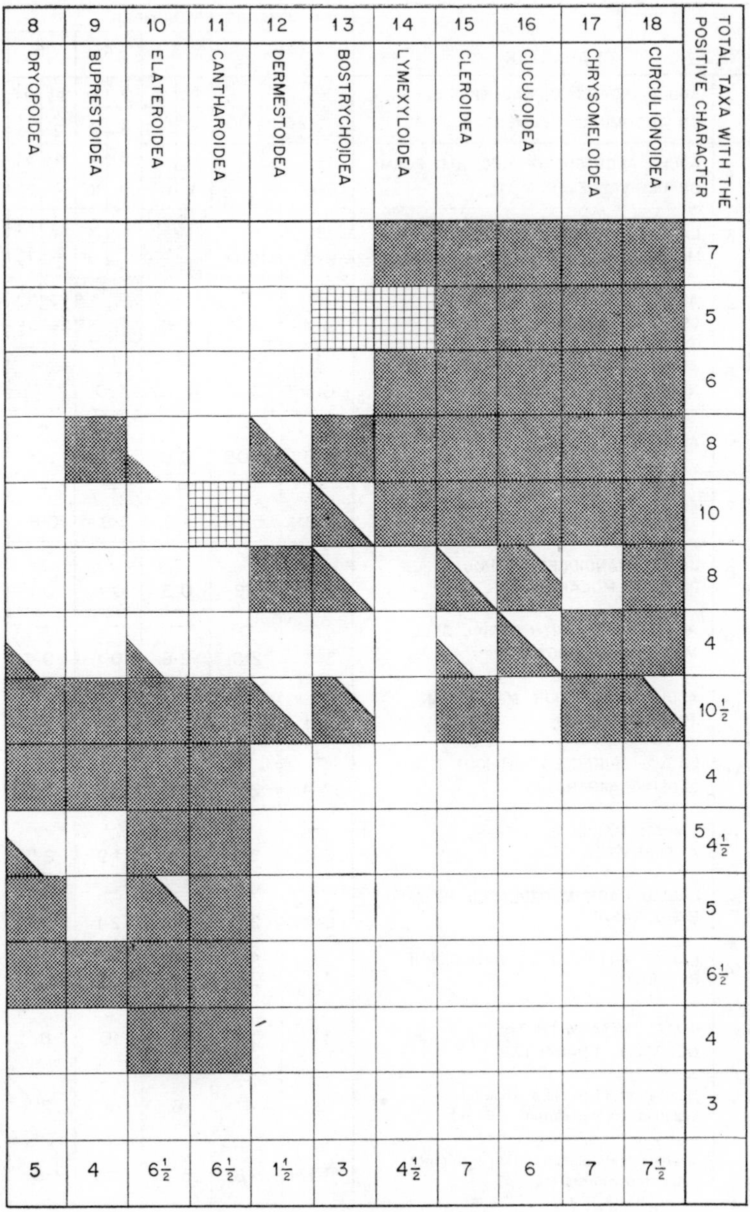

8
9
10
11
12
13
14
15
16
17
18
DRYOPOIDEA
BUPRESTOIDEA
ELATEROIDEA
CANTHAROIDEA
DERMESTOIDEA
BOSTRYCHOIDEA
LYMEXYLOIDEA
CLEROIDEA
CUCUJOIDEA
CHRYSOMELOIDEA
CURCULIONOIDEA
TOTAL TAXA WITH THE POSITIVE CHARACTER
7
5
6
8
10
8
4
10½
4
5 4½
5
6½
4
3
5
4
6½
6½
1½
3
4½
7
6
7
7½

Fig. 1b

No.	CHARACTER	CHARACTER NUMBER									
		1		2		3		4		5	
1	ADULT: NO FUNCTIONAL SPIRACLES ON ABDOMINAL SEGMENT 8	▨	▨	$4\frac{1}{2}$ / $2\frac{1}{2}$	$\frac{1}{2}$ / $10\frac{1}{2}$	5 / 2	$\frac{1}{2}$ / $10\frac{1}{2}$	5 / 2	3 / 8	$6\frac{1}{2}$ / $\frac{1}{2}$	$3\frac{1}{2}$ / $7\frac{1}{2}$
2	ADULT: AEDEAGUS OF CUCUJOID FORM OR DERIVABLE FROM IT	+ 7·6		▨	▨	$4\frac{1}{2}$ / $\frac{1}{2}$	$1\frac{1}{2}$ / $11\frac{1}{2}$	5 / 0	3 / 10	$4\frac{1}{2}$ / $\frac{1}{2}$	$5\frac{1}{2}$ / $7\frac{1}{2}$
3	LARVA: MAXILLA WITH UNDIVIDED MALA	+ 9·0		+ 10·0		▨	▨	5 / 1	3 / 9	$5\frac{1}{2}$ / $\frac{1}{2}$	$4\frac{1}{2}$ / $7\frac{1}{2}$
4	ADULT: MALPIGHIAN TUBULES CRYPTONEPHRIDIC	+ 3·4		+ 8·7		+ 5·5		▨	▨	$5\frac{1}{2}$ / $2\frac{1}{2}$	$4\frac{1}{2}$ / $5\frac{1}{2}$
5	ADULT: HIND COXAE FLAT OR NEARLY SO	+ 6·5		+ 3·3		+ 4·75		+ 1·0		▨	▨
6	ADULT: ANTENNAE CLAVATE	+ 0·3		− 0·05		− 0·0		− 0·3		+ 1·0	
7	ADULT: TARSI PSEUDOTETRAMEROUS	+ 2·8		+ 5·7		+ 4·0		+ 1·9		+ 0·6	
8	LARVA: MANDIBLES WITHOUT DISTINCT MOLAR PART	− 0·3		− 0·2		− 0·3		+ 0·1		− 0·1	
9	ADULT: WINGS WITH RADIAL CELL MARKEDLY ELONGATE	− 3·3		− 2·0		− 2·6		− 0·1		− 0·4	
10	ADULT: ONLY FOUR MALPIGHIAN TUBULES	− 2·4		− 2·5		− 1·7		− 3·3		− 0·1	
11	LARVA: SPIRACLES WITHOUT A CLOSING APPARATUS	− 2·4		− 2·5		− 3·5		− 3·3		− 1·8	
12	LARVA: MAXILLAE WITH ARTICULATED GALEA	− 2·4		− 3·9		− 5·1		− 1·9		− 2·5	
13	LARVA: LABRUM REPLACED BY A RIGID NASALE	− 0·4		− 2·0		− 1·0		− 2·1		+ 0·1	
14	LARVA: ARTICULATED UROGOMPHI PRESENT	− 0·05		− 1·4		− 0·45		− 2·9		+ 1·1	
	TOTAL TAXA WITH THE NEGATIVE CHARACTER	11		13		12		10		8	
	Number of characters showing signification correlation ($\geqq 4$)	3		4		6		2		2	
	Average level of correlation with other tabulated characters	3·1		3·8		3·7		2·6		1·8	

Fig. 2a

| CHARACTER NUMBER | | | | | | | | | | | | | | | | | | TOTAL TAXA WITH THE POSITIVE CHARACTER |
6		7		8		9		10		11		12		13		14		
3½	4	3	1	3½	7	0	4	½	4½	½	4½	1	5½	1.	3	1	2	7
3½	7	4	10	3½	4	7	7	6½	6½	6½	6½	6	5½	6	8	6	9	
2	6	3	1	2½	8	0	4	0	5	0	5	0	6½	0	4	0	3	5
3	7	2	12	2½	5	5	9	5	8	5	8	5	6½	5	9	5	10	
2½	5½	3	1	3	7½	0	4	½	4½	0	5	0	6½	½	3½	½	2½	6
3½	6½	3	11	3	4½	6	8	5½	7½	6	7	6	5½	5½	8½	5½	9½	
3	5	3	1	5	5½	1½	2½	½	4½	½	4½	1½	5	½	3½	0	3	8
5	5	5	9	3	4½	6½	7½	7½	5½	7½	5½	6½	5	7½	6½	8	7	
5½	2½	3	1	5½	5	½	3½	2½	2½	1½	3½	2	4½	2½	1½	2½	½	10
4½	5½	7	7	4½	3	9½	4½	7½	5½	8½	4½	8	3½	7½	6½	7½	7½	
▨	▨	2	2	4½	6	0	4	2	3	1	4	2	4½	1½	2½	2½	½	8
▨	▨	6	8	3½	4	8	6	6	7	7	6	6	5½	6½	7½	5½	9½	
+		▨	▨	3	7½	1	3	½	4½	½	4½	1	5½	½	3½	0	3	4
0·0		▨	▨	1	6½	3	11	3½	9½	3½	9½	3	8½	3½	10½	4	11	
−		+		▨	▨	4	0	3	2	3½	1½	6½	0	4	0	2	1	10½
0·0		0·6		▨	▨	6½	7½	7½	5½	7	6	4	7½	6½	7½	8½	6½	
−		−		+		▨	▨	2½	2½	2½	2½	4	2½	2	2	0	3	4
4·1		0·0		3·9		▨	▨	1½	11½	1½	11½	0	11½	2	12	4	11	
−		−		+		+		▨	▨	2½	2	2½	4	2½	1½	1½	1½	5
0·1		0·6		0·0		3·1		▨	▨	2½	11	2½	9	2½	11½	3½	11½	
−		−		+		+		+		▨	▨	3½	3	1½	2½	0	3	5
1·7		0·2		0·4		3·1		2·3		▨	▨	1½	10	3½	10½	5	10	
−		−		+		+		+		+		▨	▨	3½	½	1½	1½	6½
0·7		0·3		7·2		9·1		0·2		3·5		▨	▨	3	11	5	10	
−		−		+		+		+		+		+		▨	▨	2	1	4
0·1		0·3		3·9		2·3		3·1		0·25		5·9		▨	▨	2	13	
+		−		+		−		+		−		+		+		▨	▨	3
2·2		1·0		0·1		1·0		0·9		1·4		0·3		4·1		▨	▨	
10		14		7½		14		13		13		11½		14		15		
1		2		1		2		0		0		4		2		1		
0·8		1·4		1·3		2·8		1·6		2·0		3·4		2·1		1·3		

Fig. 2b

has an 'Aristotelean essence' represented by a common block of black
squares. There remain, however, superfamilies nos 4, 5, 6, 12 and 13,
whose positions are not made evident in this matrix. The problems
presented by these groups are not such as can be solved by further
mathematical analysis of the data in this matrix, the only useful
approach to them is the phylogenetic one discussed later in this chapter.

The second stage of matrix analysis of the same basic data is shown
in fig. 2. Along both axes of the matrix, the fourteen numbered characters are indicated, and a diagonal line of squares is blacked out. On
the upper right hand side of this diagonal line, each square contains
four numbers, the top left one being the number of taxa with the
'derivative' condition in both the characters corresponding to the
square, the top right having the derivative condition of the 'column'
character and the 'primitive' form of the row one, the bottom left the
primitive condition of the column character and derivative of the
row one, the bottom right those with the primitive condition in both
characters. The four numbers should add up in each square to eighteen
(the total number of taxa). At this stage of the analysis, problems are
presented by those taxa which do not have simple black or white
squares for the characters concerned. The convention I have adopted
is that a half black or a cross-hatched square counts as a half correlation
with either a white one or a black one; for part-black squares, I have
considered whether those members of the taxon which have the
'positive' form of the one character are largely or entirely those which
have the positive form of the other character compared—if they are,
a half correlation is entered, but if not, no correlation is counted; a half-
black square and a cross-hatched one are counted as no correlation.

If the top left number in a given square is called w, the top right
one x, the bottom left one y, and the bottom right figure z, then we
can obtain a figure for chi-squared, treating the whole as a 2×2
contingency table, according to this formula:

$$\chi^2 = \frac{18\,(wz - xy)^2}{(w + x)\,(y + z)\,(w + y)\,(x + z)}$$

For the satisfactory applicability of this method, it is generally stated
that the 'expected value' for w, x, y and z should be at least 5, whereas
in our example it is only $4\frac{1}{2}$ (18/4). A value of chi squared equal to
4 or more indicates a probability of less than one in 20 that the correlation between the two characters concerned is a random one. The
values of chi squared, so calculated, for each pair of characters are

entered in the appropriate squares on the lower left side of the black diagonal, together with a $+$ (when wz exceeds xy) or $-$ (when xy exceeds wz sign). Values of 4 or more are 'significant'. At the ends of each row or column three additional figures are entered, the first, the number of taxa showing the 'positive' form of that character, the second, the average value of chi squared (ignoring the $+$ and $-$ signs) between it and all the other characters, the third, the number of other characters showing a value of chi-squared of 4 or more with it.

Both of the last two figures mentioned provide indications of the classificatory 'weight' of a character. The distribution of $+$ and $-$ signs in the table is also significant. It will be noticed that characters 1–7 show almost entirely positive correlation values between themselves, and practically as exclusively negative correlations with the remaining seven; there is a second group of positively correlated characters from 8 to 14. The possible significance of this is discussed later, when we bring in phylogenetic considerations. The characters with the greatest weight as we have defined it are numbers 3, 12, 4, 2 and 1—two from larval structure and three from adult structure, illustrating the point made in Chapter 14, that the larval and imaginal structures in the higher insects are approximately of equal classificatory importance.

The third stage in matrix analysis of the same basic data is shown in fig. 3. Here there are eighteen taxa indicated by numbers along both axes of the matrix, there is a blacked-out diagonal line as in the second matrix; the squares on the upper right hand side of the diagonal record, for the corresponding pair of taxa, the number of positive (black square) characters they have in common, while the squares on the lower left side show the number of common 'negative' characters. In this matrix, the most notable feature is the group of relatively high values for 'positive' characters constituted by taxa 14 to 18, whereas the values for common negative characters for the same group of taxa do not stand out in anything like the same way. Taxa 1 to 3 also shows relatively high correlation in the positive character, as do numbers 7 to 11. Here again, the numerical analysis merely provides another way of expressing relations which were obvious to the eye in the original non-numerical matrix.

The second type of matrix provides the basis for the analysis of what the numerical taxonomists call 'R-type correlations', which they suggest may be of value in elucidating phylogeny but which have played no part in the practical work so far published by this school. The

		TAXON NUMBER								
		1	2	3	4	5	6	7	8	9
1	HYDROPHILOIDEA		$4\frac{1}{2}$	$2\frac{1}{2}$	2	$\frac{1}{2}$	$\frac{1}{2}$	$1\frac{1}{2}$	1	1
2	HISTEROIDEA	7		3	$2\frac{1}{2}$	1	$\frac{1}{2}$	$2\frac{1}{2}$	2	2
3	STAPHYLINOIDEA	$7\frac{1}{2}$	$5\frac{1}{2}$		2	$\frac{1}{2}$	$\frac{1}{2}$	1	1	$\frac{1}{2}$
4	SCARABAEOIDEA	$7\frac{1}{2}$	$5\frac{1}{2}$	$7\frac{1}{2}$		$\frac{1}{2}$	1	1	1	0
5	EUCINETOIDEA	$8\frac{1}{2}$	$6\frac{1}{2}$	$8\frac{1}{2}$	$9\frac{1}{2}$		0	0	0	0
6	DASCILLOIDEA	$8\frac{1}{2}$	6	$8\frac{1}{2}$	$9\frac{1}{2}$	11		1	1	0
7	BYRRHOIDEA	$7\frac{1}{2}$	6	7	$7\frac{1}{2}$	$9\frac{1}{2}$	10		3	2
8	DRYOPOIDEA	$5\frac{1}{2}$	4	$5\frac{1}{2}$	6	$7\frac{1}{2}$	$7\frac{1}{2}$	$8\frac{1}{2}$		3
9	BUPRESTOIDEA	$6\frac{1}{2}$	5	6	6	$8\frac{1}{2}$	$8\frac{1}{2}$	$8\frac{1}{2}$	8	
10	ELATEROIDEA	$4\frac{1}{2}$	$3\frac{1}{2}$	5	5	6	$6\frac{1}{2}$	$6\frac{1}{2}$	7	7
11	CANTHAROIDEA	5	4	$5\frac{1}{2}$	$5\frac{1}{2}$	6	$6\frac{1}{2}$	7	7	$6\frac{1}{2}$
12	DERMESTOIDEA	9	$6\frac{1}{2}$	9	9	11	$10\frac{1}{2}$	$9\frac{1}{2}$	$7\frac{1}{2}$	9
13	BOSTRYCHOIDEA	8	$5\frac{1}{2}$	8	8	10	10	$8\frac{1}{2}$	$6\frac{1}{2}$	$8\frac{1}{2}$
14	LYMEXYLOIDEA	$5\frac{1}{2}$	$4\frac{1}{2}$	$5\frac{1}{2}$	$6\frac{1}{2}$	9	8	$5\frac{1}{2}$	4	6
15	CLEROIDEA	$4\frac{1}{2}$	$3\frac{1}{2}$	$4\frac{1}{2}$	5	7	6	5	4	5
16	CUCUJOIDEA	5	$3\frac{1}{2}$	5	6	8	7	5	$3\frac{1}{2}$	5
17	CHRYSOMELOIDEA	4	3	4	$4\frac{1}{2}$	$6\frac{1}{2}$	6	$4\frac{1}{2}$	$3\frac{1}{2}$	5
18	CURCULIONOIDEA	$4\frac{1}{2}$	3	4	5	$6\frac{1}{2}$	$5\frac{1}{2}$	4	$2\frac{1}{2}$	4
	Total negative characters in common with other taxa	$108\frac{1}{4}$	$82\frac{1}{4}$	$106\frac{1}{4}$	$113\frac{1}{2}$	$139\frac{1}{4}$	$135\frac{1}{2}$	120	98	113

Fig. 3a

		TAXON NUMBER							
10	11	12	13	14	15	16	17	18	
$1\frac{1}{2}$	2	1	$1\frac{1}{2}$	1	2	$1\frac{1}{2}$	$1\frac{1}{2}$	$2\frac{1}{2}$	28
3	$3\frac{1}{2}$	$1\frac{1}{2}$	$1\frac{1}{2}$	2	$3\frac{1}{2}$	$2\frac{1}{2}$	3	$3\frac{1}{2}$	41
2	$2\frac{1}{2}$	1	1	1	$1\frac{1}{2}$	$1\frac{1}{2}$	$1\frac{1}{2}$	2	$25\frac{1}{2}$
$1\frac{1}{2}$	2	1	1	$1\frac{1}{2}$	2	2	$1\frac{1}{2}$	$2\frac{1}{2}$	25
0	0	$\frac{1}{2}$	$\frac{1}{2}$	1	1	1	1	1	$8\frac{1}{2}$
$\frac{1}{2}$	$1\frac{1}{2}$	0	$\frac{1}{2}$	$\frac{1}{2}$	$\frac{1}{2}$	$\frac{1}{2}$	$\frac{1}{2}$	$\frac{1}{2}$	$9\frac{1}{2}$
$2\frac{1}{2}$	3	1	1	0	$1\frac{1}{2}$	$\frac{1}{2}$	1	1	$23\frac{1}{2}$
4	$4\frac{1}{2}$	$\frac{1}{2}$	$\frac{1}{2}$	0	1	$\frac{1}{2}$	$1\frac{1}{2}$	1	$25\frac{1}{2}$
$3\frac{1}{2}$	3	1	$1\frac{1}{2}$	1	2	1	2	$1\frac{1}{2}$	25
	$5\frac{1}{2}$	$\frac{1}{2}$	1	$\frac{1}{2}$	1	$\frac{1}{2}$	$1\frac{1}{2}$	$1\frac{1}{2}$	$30\frac{1}{2}$
$6\frac{1}{2}$		$\frac{1}{2}$	1	$\frac{1}{2}$	$1\frac{1}{2}$	$\frac{1}{2}$	$1\frac{1}{2}$	1	34
7	6		$1\frac{1}{2}$	$\frac{1}{2}$	$1\frac{1}{2}$	1	1	2	16
$5\frac{1}{2}$	$5\frac{1}{2}$	$10\frac{1}{2}$		2	3	$2\frac{1}{2}$	$2\frac{1}{2}$	$2\frac{1}{2}$	25
3	3	8	8		$4\frac{1}{2}$	$4\frac{1}{2}$	$4\frac{1}{2}$	$4\frac{1}{2}$	$29\frac{1}{2}$
3	2	$6\frac{1}{2}$	7	7		6	$6\frac{1}{2}$	7	46
3	2	7	$7\frac{1}{2}$	8	7		$5\frac{1}{2}$	6	$37\frac{1}{2}$
$2\frac{1}{2}$	2	6	$6\frac{1}{2}$	7	$6\frac{1}{2}$	$6\frac{1}{2}$		7	$43\frac{1}{2}$
$1\frac{1}{2}$	1	6	$6\frac{1}{2}$	7	6	$6\frac{1}{2}$	6		47
83	81	138	130	$105\frac{1}{2}$	$83\frac{1}{2}$	$95\frac{1}{2}$	84	$79\frac{1}{2}$	Total positive characters in common with other taxa

Fig. 3b

third type of matrix is basic in the analysis of what Sneath and Sokal call 'Q-type correlations', on which their practice hitherto has been based.

What the numerical taxonomists stigmatise as 'phylogenetic speculations' entered into the formation of our original matrix, in that white squares were allocated to supposedly primitive conditions of the characters and black ones to the derivative conditions. That this coding is not a matter of indifference from the point of view of subsequent numerical analysis is suggested by the observation that, in the third matrix, the classificatory groups were indicated much more clearly by agreements in 'black' squares (derivative characters) than in the white ones. If phylogenetic considerations are excluded, the natural procedure is to represent the majority condition by a white square and the minority one by a black one. The adoption of this convention would have reversed the coding of only two of the fourteen characters in our original matrix, numbers 5 and 8.

It has been suggested by Sporne [178] that primitive characters should show a significant positive correlation with each other in their incidence within a given taxon, and that derivative characters should likewise show correlations with each other. Sporne's theoretical reasons for this are complex and not wholly convincing, but, at least as regards primitive characters, there would seem to be some element of truth in his theory. The effect, if it is real, is clearly a statistical one, one could easily find cases of characters which are undoubtedly primitive within a group but which are negatively correlated, such as, for example, excavate hind coxae in adult Polyphaga and the presence of a closing apparatus in the larval spiracles in the matrix we have been considering; it is even easier to find derivative characters which are negatively correlated in their incidence, for example, most of the characters 8–14 in our matrix are almost certainly derivative and almost all of them show negative correlations with the derivative forms of characters 1–7. It appears to me that within a given group, primitive characters will more often than not, but by no means invariably show, positive correlations in incidence.

Often when primitive characters are positively correlated with each other in their incidence in a given group those primitive characters are found only in a more or less small minority of the members of the group. Where this state of affairs exists, however, so that a small section of the group can be set apart by the presence in all or some of its members of two or more primitive characters not found in other members of the group, a strong presumption exists that the present classification of the group is not phylogenetically natural. Almost certainly, the section with the primitive characters ought to be separated from the rest to form one or more primary divisions of equal status to the entire residue of the group.

Correlations in characters, as recognised by Sporne himself, may have functional significance—thus in our matrix, the character 'cryptonephridic Malpighian tubules' represents an adaptation for improved water economy, whereas the lack of a closing apparatus in the larval spiracles is related to very damp or sub-aquatic habitats for the larva; it is not surprising that these two characters are negatively correlated in the Coleoptera-Polyphaga. The lack of a molar part to the larval mandible, and the replacement of the larval labrum by a rigid nasale, positively correlated in Polyphaga, are both found in and presumably adaptive to larvae which are carnivorous and practice 'extra-oral digestion'. It seems quite possible that nearly all marked correlations in the incidences of particular characters could be given some such adaptive interpretation. Thus the tendency for primitive characters to occur together—in what are called 'primitive' taxa—could indicate a tendency to persistence of at least some aspects of an ancestral mode of life. A primitive ancestor of a taxon would have been an organism with a definite mode of life, to which its characters were probably quite well adapted; evolutionary divergence from it would in the main be connected with changes in the mode of life. The study of correlations in the incidence of characters of different types may contribute not only to improved classifications but also to deeper functional understanding of the plants or animals concerned.

The phylogenetic systematist will attach comparatively little importance to matrix analysis and the various mathematical procedures developing from it; he is liable to proceed on the orthodox scientific plan of formulating hypotheses and then looking for evidence to confirm or disprove them. His hypotheses will, of course, be about ancestry, and the evidence to confirm or disprove them will include the appeal to various empirical generalisations about evolutionary change, discussed already in Chapter 9. These generalisations include one that at all times in the evolutionary history of a group, the ancestors must have been 'viable' organisms with definite habits or modes of life to which their structural and physiological characters were adapted.

A very common type of hypothesis of the phylogenetic systematist is that some particular condition of a certain character is primitive in the taxon concerned. He will then proceed to follow out the implications of this hypothesis. For example, he might suggest that the primitive condition of the andraecium in Rosaceae was to have only ten stamens in two whorls. Evidence in support of this might be drawn from (a) indications that a similar arrangement of the stamens is characteristic of the most closely allied groups outside the Rosaceae, e.g. Saxifragaceae or Leguminosae, (b) presence of only ten stamens in those Rosaceae which show other presumed primitive characters, (c) the formation of the primordia of the stamens in the developing flowers of Rosaceae, (d) fossil evidence for greater antiquity of types with ten stamens. The hypothesis might be extended to postulate that the basic multiplication of the stamens beyond the number ten was a monophyletic change, so that all Rosaceae with more than ten stamens should be more closely related to each other than any of them are to forms preserving the primitive number ten—a hypothesis which might be tested by comparative serology, or protein-chemical techniques.

We may now return to consider, from a phylogenetic point of view, the positions of superfamilies numbers 4, 5, 6, 12 and 13 in our first matrix diagram. In connection with superfamilies 12 and 13, character 4 is particularly significant. Cryptonephridic Malpighian tubules are almost certainly a derivative character, and one which seems very rarely subject to secondary loss. With a very few exceptions (in secondarily aquatic forms) it is found throughout the entire block of superfamilies 14–18 (Cucujiformia), it is found in 13, in some of 12, and in 9. There is no other evidence for a direct relationship of superfamily 9 to the Cucujiformia, or to superfamilies 12 or 13 for that matter, so that cryptonephridic Malpighian tubules seem to be at least diphyletic in Coleoptera Polyphaga. But there are other indications of a possible relationship of 12 and 13 to the Cucujiformia, and there is practically no evidence for a direct relationship of Cucujiformia to any group of Polyphaga other than superfamilies 12 and 13. If cryptonephridic Malpighian tubules in superfamilies 12 to 18 are all inherited from a common ancestor, then either the condition has been secondarily lost in some of 12, or superfamilies 13 to 18 are all directly or indirectly derived from 12. Of the families in 12 which are known not to show cryptonephridism—Derodontidae and Nosodendridae—the second has peculiar almost aquatic habits, living in sap flows of damaged trees, and might conceivably have secondarily lost cryptonephridism, but the species of the first, so far as known, have quite terrestrial habits and there is no obvious reason for doubting that the condition of their Malpighian tubules is primitive.

To judge from the matrix, character 6 might appear to represent a serious objection to the possibility of deriving superfamilies 13–18 from 12, in that a derived condition (clubbed antennae) is indicated as constant in 12, whereas at least some of 13, 15 and 16, and all of 14 and 17, are indicated as showing the primitive (filiform) condition. However, there are good reasons for believing that the derivative condition in this character is subject to secondary loss. Phylogenetically, it seems quite probable that superfamily 13 is a direct offshoot of 12, and that 14–18 all come from a common ancestor which itself arose from something with the essential characters of 12. If this is so, superfamily 12 could not be retained in its present form in a strict phylogenetic system, it would have to be divided, one part (probably the families Dermestidae and Thorictidae) going with superfamily 13, another part (probably family Jacobsoniidae) going with 14–18, and the remainder (families Derodontidae and Nosodendridae) forming a superfamily and series of their own. In my opinion further evidence will be required to confirm this phylogenetic hypothesis before adopting such a radical modification of the existing classification. Such evidence might be obtained from comparative serology (if sufficiently stable proteins could be isolated from these insects to show relationships at such high levels), or from protein chemistry, from 'hybrid nucleic acid' studies involving these groups, from new data in comparative anatomy or comparative physiology. It is just possible, though rather improbable, that crucial evidence may come from palaeontology—if it is possible to recognise, and to determine some of the important characters of, Jurassic ancestors of any of these groups.

The position of superfamilies 4, 5 and 6 is more difficult to decide. The Dascilloidea (6), with only a single black square, might be taken as the most primitive of the superfamilies included, followed by Eucinetoidea (5) with one and a half and Dermestoidea (12) with two black squares; however, superfamily 6 has only a single family, and there are individual families—Eucinetidae in 5, Derodontidae in 12, which if separately included in the matrix would emerge as even more primitive than Dascillidae. It seems probable that all Polyphaga are derivable from ancestors which, in respect of characters tabulated in the matrix, resembled Eucinetidae, Dascillidae and Derodontidae. If the origins of superfamily 13 and of 14–18 are to be sought in or near superfamily 12, it seems possible that somewhere in superfamilies

5, 6 or 12 we might find forms allied to the ancestors of the other specialised series, i.e. numbers 1–3 (Staphyliniformia), 4 (Scarabaeiformia) and 7–11 (Elateriformia). At least character 11 suggests a possible kinship of 7–11 to superfamily 6, and I know of no characters which strongly contradict this. Some characters of the larvae, mainly not tabulated in the matrix, suggest a connection of 4 with 6, even though the characters of the adults seem on the whole more suggestive of a relation to 1–3; there are reasons for thinking that relations of 4 to Staphyliniformia or to Dascilloidea are mutually exclusive.

The pursuit of research in systematics may well necessitate a good deal of travel by the botanist or zoologist—or the palaeontologist. Not only is he likely to wish to study types and other rare material in diverse museums, but also to wish for more information about particular kinds of animals, plants or fossils than can be elicited from the available material in museums. Only at the cost of the expedition described by Lewis Carroll was it established that the Snark was in fact a Boojum. A botanist or zoologist dealing with one of the groups in which the world flora or fauna is less completely known, may well have reason to suspect that hitherto undescribed species and genera, likely to be of systematic importance, will occur in one or other of the less studied regions of the world.

At the present time, when the remnants of the world's 'natural habitats' for plants and animals are being destroyed or radically altered at an unprecedented rate, studies of the rarer animals and plants in nature should be regarded as a priority task for systematists. The botanists and zoologists of the next century will be largely dependent on our records for their knowledge of many of these types, and information which we fail to collect is liable to be irretrievably lost. Furthermore, we cannot reasonably expect ecologists, economic biologists and other non-systematists to have either the knowledge or the incentive to collect and record all the field data on plants or animals which might prove to have systematic importance. The primary responsibility for this inevitably rests on the systematists, and it is they who, if the task is not adequately discharged, will bear the reproaches of posterity.

19 The Practical Problems of Phylogenetic Classification

In the writings of G. G. Simpson [168a, etc.], great stress is laid on differing rates of evolution in different phyletic lines; his ideas are enshrined in three new words, bradytelic (slow-evolving), horotelic (evolving at normal speed) and tachytelic (fast evolving). If applied in respect of particular measurable characters, these terms might be given a reasonably objective basis—they could, for example, be defined in terms of Haldane's [81] evolutionary unit, the 'darwin'. Simpson, however, is more concerned with whole organisms, and uses the currently accepted classificatory hierarchy as his measuring rod for evolutionary change; this is in effect almost a reversal of the principle of phylogenetic systematics, in that instead of using phylogeny as a standard to which classification has to be adapted, he used classification as a standard by which evolution can be measured. It is obvious, and more or less admitted by Simpson himself, that such categories as the genus have been used in very different ways in different groups of animals (Simpson does not discuss plants in this connection). He argues, however, that though such a criterion would have little meaning if we were comparing the evolution of such dissimilar forms as horses and bivalve molluscs, it should be valid when the forms compared are sufficiently similar in general type of organisation, e.g. nautiloids and ammonoids, horses and tapirs, *Xiphosura* and true crabs, Crocodilia and Dinosauria.

If, as it appears to me, there is some truth in Simpson's argument here, it will constitute a practical difficulty for phylogenetic systematics, though not in my opinion (which here contrasts with Simpson's) an insuperable one. The case considered in Chapter 9, of the dipnoan lungfishes, illustrates the sort of discrepancy between phylogenetic and Aristotelean classification that is liable to arise—the Dipnoi being decidedly bradytelic in Simpson's terminology, whereas Tetrapoda as a whole are markedly tachytelic in comparison with fishes.

I

We may consider two further cases, quoted by Simpson himself, of extremely bradytelic evolutionary lines, the Nautiloidea among Mollusca and the *Xiphosura* among Arthropoda. Similar considerations will apply to such forms as *Equisetum* and *Ginkgo* among vascular plants. In all these groups, Mesozoic fossils are scarcely to be distinguished, at the accepted generic level, from modern species. All authorities are agreed in placing Recent Nautiloidea in a single genus *Nautilus*, though the number of species recognised varies from three to ten; many Tertiary and Mesozoic species, going back at least as far as the Jurassic period, have been attributed to the genus *Nautilus*. The Recent Xiphosura comprise four generally recognised species, which by older authors were all placed in the genus *Limulus* (*Xiphosura* auctt.); it is now usual to dintinguish three genera, and some even make two sub-families among them. One of the 'subfamilies' contains only the genus *Limulus* s.str., with its single living Atlantic species; fossils going back as far as the Triassic period have also been assigned to *Limulus* in recent works. The two other Recent genera, *Tachyploeus* and *Carcinoscorpio*, which occur in the Indo-Pacific area, have not to my knowledge been recorded as fossils. The fossils of *Nautilus*, like most of the Recent specimens studied by systematists, comprise only the shell, which is of relatively simple form (considerably simpler, for instance, than shells of the Ammonita) and offers relatively few characters which can be used in defining a classificatory hierarchy. The exoskeleton of the Xiphosura is, however, much more complex, offering very much better bases for classificatory distinctions.

The greatest possible difficulties for phylogenetic classification which could arise from these cases would be if Recent species of *Nautilus* were traced to separate Mesozoic ancestors, or if separate ancestry for *Tachyploeus* and *Carcinoscorpio* could be distinguished already in Jurassic fossils. We might then find ourselves obliged, by the logic of phylogenetic systematics, to place in separate tribes or even subfamilies species which systematists have regarded as only barely distinguishable from each other, or to make separate families for what have been considered as poorly separated genera. These problems, however, have not yet arisen, and may never arise. Recent serological studies (Schuster [165]; see Chapter 13) tend to suggest that the divergence of existing Xiphosura may be of Tertiary rather than Mesozoic origin.

A striking botanical example of the effects of 'bradytely' and 'tachytely' may be drawn from comparisons of *Ginkgo* with various Angiospermae; forms scarcely distinguishable from the Recent genus

Ginkgo have been found among Jurassic fossils, and thus antedate any authenticated fossil angiosperms. Whatever the Jurassic ancestors of an Oak, a Sunflower or a Lily looked like, none would doubt their generic distinctness from *Quercus, Helianthus* or *Lilium*!

Many of the practical difficulties of strict phylogenetic classification result from the fact that phenotypic evolution, which is usually a slow, progressive process, is greatly accelerated in certain groups at particular times, most notably, of course, in the well-known 'adaptive radiations'. A major change in the mode of life will expose organisms to strong selection operating in quite new directions, in contrast to those phases during which selection operates mainly as a 'stabilizing' (to use Schmalhausen's term [164]) or 'canalizing' (in the sense of Waddington [194]) force. A whole series of characters are liable to be affected, no doubt in quick succession rather than concurrently (cf. Chapter 16), and the end result will be the establishment of a new 'Aristotelean essence' in a relatively short time, as in the case of the Tetrapoda and Crossopterygii already considered. Formal, Aristotelean classification has based its main divisions on such radical changes in the mode of life; phylogenetic systematics, on the other hand, will accord no special recognition to splittings in the family tree which were accompanied by such drastic transformations—on the contrary, it will treat them exactly as it treats contemporary splittings which did not involve any important change of mode of life and were not accompanied by striking phenotypic changes.

As we have seen already (Chapter 12), there is no reason to expect a simple relation of proportionality between phenotypic divergencies and genetic differences; we have already good evidence that some phenotypically very similar forms may differ radically in genotype, and conversely that species which are closely allied in their genotypes may show striking phenotypic differences. It seems probable that in bradytelic lines genetic differences will in general be greater than the phenotypes would lead us to suppose, and that the reverse may be true in many tachytelic lines. Modern techniques such as DNA fibre-pairing (see Chapter 13) can provide something like a direct measure of genotypic divergence; with their aid, it should be possible to distinguish bradytelic and tachytelic lines even in the absence of a fossil record. This may help greatly in overcoming the difficulties which differing evolutionary rates pose for the phylogenetic systematist.

The second major objection brought by Simpson against strict phylogenetic classification is not, however, so easily to be overcome.

It concerns the classification of fossils. The general principle of phylogenetic classification, as we have seen, is to make all groups monophyletic and to establish the category of a group by the age of its common
ancestral species—measured in years or in generations. How should we
apply this criterion when determining the categories to be used for
groups of fossils from (say) the Eocene period? At that time a group
originating from a common ancestral species in the very late Cretaceous
might be only some ten million years old, corresponding to a category
of not more than a genus according to the criteria developed later in
this chapter; its descendants today might constitute what is normally
called an order of placental mammals, with a common ancestor rather
more than sixty million years ago. If the Eocene fossils constitute a
genus, should that genus be placed in the same order (or family) as their
modern descendants, or grouped with other Eocene genera which
share common ancestors in the Upper Cretaceous, some of which may
be ancestral to other orders (or families) of mammals? There appears to
be no satisfactory objective method of settling this question if the
Eocene and Recent forms are to be placed together in the same system.

Simpson [169] cites the Eocene fossils *Eohippus* (or *Hyracotherium*) as
either direct ancestors of modern horses, or close relatives of such
ancestors, and places them in the family Equidae, of the order Perissodactyla. He notes, however, that the same fossils might serve equally
well as ancestors of modern tapirs or rhinoceroses, placed in two other
families (Tapiridae and Rhinocerotidae) of Perissodactyla. Furthermore, he points out, these Eocene 'horses' do not differ from other
Eocene fossils customarily placed in the extinct order Condylarthra to
anything like the extent which would be expected if they are placed in
a different order from these. Simpson gives no logical reason for placing
Hyracotherium in Equidae rather than in Tapiridae or Rhinocerotidae, or
in Perissodactyla rather than in Condylarthra, and there is no principle
enunciated by him which would provide any definite solution to these
problems.

If, however, we abandon the attempt to fit fossils and modern
organisms into one and the same system, we could, as Simpson himself
reluctantly admits, classify modern organisms on strict phylogenetic
principles. Theoretically speaking, the same should be possible for the
total assemblage of organisms of any one particular time in the geological past. In practice, there is no currently available or foreseeable
general method for precise time-correlation of fossils and deposits from
different parts of the world, so that we are not in a position to assemble

world-wide collection of fossils representing a strictly contemporaneous fauna or flora. The most practicable method, at the present time, of applying phylogenetic principles to the classification of fossils would probably be to divide the geological past into a number of reasonably equal periods of time and to establish a separate classification for the organisms of each of these periods. These periods would need to be defined in relation to the recognised eras and periods of the geologists, and at least initially would best be made rather long, perhaps of the order of twenty-five or thirty million years each. Thus, in addition to our classification of modern organisms, we could have one for those of the Neogene (Upper Tertiary), of the Palaeogene (Lower Tertiary), of the Upper Cretaceous and of the Lower Cretaceous, of the Jurassic, of the Upper and Lower Triassic, one for the Permian, one each for Pennsylvanian (Upper Carboniferous) and Mississipian (Lower Carboniferous), for Upper and Lower Devonian, Upper and Lower Silurian, Upper and Lower Ordovician, and Upper and Lower Cambrian. All these classifications would not, of course, be completely independent of each other. If we adopted a hierarchy containing as many categories as we have pre-Recent system, and defined each category in relation to a specified age of origin, persisting lineages could regularly be promoted one grade in the hierarchy in each successive system in which they appeared.

If we tried to apply this system to the fossil ancestors of the horse, discussed by Simpson, the result might be something like this; in the system of Palaeogene (Lower Tertiary) fossils, *Eohippus* (*Hyracotherium*) might appear as a genus in a family which included all the Ungulate ancestors and Condylarthra; in the Neogene system the Perissodactyla (including the ancestral horses) would form one subfamily of a family corresponding to the old Ungulata, and in the Recent fauna this family would be split into two, corresponding to Artiodactyla and Perissodactyla.

Such a system would necessarily involve considerable changes in our whole principles of nomenclature as applied to fossils; one solution to some of them would be to attach a characteristic number in front of the name of any pre-Recent organism, the numbers rising progressively with age, thus *Hyracotherium* might appear as (2) *Hyracotherium*, while the Jurassic plant *Caytonia* would appear as (5) *Caytonia*. The number would then provide an indicator of this status which any Recent descendants should receive.

The major difficulties in making a phylogenetic system for fossils would not however be eliminated altogether by such a system; they would recur within the system for each geological period, on a smaller scale, and mainly affecting the categories at about genus level. Uncertainties in time correlation would inevitably cause difficulties, for example, whether the fauna and flora of the Purbeck beds of southern England should be classified in the Lower Cretaceous (4) system, or in that of Jurassic (5) fossils. However, I think it would be worth while to accept some disadvantages in order to give our overall system a new (time) dimension.

As we have seen already, the category of a given group in a phylogenetic system of modern organisms should be determined by the age of its common ancestor. In this way we can provide our system of supraspecific categories with the objective basis which it has signally lacked hitherto, and put an end to the long-term inflationary tendencies which have afflicted the use of these categories during the last two centuries. A system of categories of more or less standardised ages would be very helpful to zoogeographers and phytogeographers (see Chapter 11), and as we have seen might help towards a solution of certain hitherto intractable problems in the classification of fossils. Nevertheless, this principle of phylogenetic classification has not so far evoked wide enthusiasm, and has been sharply criticised by several eminent systematists.

G. G. Simpson [170] wrote that the work of Hennig [91]:

> is certainly one of the most valuable books on taxonomy (as here defined) that has yet appeared. My approach is different and I had developed my basic views (which I do not mean to claim as original with me) before I belatedly read that book. The agreement is, of course, only partial, but it is substantial. It is unfortunate that Hennig's work is not more readily available to English-speaking students, and unfortunate, too, that it has no index or glossary to its complete terminology. The most general criticism—and it is one that cuts both ways—is that Hennig seems to be almost totally unaware of the vast body of English and American studies extremely pertinent to his theme. That also obviously underlies the opinion of Kiriakoff (1959) that 'modern phylogenetic systematics seems to be quite unknown in the United States'. Such a statement can only be based on grossly deficient knowledge. There are some neo-typologists in the United States, as in Europe, but they are far from being either the only American taxonomists or a consensus of them.

It is indeed unfortunate that the large majority of Simpson's readers are unlikely to read Hennig's book for themselves, for the American author, though admitting some disagreement between himself and Hennig, gives his readers no indication of the nature of that disagreement. In fact, they are likely to gain the impression that all that is important in Hennig is subsumed in Simpson, and that disagreements reflect only Hennig's alleged ignorance of the thought of modern American systematists. The falsity of this can be judged from the following quotations from the two authors:

> That taxa of the same rank, whether within any one higher taxon or in general, should be of the same age is absolutely impossible in any rational system of classification. It would seem conceivable that such a criterion could be contemporaneous animals only, and attempts on those lines have actually been made. They turn out to be completely impracticable, however. Within the Vertebrata the classes (as now ranked) Osteichthyes, Amphibia, Reptilia and Mammalia, are successively and markedly younger in that sequence. By a criterion of equal antiquity they should therefore have successively lower ranks, and supposing that Osteichthyes were retained as a class, Mammalia would be a super-order, at most, or a genus if numerous intercalary ranks were not used. If the whole animal kingdom were so arranged, we would find ourselves with such taxa as an 'order Latimeriae' (for what is now the fish genus *Latimeria*) and a 'genus *Primas*' for what is now the order Primates. (Simpson, loc. cit., pp. 142–4.)

> in der relativen Rangordnung der taxonomischen Kategorien . . . die relative Lage der Entstehungszeiten dieser Kategorien zum Ausdruck gebracht ist [the relative times of origin of taxonomic categories are expressed in their relative order of rank]. (Hennig, loc. cit., p. 215.)

> In der Praxis fuhrt die Festsetzung des absoluten Ranges taxonomischer Gruppen hoher Ordnung nach den Grad der morphologischen Divergenz so gut wie immer zu einer Durchbrechung des phylogenetischen Prinzips. [In practice, the determination of the absolute rank of a taxonomic group according to the degree of its morphological divergence almost always leads to a violation of the phylogenetic principle.] (Ibid., p. 215.)

The quotations from Hennig embody the main thesis of his entire book, to which the quoted passage from Simpson is in direct opposition. Incidentally, the American author might well be accused of setting up, in this passage, a 'straw man', a practice which only a page or two before he had been reproving in Bigelow. In Hennig's sense of the term, Simpson could not be described as an advocate of phylogenetic classification, any more than could Mayr [140a]; Kiriakoff's quoted remark seems perfectly justified according to my reading (which is little hampered by language difficulties) of recent American writings on these subjects.

It is true that the general application among modern organisms of Hennig's principles of phylogenetic classification must necessarily involve some more or less drastic alterations of the current applications of the categories in some groups; presumably it is these changes which are regarded by Simpson as 'absolutely impossible' and 'completely impracticable'. As I have already tried to show in this chapter, the application of Hennig's principles to the classification of fossils need not involve any absurdities which are not inherent in the system which Simpson himself uses.

It is worth while to consider, in terms of strict phylogenetic classification, the case which Simpson uses as the skeleton of his 'straw man'. Starting from the Mammalia, we can assign them to successively higher groups, the Synapsida, the Amniota (including 'Reptilia' and Aves), the Tetrapoda (including Amphibia), the Choanata (including also Crossopterygii and Dipnoi), the Gnathostomata (including Actinopterygii and Elasmobranchii), and the Vertebrata (including the Agnatha). The question is, to what categories should we assign these groups? For the Vertebrata, the category of subphylum is both customary and reasonably in line with application of this category in other phyla. At the other end of this hierarchy, the Mammalia seem to be most closely comparable with groups like Hymenoptera, Coleoptera and Diptera among the Insecta—groups which we know from direct fossil evidence to be at least as old as the Mammalia, which show at least as much structural and adaptive diversity as them, and which greatly exceed Mammalia in numbers of species. These insect groups have for a long time been universally treated as orders, so why not allot the same status to Mammalia? If we do so, Simpson's own 'complete hierarchy' (loc. cit., p. 17) allows us six categories between Vertebrata and Mammalia (only his straw man would attempt to express a phylogenetic system of this sort in terms of the 'Linnaean hierarchy'). The Gnathostomata could be a superclass, the Choanata a class (cognate with Elasmobranchii and Actinopterygii), the Tetrapoda a subclass, the Amniota a cohort, the Synapsida a super-order—and this would still leave one of Simpson's categories unused. Simpson's treatment of the Mammalia as a class leaves only one category in his extended hierarchy to represent all the evolution between the beginnings of the Vertebrata (perhaps 400 million years ago) and the initiation of the Mammalia (about 150 million years ago, while allocating no less than sixteen categories to cover the evolution of the Mammalia within the last 150 million years. This disproportion may be understandable in view of the fact that Simpson's own special studies have been entirely within the Mammalia, but is hardly to be justified in any system which claims to be phylogenetic.

On the system I have outlined, the Reptilia would cease to be regarded as a classificatory group of modern animals, though the term might be retained, as suggested by Huxley [104], to describe an evolutionary 'grade'. The Aves would presumably take the same status as Mammalia, subordinate to a super-order Archosauria with another living order, the Crocodilia; the Squamata would also be an order, which, together with Rhynchocephalia, would compose another super-order, and the muster of living Amniota would be completed by the super-order Anapsida with a single order Chelonia. On the assumption, supported by some recent studies (cf. Parsons and Williams [149]) that existing Amphibia share a common labyrinthodont ancestor distinct from any ancestors of Amniota we could give them cohort status, perhaps retaining Anura, Urodela and Gymnophiona as orders. Though Dipnoi and Crossopterygii would be assigned to independent subclasses, the existing Dipnoi might stay in a single order or super-order; *Latimeria* would represent a monotypic genus, family, order and cohort (Coelacanthii)—in normal usage, the order including *Latimeria* would have a descriptive name, not one formed on the principle of 'Latimeriae'. Incidentally, supposing that the Primates were all put in one genus, it is difficult to see how the generic name could come to be *Primas*; surely it would be *Homo*?

The simple division of the living world into plant and animal kingdoms can no longer be regarded as satisfactory even for purely practical purposes, but as yet none of the more complex schemes which have been proposed to replace it has won anything like general acceptance. A notable review of the evidence which should be taken into account in constructing any such schemes is provided in a paper by Klein and Cronquist [117]. Putting aside the viruses, whose claim to be considered as independent organisms is disputable, the primary division of the living world which is suggested by Klein and Cronquist is into Procaryota, comprising Bacteria and blue-green algae, and Eucaryota, comprising the remaining plants, fungi and animals. The authors note that the eucaryotan cell constitutes a higher level of organisation than is attained among the Procaryota, and they postulate a derivation of Eucaryota from procaryotan ancestors. Other recent authors have cited evidence which suggests that the eucaryotan cell originated as a symbiotic system of procaryotan components. The primary division Procaryota–Eucaryota is not, according to the principles of phylogenetic classification, consistent with the evolutionary relationships which Klein and Cronquist themselves suggest. According to them, Eucaryota arose from early Cyanophyceae by way of forms like *Cyanidium*; their evidence, in fact, suggests that blue-green algae are genetically nearer to Eucaryota than to ordinary bacteria, in much the same way as we saw in Chapter 9 that lung-fishes are genetically nearer to Tetrapoda than to ordinary fishes.

A more natural primary division would probably be to separate first the Bacteria from the rest, then to subdivide the rest into Procaryota (Cyanophyceae) and Eucaryota and finally to resolve the Eucaryota into several kingdoms, e.g. Rhodophyta (red algae), Plantae (including typical green and brown algae), Mycophyta (true fungi, excluding Mycetozoa and the algal Oomycetes, plus euglenoids), Protozoa (including Mycetozoa), and Animalia (= Metazoa plus sponges). For practical purposes, Bacteria and Procaryota might also be considered as kingdoms, making seven in all. Recently published information [182] on the structure of the cytochrome-*c* molecule suggests, interestingly enough, that green plants and animals may be nearer to each other than either are to the fungi.

A similar system, applied to true plants, might work something like this: Phaeophyceae and Chlorophycease as phyla, Archegoniatae as a subphylum of Chlorophyceae, vascular plants and Bryophyta as superclasses, ferns and allied forms (Filicineae and Spermatophyta) as a class Pteropsida of vascular plants, side by side with three others for the Psilotales, Equisetales and Lycopods, Spermatophyta as a cohort (side by side with Filicineae), four super-orders within Spermatophyta, one for Cycadales, one for Ginkgoales, one for Coniferae and one for Gnetales + Angiospermae. The Angiospermae, treated as an order, would be divided into two suborders Dicotyledones and Monocotyledones, the Dicotyledonae would be resolved into superfamilies (more or less corresponding to groups like Ranales, Rosales, etc., of recent systems), and these into families more or less corresponding to the old 'natural ones'.

In the accompanying table (table 1) I have set out for comparison the complete classificatory hierarchies as customarily used in vertebrate and invertebrate animals, in higher plants and in algae, together with a suggested modification of the systems of Vertebrata and seed-plants to bring them more nearly in line with those of insects and algae. The adoption of some such scheme as this by systematists of Vertebrata and Angiospermae would certainly bring us nearer to the realisation of a phylogenetic system. It may seem a little paradoxical that the groups in which the greatest changes in the application of categories will be demanded are the most familiar and best-studied ones, those in which the greatest numbers of systematists have laboured, and whose classifications are liable to be held up as models of how such things should be done. Yet what is in question is not the structures of the hierarchies which systematists have devised for birds, mammals, dicotyledons, etc.—specialists in those groups may indeed have set an admirable example to the rest of us in the naturalness of their groupings—but the manner in which they apply categories. It is not at all evident that the profundity of a systematist's knowledge of birds gives him any special authority in deciding what sort of group in Aves should be given the same category as the earthworm group Lumbricidae, or the plant family Palmaceae.

The status hitherto awarded to a classificatory group by systematists is very likely to have been influenced by the degree of isolation it manifests; Mayr [140a] even recommended as a principle that at a particular category level the size of a group should be inversely proportional to its degree of isolation. Isolation, in this connection, means the extent of the apparent gap between members of the group and forms outside it; it is implied that an isolated group differs from the nearest related forms in several classificatory characters, not just one. As is pointed out in Chapter 16, this sort of isolation is attributable to the extinction of intermediate forms, which in turn depends on various particularities of evolutionary history, which should have no bearing on the status awarded to groups in a phylogenetic system. The fact that tree-shrews (Tupaiidae) still exist, and serve to diminish the gap between Primates and Insectivora★ is an accident of evolutionary history; if the group became extinct next year, should we be obliged to transfer the Primates to a higher category? The order Rodentia is a more 'isolated' one than the Primates, there being no surviving forms which link them with other mammals in the way that Tupaiidae may link the Primates—but in a phylogenetic system I can see no grounds for giving a higher category to Rodentia than to Primates. Similarly in botany, the Compositae are a more 'isolated' group than the

★ Recent data have cast doubts on a relation of Tupaiidae to Primates.

CATEGORY	Example of current use in Invertebrate Zoology	Example of current use in Vertebrate Zoology	Vertebrata example brought in line with standard age system
KINGDOM	ANIMALIA	ANIMALIA	ANIMALIA
SUBKINGDOM	METAZOA	METAZOA	METAZOA
SUPERPHYLUM	PROTOSTOMIA	DEUTEROSTOMIA	DEUTEROSTOMIA
PHYLUM	ARTHROPODA	CHORDATA	CHORDATA
SUBPHYLUM	MANDIBULATA	VERTEBRATA	VERTEBRATA
SUPERCLASS	TRACHEATA	AMNIOTA	GNATHOSTOMATA
CLASS	INSECTA	MAMMALIA	CHOANATA
SUBCLASS	PTERYGOTA	EUTHERIA	TETRAPODA
COHORT	ENDOPTERYGOTA		AMNIOTA
SUPERORDER	'Panorpoid Complex'	UNGUICULATA	SYNAPSIDA
ORDER	DIPTERA	PRIMATES	MAMMALIA
SUBORDER	NEMATOCERA	ANTHROPOIDEA	TRITUBERCULATA
SECTION	CULICOMORPHA	CATARRHINA	THERIA
SUPERFAMILY	CULICOIDEA	HOMINOIDEA	EUTHERIA
FAMILY	CULICIDAE	HOMINIDAE	HOMINIDAE
SUBFAMILY	CULICINAE	HOMININAE	HOMININAE
TRIBE	CULICINI	HOMININI	HOMININI
SUBTRIBE	CULICINA	HOMININA	HOMININA
GENUS	*Culex* L.	*Homo* L.	*Homo* L.
SUBGENUS	*Culex* in sp.	*Homo* in sp.	*Homo* in sp.
SUPRASPECIES	*pipiens*-group	*sapiens* group	*sapiens* group
SPECIES	*pipiens* L.	*sapiens* L.	*sapiens* L.

Note:– The oblique lines connect groups of equivalent constitutions

Table 1a

Example of current use in Cryptogamic Botany	Example of current use, Phanerogamic Botany	Phanerogamic example brought in line with standard age system	Suggested geological age for ancestral species of group
PLANTAE	PLANTAE	PLANTAE	Pre-Cambrian
THALLOPHYTA		CHLOROPHYCEAE	Pre-Cambrian
ALGAE			Pre-Cambrian
PHAEOPHYTA	TRACHEOPHYTA		Late Pre-Cambrian
	PTEROPSIDA		Cambrian to early Ordovician
	SPERMATOPHYTA	ARCHEGONIATAE	Lower Silurian
CYCLOSPOREAE	ANGIOSPERMAE	TRACHEOPHYTA	Late Silurian to Devonian early
	DICOTYLEDONAE	PTEROPSIDA	Late Devonian to Carboniferous
	ARCHICHLAMYDEAE	SPERMATOPHYTA	Upper Carboniferous to Permian
		CYCADOPHYTA	Permian to Lower Triassic
FUCALES	RANALES		Rhaetic
			Lower Jurassic
		ANGIOSPERMAE	Late Jurassic to early Cretaceous
		DICOTYLEDONAE	Cretaceous
FUCACEAE	RANUNCULACEAE	RANUNCULACEAE (s.n.)	Upper Cretaceous
	RANUNCULOIDEAE	RANUNCULOIDEAE (s.n)	Paleocene or Lower Eocene
	RANUNCULINEAE	RANUNCULINEAE	Upper Eocene
			Lower Oligocene
Fucus L.	*Ranunculus* L.	*Ranunculus* L.	Upper Oligocene
Fucus s. str.	*Ranunculus* s. str.	*Ranunculus* s.str.	Lower Miocene
vesiculosus – group	*repens* – group	*repens* group	Upper Miocene
vesiculosus L.	*repens* L.	*repens* L.	Pliocene
Note:– The oblique lines connect groups of equivalent constitutions			

Table 1b

Magnoliaceae—the definition and extent of the former family has remained essentially unchanged for a century and more, whereas hardly any two major systematists have been agreed about the extent of Magnoliaceae—but this would not justify assigning a higher category to the Compositae.

If phylogenetic classification proceeds usually by dichotomous divisions, and very unequal ones at that, it will necessitate the use of many more categories than were needed for older, 'formal' systems. In the older classifications, each group was usually resolved into a number of subgroups, perhaps of the order of five, and the subgroups were more equal in size than phylogenetic divisions tend to be. Furthermore, the number of recognisable levels of phyletic branching in two related groups of the same category is liable to be very different. For these reasons, those who look on classification as a convenient system for pigeon-holing knowledge will consider phylogenetic systems to be untidy and inefficient.

The most general and fundamental difficulty confronting those who would construct phylogenetic classifications is best explained by reference to a diagram (fig. 4). In this dendrogram, the common ancestral species of an (imaginary) family is represented at A; B and C indicate two distinct daughter species each of which is shown as giving rise to a series of later species and genera extending up to the present day (indicated by the horizontal line at the top). Existing species of the family are numbered. If the progeny of A form a natural group, the descendants of B and of C represent its natural sub-groups. As good species, B and C were doubtless phenotypically distinguishable, there were definite characters by which they could have been separated in a key. Unfortunately, *there is no law of nature which states that the distinguishing features of B and C must be preserved in their respective progeny.* If A was a flowering plant, we might suppose that B and C differed chiefly in their degree of adaptation to dry habitats; if B was originally distinguished by dry-habitat adaptations (affecting leaf-form, stomata, vestiture etc.), it is quite possible that among the progeny of C similar adaptations occurred later, and among the progeny of B some might well have re-adapted to damper habitats and lost the original distinguishing features of B (this would probably not involve any serious contravention of 'Dollo's Law'). No doubt, at some stages in the diversification of the progeny of A, evolutionary changes took place whose effects were lasting and not duplicated in any other line, e.g. at points 'x', 'y' and 'z' in the dendrogram. Perhaps at x the flowers became zygomor-

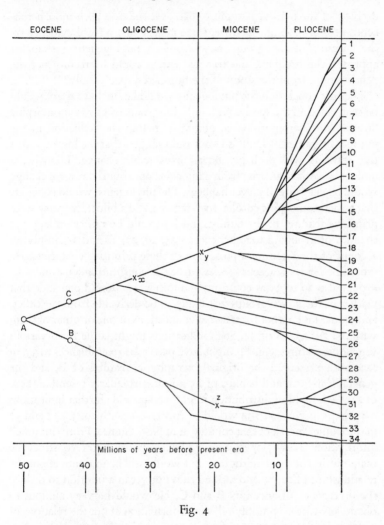

EOCENE	OLIGOCENE	MIOCENE	PLIOCENE

Millions of years before present era

50   40   30   20   10

Fig. 4

phic, at y one of the stamens was lost, and at z the gynaecium became unilocular, all three characters persisting in the progeny up to the present day. Now these are characters of types which systematists would be likely to seize on in order to establish major subdivisions of the family. We can imagine controversies developing among herbarium systematists, as to whether the zygomorphic corolla, the number of stamens, or of carpels, provided the most natural basic division of the family. Assuming that, as is the case with most herbaceous groups of Angiospermae, there is little or no fossil evidence on the ancestry of the group, how would a phylogenetic systematist approach the problem, and to what extent could he reconstruct the dendrogram, from the study of the numbered species only?

The first conclusion the phylogenist would be likely to draw would be that, since all the species with only four stamens have zygomorphic flowers, the zygomorphic corolla arose before the reduction in the number of stamens—Dollo's Law would suggest that the loss of one of five stamens will probably be an irreversible change. Likewise, a unilocular ovary can arise from a bilocular one, but the reverse change would be very rare, if it ever happens. The phylogenist would conclude that an actinomorphic corolla, five stamens, and a bilocular ovary were primitive features in the family, and he would be rather unhappy at the idea of grouping together species 25, 26, 27, 28 and 34 simply on account of their common possession of these primitive characters. He would be ready to accept species 1 to 24 as constituting a natural subgroup, also 29 to 33 as constituting another. He would conclude that neither of these two groups was likely to be derivable from the other, and would look rather for traces of connections of one or other of them with 25, 26, 27, 28 or 34. Such indications might come from various vegetative features, e.g. 34 might have particular resemblances to 33, in that both preserved the original 'xerophytic' features of B, and the general leaf-form and habitus of 25 might particularly resemble those of 24. Geographical distribution might well provide further indications —perhaps species 29 to 34 would be Australian, with 29, 30, 31 and 32 in the tropical north of that country or in New Guinea. From this much information, the phylogenist would be in a position to reconstruct the main outline of the dendrogram, but would still be in doubt about the relationships of 26, 27 and 28, and thus not yet in a position to deduce the divergence of ancestors B and C. He would be very anxious to obtain any further possibly relevant information about the relations of numbers 26, 27 and 28, e.g. from biochemical and serological studies, also perhaps from herbivorous insects.

This hypothetical example is, I think, a reasonable representation of the actual problems faced by, and the lines of reasoning used by, phylogenetic systematists. Supposing that some new lines of evidence—perhaps serological—eventually established the relationship of species 26, 27 and 28 to 25, and thus to the stem of 24, etc., the phylogenist would then be in a position to reconstruct the entire dendrogram in its main features, and to hazard a good guess at the characters of ancestors B and C. He would presumably establish two subfamilies, one for species 1 to 28, the other for 29 to 34. These two would be very difficult to distinguish by simple definitions or analytical keys; the main characters used in practice would be the zygomorphic corolla and the unilocular ovary; the definition of the 'B' subfamily would incorporate a special clause to cover the case of species 34, and that of the 'C' one would include corresponding special clauses to cover species 25 to 28. These complicating exception-clauses would annoy tidy-minded formal systematists but would probably not make for very serious practical difficulties. The further subdivision of the two subfamilies would probably be on these lines: in the 'B' subfamily, two tribes, one for species 34 (with bilocular ovary), and one for 29 to 33 with unilocular ovary; in the 'C' subfamily, a tribe for species 1 to 21, with four stamens, one for 22 to 24, with five stamens and a zygomorphic corolla, one for species 25, and one for 26 to 28. The ages of the separate ancestries of the tribes would thus vary, the youngest (that containing species 34) going back to the late Oligocene period and the oldest (the tribes for 25, and 26 to 28) being traceable to the Lower Oligocene period. Inequalities of this sort could hardly be avoided in any system employing a limited number of categories.

20 Zoological and Botanical Nomenclature

Say, what's the rest but empty boast
The pedant's idle claim
Who, having all the substance lost
Attempts to grasp a name?
(Solomon, in Handel's oratorio of the name)

The main avowed aim of the international codes of nomenclature, botanical, zoological and microbiological, is the promotion of stability and uniformity in the scientific naming of organisms. A classificatory name, however, does not really denote an organism (Kenneth Grahame's Mr Toad not withstanding) or even a specific group of organisms, it embodies a concept; it is analogous, in ordinary grammar, to terms like Windsor chair or digital computer rather than to proper names. The generic name *Campanula*, for example, embodies the concept that a whole range of plants, among which we may distinguish many species, can be recognised as a group by the common possession of certain specified characters, and that this group is distinct enough, diverse enough, or old enough, to qualify for the category of a genus. The interests of stability and effective communication would best be served by measures to ensure that as far as possible a given classificatory name will always embody the same concept. The current codes of nomenclature in botany (1956) and zoology (1961) have not been formulated with this as their main aim, no doubt because it was considered impracticable by those who drew up the latest versions of these codes. Instead, both codes are concerned to provide the most objective possible set of rules by which a 'correct' application of any name can be discovered no matter how much classificatory concepts may change. The manner in which both these august international committees have in effect passed votes of no confidence in themselves and their colleagues as systematists contrasts oddly with the faith they both display in the durability of their achievements as legislators. For example, the latest zoological code states (Article 61) 'Nucleus of a taxon, and foundation of its name, the type is objective and does not change, whereas the limits of the taxon are subjective and liable to change'. Neither code attaches any legal force whatever to the content of the definition of a taxon, i.e. to the actual concept of the systematist. The kind of image one senses in the minds of these legislators is of classification as an anarchic sea in which the only way to confer any sort of stability on names is to attach them by systems of threads to anchors (the holotypes) on the bottom.

The two basic principles of the botanical and zoological codes today are Priority and Typification. These provisions apply for taxa up to the category of superfamily (in the zoological code) or order (in the botanical code). The principle of priority is that, where more than one 'available' (see later discussion of this term) name is found to be attached according to the current code to the same taxon, the first bestowed of them should be used and the other(s) relegated to the status of 'junior synonym'. For names of the species and genus groups, where the same name has been legally published for two different taxa of the same or of 'co-ordinate' rank, the latter application of it is to be irrevocably suppressed as a 'junior homonym'. For the specific name (i.e. the second part of the Linnaean 'binomen'), the requirement is that it must never have been used before with the generic name in question. For nomenclatorial purposes, subspecific names are 'co-ordinate' with specific ones and subgeneric names with generic ones— the 'type' subspecies of a species must, however, have the same subspecific as specific name and the type subgenus the same subgeneric name as the generic one. In the case of specific names, a name which was not a homonym when proposed may become one later (either through the transference of additional species into the original genus from other ones, or through transference of the species itself into a new genus), a phenomenon known as 'secondary homonymy'. A secondary homonym has to be replaced while the species concerned remains in that genus, but its rejection is not irrevocable; if the species is later transferred to some other genus, the original name may be reinstated if it is not then homonymous.

For example, if the name *niger* published in 1820 for a species of the genus *Animalus* were found to be based on the same species as *atratus* bestowed in 1803, not necessarily with the same generic name, *niger* would be rejected as a 'junior synonym' but would not thereby lose 'availability'. The binomen *Animalus atratus* would be used for the species, but if someone else in 1885 described a new species as *Animalus niger* that name would be permanently rejected as a 'primary homonym' of *Animalus niger* of 1820—the species name niger could not be restored for it even if it were transferred to a new genus. If *Animalus atratus* were transferred in 1890 to another genus *Brutus* in which there was already a species *atratus* dating from 1870, the name *atratus* of 1803 would have to be rejected as a 'secondary homonym' and the name *niger* of 1820 would probably be reinstated for the species. If yet another classificatory change removed this species to a third genus *Novus*, in which there was no previously defined species *atratus*, the species would revert to that name and be called *Novus atratus*. Whoever discovered the homonymy of the *Animalus niger* of 1885 would have the right to propose a replacement name for it; in such cases it is a frequent courtesy to rename the species after the bestower of the replaced name. e.g. *Animalus smithi* (*Animalus niger* Smith 1885, nec Schmied 1820). The replacement name would take priority as from its date of publication, not from 1885.

The zoological code expressly states that, with a few exceptions, a difference of a single letter is sufficient to make two species or generic names non-homonymous; it is even stated in the code that *londonensis* and *londinensis* could be accepted as names for different species of the same genus. The botanical code, however, rules that what it called 'orthographic variants of the same name' are to be treated as homonyms. Here, as in most other differences between the two codes, the botanical one shows greater respect for tradition and common sense.

Though the principle of priority in nomenclature is an old one, having been incorporated already in the 'Strickland Code' drawn up for the British Association in 1842, the principle of typification as now understood is quite modern. Originally, the main purpose of the priority principle was to settle the acromonious disputes which arose when Mr A in London and Herr Dr B in Berlin were found, in ignorance of each other's activities, to have described as new the same species under two different names. During the first half of last century the principle was rarely, if ever, invoked to change the accepted applications of names given by dead authors—the enlightened British of 1842 did not like retro-active legislation. When, as sometimes happened, authors of this period mentioned a species as type of a genus, the term did not have its modern nomenclatorial connotation. A species mentioned as a type a hundred or more years ago was normally the commonest, best known or most widely distributed, preferably European member of its genus with the idea that other collectors would be likely to have specimens of it in their collections and thus be able to use it as an example for checking the generic characters on. The founders of modern plant and animal nomenclature, such as Linnaeaus and Fabricius, did not specify types for genera and it is doubtful whether the majority of plant and animal genera had types specified for them before 1910. The citation of a type species for a new genus did not become legally obligatory in zoology till 1931, and in botany not till 1958. During the last fifty years, various systematists have taken it upon themselves to propose types wholesale for old genera which seemed to lack them. In spite of the fact that the current rules make type designation the all-important act as far as nomenclature is concerned, such designation has only too often been made in a casual manner with no serious study of the issues involved; hence spring many of our present nomenclatorial troubles.

The current codes incorporate provisions to the effect that the original specifier of a species name as type for a genus is assumed to have used that name correctly (i.e. in accordance with our latest code of rules!) unless there is evidence to the contrary *in the actual publication in which the citation is made*. This means, in effect, that the type is the name itself rather than an actual species—if later discoveries or legislation lead to a change in the application of the species name, this will probably mean that a different species becomes the type of the genus. The zoological code uses the phrase 'nominal type species' in this connection; a nominal species seems to be the accepted gobbledygook equivalent of a species name.

For names in the 'Family group', i.e. from subtribes up to superfamilies, and for orders in the botanical code, the rule is that the group name is to be formed by adding the appropriate ending to the stem of that included generic name which was first used to form a supra-generic group name. Thus, a systematist proposing a new taxon within this range of categories is rarely free to select a suitable type for it; if any of the included generic names (including synonyms as well as valid names) has previously been used to form a supra-generic group name, he has no freedom at all in this respect. As both codes permit the use for different genera of names differing only in their terminations and not in the 'stem' part, for example, in the animal kingdom *Sphaerius* for a beetle and *Sphaeria* for a mollusc, homonymy is liable to appear in names of the family group. The 1956 botanical code has no specific provisions to deal with such situations, but the zoological code of 1961 prescribes that such homonymy in family-group names in the animal kingdom must be eliminated by altering one of the names concerned, normally the 'junior' one, usually by inserting one or two additional letters between the stem and the standard termination. The issue between the entomologists and the malacologists over the possession of the family name Sphaeriidae has not yet been resolved, however.

A family-group name is now officially based on a 'nominal type genus', much as a genus name is based on a nominal type species; here again, this means that the type is in effect a name rather than an actual genus. The law assumes that the designator of a new taxon in the family group 'correctly' identified the genus from which he formed its name, unless (as is rarely the case) there is definite evidence in the publication itself that he was applying the name 'wrongly' (that is, wrongly according to our retroactive rules).

The only category of name whose type is not another name is the species, for which the type is an actual museum specimen of an organism, the Holotype.

The zoologists go even further than the botanists in following the logic of the principle of typification, in that their latest code permits the availability of a generic name provided only that a species name was mentioned in connection with it, in the absence of any other definition. This has led to the re-instatement of numerous names long rejected as 'nomina nuda'. The 1956 botanical code, however, states that the citation of a species name does not by itself make 'available' a new generic name in the plant kingdom—something which purports to be a definition, in Latin, must be provided as well, though there is, of course, no requirement that this definition should be sufficient to distinguish the group or even that its terms should apply to the forms included in the group.

The mandatory provisions of the present code in botany deal with names of taxa up to the category of orders, and the zoological code similarly covers the hierarchy up to the level of superfamilies. The names of taxa above these levels are not subject as yet to any laws but those of custom. The higher categories officially recognised in the current botanical code are: Regnum (Vegetabile), Divisio, Subdivisio, Classis, Subclassis; in the zoological code no categories above the 'family group' are specifically named.

Standardized endings for taxa in botany and zoology

Category	Botany (and Microbiology)	Zoology
Subtribe	-inae	(-ina)
Tribe	-eae	-ini★
Subfamily	-oideae	-inae
Family	-aceae	-idae
Superfamily	—	-oidea★
Order	-ales	—

Ending in parentheses is customary but not specifically cited in the code; endings with asterisk are recommended but not mandatory in the codes.

The current codes include clauses permitting the retention of long-used and familiar names which would be invalidated by their other provisions; the Botanical Code has as an Appendix a list of Nomina Conservanda et Rejectanda of genera, and provides for additions to this list by decision of the General Nomenclature Committee of the International Association for Plant Taxonomy, which require, however, to be ratified at the next subsequent International Congress of Botany. The International Commission on Zoological Nomenclature has published Official Lists of accepted and of rejected names in Zoology; both lists are being extended continually through decisions of the International Commission on Zoological Nomenclature. When proposals for the acceptance or rejection of particular names are being considered by the Commission, they are published in the Bulletin of Zoological Nomenclature and any interested systematist may submit relevant considerations before the Commission formally votes on the proposals. Once the Commission has voted, its decisions do not require ratification at a subsequent International Congress of Zoology.

The codes require that a generic name should be treated formally as if it were a Latin noun (substantive) in the nominative singular case, with a definite gender, masculine, feminine or neuter. Neither code, however, makes it mandatory for the author of a new name in the genus group to specify its gender, and the botanical code provides only 'Recommendations' instead of 'Rules' concerning the genders of already published generic names. The zoological code, however, specifically rules that generic names formed from words included in the standard Greek or Latin dictionaries take the genders of those words, unless the ending is altered so as specifically to indicate a different gender, in which case the name takes the gender appropriate to its ending. Species names may take the form of a (Latin) adjective, in which case their endings have to take the gender form appropriate to the genus name with which they are used—if such a species name is transferred from one genus to another with a different gender, the ending of the species name may need to be altered correspondingly. Difficulties of gender in species names may be, and now often are, evaded by giving the name the form of a (Latin) noun in the genitive case, or of a noun in the nominative singular case 'in apposition'. Thus *Cedrus deodara*, where the generic name is clearly masculine while the species name looks as though it should be feminine, is accepted on the assumption that *deodara* is another noun in apposition and not an adjective.

Neither code imposes any limits on the number of letters in the name of a taxon, nor do they include mandatory provisions governing permissible combinations of letters in names. As far as can be seen, if someone published *Xqgofwlus* as a generic name in either the plant or the animal kingdom, there is no specific rule which would ensure its invalidation; presumably this could only be achieved through a special application for a Plenary Powers decision of the appropriate International Committee on Nomenclature.

The starting point of modern scientific naming of plants (other than fungi) is accepted as Linnaeus's "Species Plantarum" of 1753; plant species and generic names used in that work have priority over all others. For fungi, the accepted nomenclatorial starting points are the works of Fries [67a] of 1821–32, and Persoon [149a] of 1801. Zoologists accord a comparable position to the tenth edition of Linnaeus's *Systema Naturae* of 1758; for animals, generic and specific names published in that work have priority over all others save those of spiders published by Linnaeus's disciple Clerck in 1757. As may be imagined, the arachnologists have had a difficult fight to maintain this single exception against the objections of consistency-loving Americans and others.

Linné both standardised the binomial system of naming plants and animals, and used it in a more complete Aristotelean classification of the two kingdoms than had ever before been attempted. Species for Linné were grouped in genera, and designated by a combination of two names, a first (generic) name which had to be different from the name of any other genus (this requirement was later modified to permit the same generic name to be used for a plant and an animal), and a second (specific, of 'trivial') name which could not be used for any other species within that genus. The combination of the two was thus necessarily unique. The total of Linnaean genera was not very great, of the order of hundreds in each kingdom, and an average genus contained a number of species running into double figures. It was reasonably practicable for an eighteenth-century naturalist to know the significance of all the generic names used in botany or zoology or even in both. The genera were grouped by Linné into orders, and these into classes, both categories designated by single names for which no clear rules were laid down—Linné probably thought these names of less practical importance. The convenience of this system must have been a major factor in winning it wide and speedy acceptance.

The practicality of the binomial system has however been seriously undermined by subsequent changes, in a way which was not inevitable and for which Linné himself could hardly be blamed. Enormous numbers of species, both in plants and animals, have been added to those recognised by Linné; there are now more than 300,000 recognised species of plants and something approaching a million of animals, compared with a few thousands of each kingdom in Linné's time. The binomial system could have absorbed this enormous increase without serious loss of utility if most of the additional species had been incorporated in already established genera, so that the number of the latter did not rise very greatly. In practice, however, the multiplication of species since Linné's time has gone hand in hand with the splitting of his genera, and the average number of species per genius has actually declined—a figure of about five would probably be a reasonable estimate today. The days when a properly educated botanist (or even more so, zoologist) could be expected to know all the generic names used in his kingdom are very long past; unless you are dealing with the most familiar genera, it has become a practical necessity of communication to give an indication of the higher group an organism belongs to as well as its binomial designation. In some groups (e.g. birds) generic splitting has gone to such lengths that the average number of valid species per genus is down to something between two and three, and among modern ornithologists (e.g. Cain [25]) there is serious question whether the generic names serve a practically useful purpose at all.

What has happened has been the result of looking on the genus as the next higher recognisable classificatory grade to the species, together with the ever closer anatomical study of plants and animals, whereby significance came to be attributed to smaller and smaller differences.

Linné himself having been mainly a naked eye or hand lens anatomist, it is not surprising that the extent to which his genera have been subsequently split is greatest in the smallest and most obscure forms and least in the large ones. The disparity in Linné's treatment of large and small organisms would have had to be rectified in any case, but it should have been possible to retain something like the broad Linnean conception of the genus. Later refinements of discrimination could have been accommodated by admitting, for specialist purposes, more and more intermediate categories between the genus and the species. If systematists had adopted this policy, they would have manifested a degree of public responsibility and foresight which experience shows it is unrealistic to expect of ordinary human beings; what happened was doubtless historically inevitable. There is no reason to suppose that contemporary biologists, even as represented by our august International Commissions of Nomenclature, will be any more successful than their predecessors in anticipating the difficulties which will face posterity.

Experience suggests that a human systematist may be expected to carry in his head knowledge of the significance of generic names of the order of a thousand, or at most a few thousands. On this basis we might conclude that Mammalia or ferns, monocotyledons and perhaps birds, represent groups which might each be dealt with on a world-wide basis by single systematists—though each of these groups has too many species for a single systematist to be able to recognise them all reliably on sight. Such groups as dicotyledons or the major orders of insects (Coleoptera, Diptera, Hymenoptera, Lepidoptera) would be too large to be treated by single systematists on this criterion. It must be remembered that the practical systematist needs to remember more generic names than there are valid genera in his group; modern codes of nomenclature have ensured that numerous synonyms have to be known in all groups.

It is an interesting paradox that the present century, which has more and more professed the unity of Biology as against the separatism of former botany and zoology, has also witnessed the final and apparently irrevocable splitting of the original unified Linnean code of nomenclature into completely independent botanical and zoological codes, and has recently seen a 'unilateral' declaration of nomenclatorial independence by bacteriologists and virologists. The zoologists, as now seems usual, were in the forefront of this fashionable development, their nomenclatorial independence having been declared as early as 1901;

the botanists followed suit in 1905, and the micro-biologists in 1958. Unfortunately, at no time has any rule been laid down to determine, in the disputed territory of the Protista, which are to be regarded as nomenclatorially plants and which as animals—groups like the Mycetozoa (Myxomycetes) have caused confusion in the generic naming of both kingdoms.

For Linné and nearly all the older systematists, right up to the beginning of the present century, every classificatory group was based essentially on a set of characters, or perhaps on an assemblage of organisms manifesting them. Classificatory names were regarded as means, not ends—they were not treated as entities existing in their own right so to speak, but as vehicles for classificatory concepts. Today, however, 'nous avons changé tout cela'. As we have seen, according to both codes the application of a supra-specific name is determined solely by what is called a 'nomenclatural type' or 'nominal type species'. Neither code accords any legal weight to the content of the definition of a taxon, nor is there any clause in either specifically designed to tie names to classificatory concepts. Neither code even has any requirement that any or all of the characters specified in the definition are present in the type of the group or in any of its members.

Present rules in both botany and zoology require that, in order to be 'available', a name for a new taxon should have been validly published, with something purporting to be a definition or indication, though neither of the International Committees has succeeded in formulating a simple and satisfying criterion of valid publication. The generally accepted interpretation of this requirement is that the name in question should have been reproduced in some numbers in permanent print, and that the printed work should have been distributed in such a way as to have made it at least potentially available to anyone wishing to acquire a copy of it. The established norm is provided by the journals of reputable scientific societies, which are usually printed at least in hundreds, and for which a subscription rate is commonly quoted for non-members. Duplication of typewritten material by ordinary office methods has been ruled not to constitute publication, even though such duplicated copies may be available to anyone asking for them, and such modern techniques as micro-films and micro-cards have also been disallowed. On the other hand, the modern 'offset' method of producing books by direct electrical reproduction from typescripts has been tacitly accepted as valid publication.

The question of distribution also raises difficulties. Certain old works, which were privately printed in very small editions and sent personally by the author to a few friends only, have been officially ruled as 'not validly published', for example, Hübner's *Tentamen* of 1806 (Opinion 97 of the Zoological Commission) so that names appearing in them do not thence become 'available'. In recent times, descriptions of new taxa of both plants and animals by many established systematists have appeared in print in glossy brochures sent out by the Companhia de Diamantes di Angola; no price or subscription rate is quoted for these, which it seems are sent out unsolicited to those fortunate enough to be on the Company's mailing list. There might be serious doubt as to whether this constitutes legally valid publication, but as far as I am aware no one has yet raised the matter with either of the international committees on nomenclature.

The present rules for both kingdoms place a tremendous legal weight on the holotypes, which constitute the only points at which the hierarchy of names and the hierarchy of classificatory groups are tied together. The theory of this is that a single specimen can only belong to one species, whereas it always remains possible that a group of specimens might prove to contain more than one species. If we ignore grafts and hybrids, and define a specimen as comprising the whole or part of a single organism, the postulate may be conceded. However, neither code does, in fact, define a specimen in this way; a botanical type for instance often consists of a herbarium sheet bearing detached portions of foliage, flowers, fruits, etc., and there is no proof that they all came from the same individual plant. Conversely, a single individual plant, e.g. a tree, might provide material for large numbers of herbarium sheets, but only one of these could be a holotype. As we have seen in Chapter 3, a single specimen does not by itself provide a sufficient criterion to define a species; there is no general method by which, in comparing one specimen with another, it can be determined whether or not they belong to the same species. A further objection to our exclusive reliance on the type system is that holotypes are all too perishable, and if they receive the frequent and intensive study which the logic of our present system demands, their useful life will not be very long. In practice, the use of holotypes is restricted because, in most groups they are scattered widely in the museums of the world, which as a rule are very reluctant to send them outside their own doors for study. A substantial part of our present collective expenditure on taxonomy goes to finance the consequent globe-trotting of conscientious typological systematists.

The complete divorce between classification and naming which is implicit in the present rules of nomenclature may seem a little surprising in an era which has seen the widespread popularity of linguistic and semanticist philosophies. The exponents of these recently fashionable modes of thought maintain that words are, in the development both of our species and of individual human beings, prior to concepts. Conceptual thinking, it is asserted, arises from the use of words, and not *vice versa*; if the Greeks did not have a word for it, then they could not have had the idea either. The logic of this theory, as applied to the naming of classificatory groups in biology, would surely be that the actual application of nomenclature is likely to have a predominant influence in shaping the classificatory concepts of botanists and zoologists. If this is so, should not every effort be made to tie classificatory group-names to the concepts they are intended to embody? And should not our codes of nomenclature be formulated so as to correspond as closely as possible to the actual procedures of good systematists? The present codes have abandoned any such aims in order to pursue an elusive goal known as 'objectivity'—concepts are by definition subjective, and in the vocabulary of our legislators this word has acquired a pejorative significance.

Classification, as we have seen, can aspire to be natural, a perfectly natural classification of plants and animals might even be considered as objectively existing, and thus requiring to be discovered rather than invented. Nomenclature, on the other hand, is generally agreed to be essentially human and artificial, the names being merely tools for the purpose of human communication, and the rules are merely scientific conventions. There is a body of opinion which sees the ordinary civil laws of states as more or less imperfect derivates of an ideal 'natural' or 'divine' law which could be regarded as objectively existing like the natural system of plants and animals. However, not even the most ardent advocates of the various codes of nomenclature which have been promulgated during the last century have claimed this sort of authority for them; the argument has always been that *if* all biologists would regulate their naming of organisms strictly by the code, we should soon achieve the desired end of uniform nomenclature. Now the naming of plants and animals is a field in which by now the past is likely to have played a bigger part than the future will—the amount of essential naming done by our predecessors is not likely to be matched by the activities of ourselves or our successors. Botanical and zoological libraries and museums are filled with books and collections in which

organisms are referred to by names current before the latest codes of
nomenclature were formulated. For some reason which I have never
understood, the legislators for botanical and zoological nomenclature
have throughout insisted that their codes should be retroactively
applied—the nomenclatorial activities of their predecessors have to be
forced into compliance with the latest set of rules. This may seem a
strange attitude to be adopted by men coming as a rule from socially
advanced and enlightened societies, in whose civil laws retroactive
legislation is universally regarded as a very undesirable thing. Admittedly, the first international codes of nomenclature, both for animals
and plants, incorporated 'escape clauses' which were intended to
permit the suspension of retro-activity of the rules where this seemed
likely to alter old and widely used names. These escape clauses were,
however, so vaguely worded that people have felt unable to decide
when they could properly be invoked; as far as I am aware, there have
been very few occasions when either the botanical or the zoological
committees have ever pronounced officially on the applicability or
inapplicability of the escape clauses to particular cases. The latest
zoological code includes a rather more precisely formulated escape
clause, in Article 23, clause (b), 'A name that has remained unused as
a senior synonym in the primary zoological literature for more than
fifty years is to be considered as a nomen oblitum', with the additional
provisions that (i) after 1960, a zoologist who discovers such a name
is to refer it to the Commission, to be placed on either the Official
Index of Rejected Names, or if such action better serves the stability and universality of nomenclature, on the appropriate Official
List; (ii) a nomen oblitum is not to be used unless the Commission
so directs; (iii) this provision does not preclude application to the
Commission for the preservation of names, important in applied
zoology, of which the period of general use has been less than fifty
years.

Although the International Colloquim which formulated this code
had been specifically instructed by the International Congress of
Zoology to draw up a new code incorporating a stronger 'conservation' clause, the one in question was only adopted against strong
opposition, particularly from American systematists; it seems to me
that many of these same people are now endeavouring to apply
sabotage techniques in an attempt to demonstrate the practical un-
workability of the clause, with a view to getting it deleted at the next
revision of the code in nomenclature, as in some other spheres of

human life, the American 'consensus' seems determined to impose its views on the world.

It might seem that if we are to have a code of rules which is retro-active, a code to determine the proper application of names the large majority of which were bestowed by our predecessors, we should try to make the code conform with the practice of systematists of the past rather than with the predilections of those of the present day. This sort of 'democracy of the generations' is not one of the most conspicuous virtues of modern, progressive man.

One of the effects of the retro-active application of new codes of nomenclature has been the paradoxical one that vernacular names of animals and plants have become a much more stable and unambiguous means of denoting species than are the Linnaean binomials. This is particularly clear in such groups as the macro-Lepidoptera, the birds and the Orchidaceae. It is unfair to use this argument, as has been done, e.g. by Cain [26], in order to discredit the principle of Linnaean nomenclature; the blame should be shared between the 'splitters' among systematists, and those who preach and practice the principle of retro-active legislation in nomenclature.

The adoption of the 'principle of typification' as the basis of naming has probably had even more undesirable effects on the nomenclature of taxa of the 'family group' (superfamilies, families, tribes and sub-tribes) than it has on generic names. In order to determine the 'correct' appellation of a taxon in this group, it is necessary to find out which of the included generic names was the first to be used in forming the name of a supra-generic taxon, quite irrespective of the content of that taxon. As a rule, a modern systematist naming a new taxon in the family group of categories has no freedom in naming it or in establishing a suitable type for it.

A hundred years and more ago, great numbers of small and often not very professionally produced publications were appearing, filled with definitions of new genera and species of plants and animals. With authors often amateurs and not very punctilious about proof-reading, and with not very literate typesetters confronted with strange and often long hand-written graeco-latin names, it is hardly to be wondered at that misspellings in the original publication of names were rather common. Such mistakes were normally corrected in the next publica-tion to deal with the taxon in question, either by the author or by someone else; in cases where the correction was by someone else, it was very rare for the original author to register any objection. The corrected forms of names normally passed immediately and without question into general use, and in most cases have stayed current up to recent times, but most of them are invalidated under the latest codes.

Successive revisions of the International Code of Zoological Nomenclature (the botanists have been far less guilty in this as in some other respects) have progressively eroded the principle that the scientific naming of animals has anything whatever to do with linguistics; the latest zoological code treats scientific names of animals as to all intents and purposes arbitrary combinations of letters. A spelling difference in a single letter is now considered to render two names non-homonymous. It may not be altogether a coincidence that our legislators in nomenclature have discovered the single letter just when musical composers were 'discovering' the single note and abstract-expressionist painters the single brush-stroke; all these phenomena reflect the special cultural climate of our age.

As a final illustrative example of the operation of the current zoological rules I will describe an instance from my own experience in the family Chrysomelidae of the insect order Coleoptera. The genus *Lema* of Fabricius (1798) as originally proposed was completely equivalent to *Crioceris* of Geoffroy (1762), of which it included all the original species. The nomenclature in Geoffroy's work was not strictly binomial in the Linnean sense, as he used french words for the second (specific) names; the work has recently been declared invalid for nomenclatorial purposes (Opinion 228) by the International Commission. However, in 1785 many of Geoffroy's names, *Crioceris* among them, were re-published in correct Linnean form. The name *Crioceris* thus has priority from 1785, and should be attributed to Fourcroy. Under the present rules, the fact that *Lema* was originally a synonym of *Crioceris* does not preclude the maintenance of both as valid generic names, if suitable type designations have been made. Lacordaire (1845) split Fabricius's genus *Lema* into two parts, and restored the name *Crioceris* for one of them; he also mentioned a species, *C. merdigera* L., as a type of *Crioceris*, a designation which would make *Crioceris* synonymous with the much later generic name *Lilioceris* Reitter (1912). However Curtis had previously (1830) cited *C. asparagi* L. as the type of *Crioceris*, which name he of course still treated as synonymous with *Lema*. Lacordaire mentioned three European species as examples of *Lema* in his sense, without specifying any one of them as the type. Jacoby (1908) seems to have been the first to specify a type for *Lema*, and he cited *cyanella* L., a name which had appeared among Lacordaire's three. Unfortunately two quite distinct species had been confused under this name by Fabricius, those subsequently named as *lichenis* Voet (1806) and *puncticollis* Curtis (1930). Voet presumably named his species under the impression that the true *cyanella* L. was what Curtis later called *puncticollis* and Curtis would have assumed that *lichenis* Voet was the true *cyanella* L. Both species appear over the name *cyanella* in the Banks collection, which has been named by Fabricius, and Gyllenhall (1813) explicitly referred to the two forms as opposite sexes of one species. There can be little doubt that Jacoby took *cyanella* to be synonymous with *puncticollis* Curtis, as could be seen in the British Museum collection of the group which he had arranged. On the other hand, there is only one specimen now standing over the name *cyanella* in the Linnean collection in London, and it proves to belong to *lichenis* Voet.

In 1886 the French Des Gozis published a privately distributed pamphlet [78*a*] which startlingly foreshadowed the attitudes of nomenclatorial legislators of sixty or seventy years later, and which encountered correspondingly sharp criticism from his contemporaries (see, for example, D. Sharp in the *Transactions of the Entomological Society of London* of the same year). Des Gozis, arguing that it was not possible to establish *Lema* F. as a genus separate from *Crioceris*, proposed a new name *Oulema* to replace Lacordaire's *Lema* and designated for it the type *melanopa* L. In 1927 the German Heinze defined a new genus *Hapsidolema* for a group of species formerly included in *Lema*, among them *melanopa* which Heinze indeed designated as the type of his new genus, in apparent ignorance of Des Gozis. This genus will certainly include also *lichenis* Voet. (*cyanella* L.) The current zoological rules state (Article 70): 'It is to be assumed that an author correctly identifies the nominal type species that he either (1) refers to a new genus when he establishes it or (2) designates as the type species of a new or of an established genus', so presumably we must take Jacoby as having designated *cyanella* L. (*lichenis* Voet.) as the type of *Lema* F., so that the name would become synonymous with *Oulema* and *Hapsidolema*; if Heinze's generic distinction is sustained, then what he called *Lema* will be left without a well-established name. In looking for an 'available' name to replace the *Lema* of Heinze (1927), there are one or two old and disused names which we may need to consider, for example *Auchenia* Thunb. (1791) and *Petauristes* Guerin (1944), both of which as originally published included species of Lacordaire's *Lema*.

The most satisfactory resolution of this situation would probably be for the International Commission to set aside Jacoby's type designation for *Lema* and all other previous ones, and to establish another, unambiguous Fabrician species as type of the genus, for example *Lema cyanea* F.. This would firmly establish the name *Lema* in its traditional sense, and Gozis's name *Oulema* could prevail for Heinze's *Hapsidolema*. An application on these lines is at present before the International Commission.

21 The Practical Work of the Present Day Systematist

If a representative committee of modern biologists were faced with the problem of designating a holotype for the word systematist, there can be little doubt that the person finally selected, whether a botanist or a zoologist, would be a museum worker. The scientific staffs of museums and herbaria include the majority of those biologists who would normally be referred to as systematists *tout court*. We need to consider the functions which such people are expected to perform, and those which they actually fulfil. In the United Kingdom, the largest institutions of this kind are the British Museum (Natural History) and the herbarium of the Royal Botanical Gardens, Kew, and their staffs will provide typical examples of professional systematists.

The staffs of such institutions are expected to devote the major part of their time to curatorial work on the collections, and to the determination of specimens sent in from outside for that purpose. In return for providing this identification service for outside bodies and persons, most museums and herbaria claim the right to retain a proportion of the material identified by them, and this usually provides a major source of material for the expansion and improvement of the collections. Most museums have a certain amount of money available annually for the purchase of material for the collections, and are liable at times to receive gifts or bequests; the larger institutions may also from time to time send out scientific expeditions to collect material for their collections, and members of their scientific staffs may have the opportunity of participating in such expeditions. However, the emphasis on such expeditions is likely to be overwhelmingly on the collection of specimens, members of them are unlikely to have much opportunity for study of habits, life histories, ecological relations, etc. of the organisms which they are collecting. The success of a museum expedition is usually measured by the number of specimens it brings back, and resulting publications are likely to comprise little more than lists of species from particular localities and descriptions of occasional new species.

In the course of his routine determination work, the museum systematist is liable at times to be confronted with plants or animals which appear not to be assignable to any previously defined species. He then has the option, which many museums would regard as a duty, of describing new species. By such publication he may establish 'holotypes' in the museum collection, and these represent an increasingly valuable form of 'capital' for the institution. The more holotypes there are in a museum, the more likely it is to attract the most eminent systematists to work within its walls, the more material is likely to be sent to it for determination, and hence the more new holotypes and paratypes are likely to accrue to its collections. Of course, if the museum systematist's descriptions are really full and accurate, others may be enabled to recognise his species without the need to send specimens to the museum for checking—such thorough descriptions, it seems, though they may help to build the reputation of the systematist himself, may not serve the material interests of the institution employing him. It may not be unconnected with this circumstance that, as we saw in Chapter 20, the current codes of botanical and zoological nomenclature attach no legal force whatever to the content of the definition of a new taxon; the entire legal weight rests on the types themselves.

It is generally agreed among zoologists and botanists outside museums that one of the most valuable services to science which can be performed by systematists is the preparation of good systematic monographs of more or less extensive groups—works whereby the non-specialist may be enabled to identify organisms for himself and in which he can find a compendium of information and bibliographic references concerning them. Such works necessarily involve a great deal of time and trouble in their preparation; most museums do not encourage their staffs to undertake works of this kind, nor are they willing as a rule, to permit members of their staffs to spend any important part of their official working hours in such an activity. Among museum staffs, writing such books is generally regarded as a spare-time activity and is most commonly undertaken only after the systematist has retired from his salaried post, by which time his work is apt to have deteriorated in quality.

Real systematic research, such as we have discussed in Chapter 18, constitutes as a rule a very small and strictly unofficial part of the work of a member of a museum or herbarium staff. His major occupation will be identification, and his professional skill in performing this function will largely depend on his ability, confronted with a plant or animal specimen, to know just what cabinet, drawer or bottle in the museum collection is likely to contain similar specimens. The more efficient he is at this, the less real intellectual, systematic content there will be in his routine activities.

The museum systematist of the present day is overwhelmingly preoccupied with problems of discrimination among closely allied forms. The experience of being confronted with representatives of a genuinely new supra-generic taxon is exceedingly rare for most of them, and in most groups of higher animals and plants, even the discovery of a new genus would be a notable event. If the museum worker of today has any novelties to describe, these will usually be new species and subspecies in 'difficult' genera. As a result, his attention will be overwhelmingly concentrated on the search for differences, similarities being taken for granted and scarcely thought about. Hence he is liable, not only to be biassed in the direction of 'splitting', but also to have an inadequate appreciation of the bases of the higher classification of the groups with which he is concerned.

There are also members of the staffs of academic botany, zoology and other departments who may be referred to as systematists, a small number of them having been explicitly appointed as such. The last mentioned group tend to be preoccupied with the philosophical rather than the botanical or zoological problems of systematics, their publication often take the form of learned books and papers on the Theory of Classification, Numerical Taxonomy, etc. Academic systematists have usually far greater opportunities for the pursuit of systematic research properly so called than have their museum colleagues, but are apt to be handicapped particularly by not having large collections at their disposal. Some of them try to carry out detailed revisional studies of limited groups, and in the process become involved in protracted visits to, or large loans from, relevant museums or herbaria. A few of them occupy themselves mainly with classificatory problems at the higher levels, e.g. families, orders, etc.

Classification at these higher levels is a sphere of research which museums rarely encourage their staffs to devote much time to, but one which is well suited to be pursued in universities. A division of scientific labour by which the main responsibility for identification and species systematics devolves on the museum and herbarium staffs, while leaving the main responsibility for supra-generic systematics to their academic colleagues, would be a sensible rationalisation of present trends. To work satisfactorily, however, this division of function would require a degree of mutual understanding and respect which may be rather difficult to achieve.

The academic systematist is likely to be expected by his colleagues to identify organisms on demand, and would be well advised to provide himself with the basic requisites for doing this in at least some groups. Even a small reference collection of reliably identified specimens will be a very valuable adjunct to the systematic monographs, etc., which may be available for this purpose. Good personal relations with museum specialists in particular groups are also worth cultivating.

Academic systematists often succumb to the temptation to specialise in techniques or approaches rather than in particular taxa, they are liable to preach and practise numerical taxonomy, comparative serology, stem anatomy, chromosome studies, etc. From the point of view of conducting original research, there is a good deal to be said in favour of this; once you have mastered a more or less difficult technique, it seems a pity to waste your hard-earned skill. There would be little harm in such specialisation as far as practical research is concerned, if the academic systematists pursuing it also accepted full scholarly responsibility for the co-ordination of all other types of relevant information about the groups concerned—this, however, many of them are reluctant to do. Our present academic organisations attribute little value to scholarship in the traditional sense.

In addition to identification work and the occasional description of new species or subspecies, many museum systematists from time to time produce 'revisions' and catalogues of genera or higher taxa, which are undoubtedly useful even if most of them fall far short of the scholarly monographs which the rest of us would like to possess. In a 'revision', we may hope to find a redefinition of the taxon, and keys whereby to distinguish the included genera and species, together with some distributional information—either for the entire world fauna or for some clearly marked part of it such as the New World species. In a catalogue, there will usually be no definitions or keys, but full synonymy and bibliographic citations to all published references to members of the taxon in systematic literature may be expected. Catalogues prepared by museum systematic zoologists rarely cite references to included species in physiological, ecological and other non-classificatory literature, where these references are not cited in the systematic part of *Zoological Record*. The last-mentioned work, providing an annual author, systematic and subject classified index to zoological publications, is indispensible to the serious zoological systematist; the absence of a really comparable botanical work is deplored by most systematic botanists.

Unfortunately the typical museum systematist, even when he is able to find the time, the energy and the publisher for preparing a determination work for the use of non-specialists, is apt to be not very good at the task. For one thing, in books for the practical identification of plants and animals, the essential part is apt to be a series of analytical keys, and determination by means of such keys is something which most museums systematists rarely practice. They usually operate by direct comparisons with a large reference collection, and are apt to rely heavily—sometimes too heavily, to the extent of making gross mistakes—on that elusive quality, 'general facies'. In keys constructed by museum systematists, the beginner is only too often confronted with a couplet such as this:

18. Legs very long and slender   19
    Legs of normal length   26

The character used here may be a perfectly good one, and the one which the professional systematist himself relies on to distinguish the groups concerned, but expressed in this way is apt to cause headaches to the non-specialist trying to identify a single species in the absence of a reference collection. The idea of 'normal length' is one which the museum specialist has derived from the reference collection and from his long experience of the group. For the beginner, the best substitute for these things is lavish illustration—but, whether the systematist prepares illustrations himself or someone else is professionally employed to do so, they will add greatly to the effective cost of the work as well as to its value. Another frequent drawback to determination works prepared by museum specialists is the use of an unnecessarily technical vocabulary, which the specialist has been using for so long that he finds it difficult to dispense with or even to recognise as such; this however can usually be surmounted with judicious editorial assistance and the provision of a glossary.

The classical form of systematic monograph begins with a historical account of previous treatments of the taxon concerned, then provides a more or less exhaustive definition of it, discusses reasons for inclusion or exclusion of doubtful forms, the relations of the taxon concerned to other ones, and perhaps its phylogenetic ancestry. Then follow a key to the primary divisions of the taxon, definitions and general considerations for each of these divisions in turn, similar keys and definitions to their subdivisions, and so on down to the level of species. The species are described individually, with the inclusion of any available information about geographical distribution, young stages, life cycles, habits etc. All characters which are used to distinguish groups within the taxon and which are not readily and unambiguously expressible in words should be figured. Habitus figures of whole organisms are desirable but not usually essential. It is a great help in practice if page references to subsequent detailed treatment of subordinate taxa are given at the points where these subordinate taxa are keyed out. Finally, the bibliography should be as full as possible and preferably not limited to purely 'taxonomic' works.

The construction of the keys which form the core of most such monographs is a matter on which there are various schools of thought, discussed by Osborne [147]. It is generally agreed today that, for any but the simplest and shortest keys, numerical forms are preferable to the 'hieroglyphic' or 'spatial' kinds used in many old works—a 'spatial' key is one in which pairs of alternatives are indicated by varying degrees of indentation from the margin of the page, and the 'hieroglyphic' type uses all manner of signs, other than numbers, to indicate pairs of alternatives, such as single or multiple letters (a, aa, aaa, etc.) asterisks, =, +, % and similar symbols. In numerical keys the alternatives are numbered in sequence; each pair of alternatives may be presented as a numbered couplet, the halves of which lead either to taxa or to further numbered couplets—or the two alternatives may appear separately, with a number reference to each other, in this manner:

1(6) .	.	.	.	.	.	.	.	.	.
2(3) .	.	.	.	.	.	.	.	Group A
3(2) .	.	.	.	.	.	.	.	.	.
4(5) .	.	.	.	.	.	.	.	Group B
5(4) .	.	.	.	.	.	.	.	Group C
6(1) .	.	.	.	.	.	.	.	Group D

The same key, in the other numerical form, would appear like this:

1	.	.	.	.	.	.	.	.	Group D
	.	.	.	.	.	.	.	.	.	2

2	.	.	.	.	.	.	.	.	Group A
	.	.	.	.	.	.	.	.	3
3	.	.	.	.	.	.	.	.	Group B
	.	.	.	.	.	.	.	.	Group C

The first illustrated type of numerical key is similar in principle to the hieroglyphic or spatial kinds, and is apt to be tiresome for a lengthy key, as alternatives are liable to be widely separated; you may well start off with something like this:

1(168) Stamens 5

and may have to turn over numerous pages before you find the alternative

168(1) Stamens 4 or less

Apart from the mechanics of their construction, keys differ in another important way. They may be 'short' in the sense that each alternative is presented in terms of a single character (and usually a single phrase) as in the example just quoted, or 'long' in which case each couplet is liable to refer to a number of characters and to make more or less complicated provisions for exceptions in respect of particular ones among them. 'Long' keys are apt to look forbidding at the outset, and to be cumbrous in use, but have the advantage that they may make possible determinations when some of the characters cannot be observed in the specimen(s) at hand; they may also permit the keying out of groups in ways which correspond more or less to their natural relationships. The short type of key is probably preferable on the whole for quick routine determination work; it should always be supplemented by descriptions or definitions of the taxa distinguished in it, which may not be necessary when the key is of the 'long' type.

In short-type keys it is best to deal with exceptions and ambiguities by keying out groups showing them at two or more places in the key; for example, in the last quoted key, a genus in which there are four perfect stamens and a large staminode could be included under both halves of the couplet, as could another in which some species have four, some five stamens.

Even when the long type of key is used, it may sometimes be desirable to key an occasional group out in more than one place, in order to avoid the use of excessively complex series of alternatives in the couplets.

The amount of information, in addition to keys and formal definitions, which is put into systematic monographs in practice is very variable. The majority of available works of this kind limit themselves almost entirely to the recording of such characters as can be observed in preserved museum specimens with a minimum of dissection; ecological, developmental, physiological and internal-anatomical characters rarely receive more than cursory mention. There are, however, some few published monographs which genuinely attempt to present the whole available knowledge of the group in systematic form—a classical prototype of these is Charles Darwin's *Monograph of the Cirripedia*, and more recent ones include Stephenson's *The Oligochaeta* [181] and Ame's *Monograph of the Chaetomiaceae* [4]. In the circumstances obtaining at present, it is hardly fair to expect museum systematists to produce works on this model, but I think it ought to be the ambition of every academic systematist to leave at least one such monograph as the legacy to posterity of his life's work.

Unfortunately, even when a systematist in a university department has the will and the facilities to set about writing such a monograph, he may be deterred from the enterprise by the expectation of grave difficulties in getting his work published. The anti-systematic bias which so many botany and zoology departments today instil, deliberately or not, in their students, has the effect that these students are effectively discouraged from thinking of themselves as specialists in some particular group of organisms; as a result, not merely are some potentially good systematists likely to become mediocre experimentalists instead, but also our university graduates are discouraged from buying systematic monographs for themselves. The potential sales of such works are thus being seriously curtailed at the same time as technological and social changes are making it more and more difficult and expensive to publish works with small circulations.

The main service which the present day world expects of its systematists remains, however, the speedy and reliable identification of organisms, and we need to consider briefly the problems and methods involved in doing this.

As already indicated, if the systematist works in or has access to one of the great museums, his main recourse will be to comparison with specimens in the collection; in the interests of speed and efficiency, he should familiarise himself as soon as possible with the layout of these collections in as much detail as possible. Difficulties are liable to arise in this process for two reasons, either the incompleteness of the collection or the unreliable determination of parts of them. Incompleteness of the collections may involve the absence of any specimens of certain species, and the representation of others by too few specimens to give a proper idea of the range of phenotypic variability (cf. Chapter 3).

The systematist will attempt to make good deficiencies in his reference collection by consulting published works in which the missing species are defined, or by communicating with specialists in other museums. In dealing with published work, he is likely to find it written in various languages; it is hardly possible to be an effective world specialist in a particular systematic group without a working knowledge of at least French and German as well as the English language. An index of the addresses of relevant systematists and museums throughout the world is a valuable possession for any serious systematist.

The museum systematist will probably find himself with full responsibility for the arrangement and upkeep of some part of the institution's collections. When he takes over this responsibility, it is quite likely that he will find this part of the collection arranged according to some long outmoded classification. He then has to decide, whether to leave the general arrangement as it is, or to undertake a total reorganisation of this section, to arrange it in accordance with current ideas of natural relationships. Such an undertaking might well occupy most of the working time of himself (and any technical assistance which may be available) for several years, and will effectively devalue the expertise of any persons who had been accustomed to determining specimens from the collection as formerly arranged.

A further problem which is likely to face the museums or herbarium systematist is that of dealing with numerous requests which he may receive from outside for loans of type material. As we have seen, the current rules of nomenclature make it very important for the systematist to study the types of species with which he may be concerned, and also make the holotypes the most valuable form of capital a museum possesses. It is vastly cheaper to transport a holotype from Paris to Washington and back than to transport a systematist from Washington to Paris and back. In practice, most of the major museums do not normally permit unique holotypes (i.e. species of which the only specimen they possess is the holotype) to be borrowed for study outside the museum, but many of them permit established specialists to borrow holotypes of species of which they possess a type series, including paratypes. In such cases, if the holotype is lost or destroyed, one of the paratypes could be designated as a Neotype in its place.

Another duty which is expected of many museum systematists is the arranging of displays for the benefit of the general public. The large majority of museums and herbaria are financed from public funds, and are expected to provide displays which will be appreciated by the taxpayers who finance them. The annual public attendance figures of a museum are commonly taken as an index of its success; if they fail to increase, or even decline, over a period of years, strong pressure may be expected for radical changes in the policy of the museum.

Marshall McLuhan's doctrine, that 'the medium is the message', is at least as true in relation to modern museum displays as it is to the mass media of our day. The popular appeal of a museum display of natural history today depends far more on the display techniques used than on its basic scientific content. The keys to success are sensational and eye-catching methods of presentation, and as frequent as possible changes in the exhibits. In the last century, the natural history displays in local museums were directed at the local amateur naturalist rather than the general public, all that was required was to put an adequate amount of intrinsically interesting material out where it could be seen, and to provide simple hand-written labels. Visitors were comparatively few but most of them brought positive interests with them; if curiosity was not satisfied with the information provided in the labels, the curator was generally accessible. The museum provided for the local natural history society something analogous to a church.

The difficulty in those days, of course, was that the curator of the museum was expected to know about everything, botany, zoology and geology, so that he hardly ever had the time or the energy left to pursue systematic studies to a very high level in any one group. The main legacy we in Britain owe to these nineteenth-century natural history museums and societies is our detailed knowledge of the distribution of the British fauna and flora, and of the fossils of our sedimentary rocks; the last fifty years which have seen the decay of this type of museum have also seen notable changes in the distribution of many species of plants and animals. The smaller local museums of today are becoming more and more predominantly concerned with human history and archaeology, and their curators rarely have much knowledge of or interest in natural history. The larger museums maintained by the more important cities however do, as a rule, try to maintain at least one natural historian on the curatorial staff. Like the old-time curator, he may well be expected to take responsibility for collections and exhibits in botany, zoology and geology; additional duties which are liable to devolve on him are those of giving or at least organising regular lectures to school classes, and of operating a system of loans of natural history material to schools for class use. It is hardly to be wondered at that few such people are able to pursue serious studies in systematics.

The number of British museums whose staffs include more than one full-time natural historian is very small—apart from the State-financed ones in London, Edinburgh and Cardiff, only Manchester and Leicester really qualify under this heading. The British Museum, the Kew and Edinburgh Herbaria and the National Museums of Scotland and Wales, are about the only British institutions, other than universities liable to employ full-time specialist systematists.

The British Museum (Natural History) has recently established a 'Unit of Experimental Taxonomy', equipped with facilities for comparative serology, electron microscopy, etc., which are at least theoretically available for the investigation of systematic problems arising in any group of organisms represented in the museum's collections. This is in principle a desirable development, but seems a little illogical in an institution which hardly recognises the right of its staff members to pursue systematic researches to any significant extent in more old-fashioned ways.

22 Epilogue: The Future of Systematics

What's to come is still unsure.

SHAKESPEARE

It has been pointed out, e.g. by Bertrand de Jouvenel [112] that public institutions and/or private corporations in our advanced technological societies need continually to forecast future developments if they are to decide rationally between various alternatives in making major capital investments at any particular time, while at the same time the increasing speed and unpredictability of technological developments are making such forecasting ever more hazardous. Professor J. K. Galbraith [68] suggests that once a really big capital investment is made, the institution or corporation concerned will exert all possible influence to ensure that consumer choice, etc., develop in the way which was assumed in making the forecast on which the investment was based; such influence, Galbraith suggests, may be strong enough for a considerable period to outweigh 'spontaneous' consumer or market choice.

Planning problems of a similar kind confront those who direct university departments and faculties, research institutes, publishing houses, and even to some extent, museums. To a considerable extent, the development of systematic botany and zoology over the next twenty years or more will be determined by capital allocations now being decided. As far as this author is able to judge, these decisions are only too often ones which decree a very meagre and narrowly restricted future for systematists, reflecting as they do the ideological bias of administrating scientists now in the fifties and sixties, whose mental attitudes were formed during the inter-war years.

One symptom of this mental attitude has been the disappearance from the cover of the scientific magazine *Nature* of the picture of the 'great globe itself' and of the Wordsworthian quotation 'To the solid ground of Nature trusts the mind that builds for aye'. The idea of Nature as 'solid ground' has been undermined by twentieth-century physical theories like wave mechanics, as well as by the prevalence of various forms of Positivist philosophy. Among contemporary intellects, as among the architects who design the new institutes in which so many of them operate, the idea of 'building for aye' is hardly respectable. One should, it is felt, be content with the esteem of one's contemporaries and the more tangible rewards that go with it, and take no thought for the centuries to come. To the typical contemporary mind, the very idea of planning for posterity smacks of impiety, even of blasphemy against the religion of progress. Future generations, it is felt, will be so enormously superior to us in all departments of technology that we cannot presume to anticipate or provide for their requirements. In our age, the short-term view predominates to a degree unprecedented in developed civilisations.

Classical, systematic botany and zoology are the products of the cumulative labours of many generations, like Milan Cathedral but unlike the United Nations building at Lake Success. In order to be in a position to make permanently valuable additions to systematic botany and zoology, each generation has need first to assimilate and evaluate all that its predecessors have achieved. Real scientific progress has meant that this task has become more and more onerous for each succeeding generation. There is a powerful urge in 'forward-looking' circles today to reject such responsibilities outright, and to set about the construction of a 'new systematics' which will be as technological and mechanical as the aforementioned U.N. building. The problem, it is argued, is no different in principle from that of classifying the books in a large library, and should be tackled with the aid of suitably programmed computers.

Modern Biology stands to classical Botany and Zoology in much the same relation as that of an 'action painting' by Jackson Pollock to a landscape by the younger Turner. Instead of a systematically ordered and harmonious whole, we find a chaos of 'brush-strokes', each made 'for its own sake'. To quote a recent poet: 'The age demanded an image of its accelerated grimace'—and this demand has proved effective in biology as well as in the fine arts. Unfortunately, the 'Hugh Selwyn Mauberly' of botany and zoology has no poet like Ezra Pound to speak up for him. We may, however, entertain a reasonable hope that the twenty-first century will have little more respect for the intellectual traditions of its predecessor than our century has shown for those of the last one. There are, indeed, signs already visible of a developing challenge to the conventional twentieth-century attitudes in the fine arts.

The big question concerning the immediate future of botanical and zoological systematics is whether this contemporary-minded techno-logical approach will come to dominate everywhere as it has recently threatened to do in the U.S.A. Even in that great country, the predic-tions of Ehrlich [155a] for 1970 seem unlikely to be fulfilled; another advocate of 'numerical taxonomy', Michener [143a], is more con-cerned, despite his title, to say what he thinks should happen than to predict what will actually come to pass. The main effective rival to the technological approach is the theory of phylogenetic classification, as expounded in Chapter 9 of this work; influential in Germany and the U.S.S.R., it has substantial bodies of support also in France and Britain, though regrettably weak in the U.S.A. It is a lesson of history that, when fashionable intellectual attitudes appear to prevail against endur-ing cultural values, their victory is likely to be temporary, and I am optimistic enough to believe that this will be true of the technological approach to systematics.

The usual procedure in predicting the future is to single out the observable trends which have led from the more or less recent past to the present, and to extrapolate them. Prophecies made on this basis in the past have often proved essentially correct, but at times they have failed grossly. Almost all historical trends are liable to stop or even to go into reverse at some time or other, and at any instant in history a new and highly significant trend may be in the process of initiation.

The academic trend over the last fifty years or more has been one of a progressive depreciation of the importance of systematics, and it has produced the effect that young botanists and zoologists graduating from our universities today do so probably with less real systematic knowledge than at any time in the last hundred years. A simple extra-polation of this trend into the future would suggest the virtual dis-appearance of systematic content from academic botany and zoology courses before the end of this century. There are, however, reasons for expecting this trend to be stopped and even reversed in the near future.

The first is, of course, the practical indispensability of classification, of plants and animals for the purposes of food-production and of micro-organisms for medical purposes. The discrimination of species is a necessity for those who deal in any way with wild animals and plants, and the generalising power of a fully developed hierarchical classification has repeatedly proved its practical value. Of course, large numbers of existing species of plants and animals are likely to become extinct in the near future as a result of ever increasing human pressure on their habitats, but even at the end of this century there will almost certainly be sufficient species in both kingdoms surviving to make fully developed classificatory systems practically desirable.

Another reason for doubting whether the anti-systematic trend which has affected botany and zoology for the last fifty years will continue to do so for the next half century concerns the general movement of cultural fashions. Until very recently, in almost all branches of high culture, the movement in the present century has been further and further away from the ideals and principles of the nineteenth century. In the last ten years or so, there have been clear signs that this trend is coming to an end, and we may even be witnessing the start of a significant reverse movement. Tennyson and Matthew Arnold, Donizetti and Mendelssohn, Dickens and Hugo, Ruskin and Millais, Butterfield and Gilbert Scott, have all recently been restored to more or less respected positions in our cultural heritage, after protracted oblivion or denigration. The great edifices of systematic botany and zoology are also very largely products of nineteenth-century civilisation, and in a society which is moving in the direction of greater and greater appreciation of its cultural heritage from that period, they can look forward to rehabilitation, perhaps even to an era of excited rediscovery by a younger generation rebelling against the intellectual traditions of its teachers.

A new generation of botanists and zoologists has the opportunity, not merely of re-discovering the rich diversity of the plant and animal kingdoms and the complexity of their evolutionary history, but also of improving the classifications of the past, in the light of new information and better evolutionary understanding—and the resulting improved systems should in their turn provide better bases for generalisation from observations based on limited numbers of species.

As already noted, to be in a position to improve on a previous classification, you need first to understand fully the bases of that classification, and then to bring into consideration further comparative data, unused by previous systematists. It is necessary, and difficult, to maintain a balance in your appreciation of new and old information. The systematic importance of information is in no way dependent on its novelty, or on the sophistication or complexity of the techniques by which it was obtained. One of the objections to the conventionally 'progressive' outlook of today is that it makes it harder to maintain a proper balance in such matters. A mental attitude of despising or patronising your predecessors is no more likely to inspire real improvements on their achievements in systematics than it is in the fine arts. I have suggested that there are reasons for expecting that, if and when a major revival of interest in systematic botany and zoology does become manifest, the most favoured approach will be the phylogenetic one, rather than the currently fashionable 'numerical taxonomy'. The 'biometry' of Pearson and others was a similarly fashionable trend in advanced biological circles at the beginning of this century, but fell into some disrepute when results achieved through it failed to justify the expectations aroused by enthusiastic advocates; a very similar situation seems likely to develop in relation to numerical taxonomy. In this, as in other fields of biology, the classical Darwinian approach seems the best adapted to long-term survival.

If phylogeny finally comes to be accepted as the only stable and objective basis for classification, we can expect more and more importance to be placed on the fossil evidence. A serious systematist working on any group for which a fossil record exists will probably be expected to acquaint himself at first hand with the fossils of his group, and palaeontology will be recognised as an essential element in the academic training of prospective systematists. It may seem paradoxical if, at the same time, as I have suggested, the classifications of fossils become formally separated from those of modern organisms.

Studies of protein structure, nucleic acid base sequences, and comparative serology will probably also come to be more and more integrated with other lines of systematic research; we may hope to see the establishment at a number of centres of units of 'experimental systematics' directed by qualified systematists. Such units would probably operate in close connection with (and proximity to) major museums and university departments.

The description of new species will inevitably form a continually decreasing proportion of the activity of systematists; it is already comparatively rare in Vertebrata and Angiospermae, though undoubtedly large numbers of existing species remain to be described in such groups as Nematoda, the lower fungi, Protozoa, and some orders of insects. In recent times, the reaction of systematists specialising in groups in which real new species are rarely found has commonly been to pursue what has been called 'ingrown taxonomy'—to concentrate their interest on infra-specific classification, with a tendency to publish sub-specific names for geographical and other variants which are in the slightest degree distinguishable. Many ordinary botanists and zoologists feel that this sort of activity leads to a proliferation of nomenclature with little practical advantage to science, and we may expect this widely held view to have eventual effect in discouraging such activity among the systematists themselves. There is no objection, of course, to the serious study of variation within species, but as a rule, such a study is better approached from the point of view of population genetics than in terms of formal classification.

Eventually, it seems likely that the main interest of practical system-atists will return to the improvement of the existing system at all levels from the species level upwards, with the incorporation of ever more comparative data in their systems; their stock in trade will come to include, in addition to the established types of museum collections of preserved organisms, preparations showing developmental features, Ouchterlony plates and immuno-diffusion patterns, chromatograms, perhaps even 'banks' of anti-sera stored at low temperatures, casts or photographs of relevant fossils, records of physiological and be-havioural experiments, and so on. The systematists might then regain their rightful place as upholders of the traditions of scholarship in botany and zoology, and as the final co-ordinators of all kinds of information about organisms.

The publication of descriptions of isolated new species of organisms is already becoming a slightly disreputable procedure, at least in the eyes of most botanists and zoologists who do not have a vested interest in doing it, and the editors of more and more scientific journals are becoming reluctant to accept papers of this kind. In most cases, the best interests of the scientific world would be served if a systematist who, in the course of routine determination work, encountered speci-mens which seem to represent a hitherto unnamed species, would merely add a note to this effect to their labels, with the intention to bring them to the attention of the next person to undertake a serious monographic revision of the group. There are instances, particularly where economically important species are concerned, where it may really be desirable to publish a name for a particular new species. In any case, it is likely to become the rule in the not very distant future that anyone publishing a definition of a new species should provide a small key whereby the new species can be distinguished from its nearest relatives.

The discovery of new genera based on new species, rather than the formation of new genera by splitting old ones, is becoming quite a rare event in many groups of plants and animals, and may be expected to become even more so in the future, so that pressure for a ban on the publication of isolated new genera is not likely to be very strong. We may hope, however, to see an end to the general trend, operative for a century and more, of splitting of older genera. The reduction in the average number of species per genus has now gone so far as seriously to threaten the practical utility of the Linnaean binomial system (cf. Cain [25]). The desirable trend now would be to reduce large numbers of currently accepted genera to the level of subgenera or even species-groups (cf. Chapter 5), and at least the idealists among us may hope that a change so clearly in the interests of the scientific majority is almost bound to come about.

We can be certain that the existing codes of nomenclature for plants and animals will be subject to revision and amendment in future, but whether future revisions will be in the direction of strengthening, or restricting, the application of the 'principle of typification' is less easy to determine. The principle itself is a relatively new one, which has come to occupy an increasingly important place in both codes during the last fifty years, and it seems quite possible that its predominance will be carried still further in the next generation, so that it comes to be applied to taxa at all levels, not merely up to the family (in animals) or order (in plants) level of nomenclature. If so, this will constitute a manifestation of American predominance in yet another field of human culture. On the other hand, the name-changing which has affected so many groups of animals and plants in recent times, and which has been so widely deplored by the generality of botanists and zoologists, has to a very considerable extent been the result of applying the typification principle. If the systematists ignore the rising ground-swell of discontent on this score, and go on to enact laws which are likely to increase its causes, they may seriously damage their own general standing in the scientific world. Furthermore, even from the point of view of pure systematists, the principle of typification has the effect of making nomenclature rest on principles completely different from those of classification, and hampers the use of names as vehicles for real classificatory concepts. It may well be that the next revisions of the codes will include further restrictions, rather than extensions of the application of the typification principle.

A serious problem confronting systematists today, and one likely to become even more acute in future, is that of publication. Systematic monographs of particular groups tend to become more and more technical and complex, and less and less comprehensible to the ordinary botanist or zoologist who is not a specialist in the group concerned. The immediate potential market for serious taxonomic works is tending to shrink just when social and technological changes are making it more and more impossibly expensive to publish books with small circulations. Of course, really 'solid' systematic works are likely to continue in demand for a very much longer time than most current scientific publications, but a 'dynamic' and inflationary era does not encourage publishers to lock up much capital in forms which are unlikely to yield a profit within ten years.

In order to achieve publication of a scholarly systematic monograph today, it is nearly always necessary to obtain financial backing from some institution or foundation. It would probably be better if governments or government agencies undertook the responsibility of sponsoring the publication of such works, and perhaps also disposed of some funds which could be used to assist authors in the preparation of works of this kind. In Britain, such a function might be vested in the National Environment Research Council or the Science Research Council. A public body of this kind could establish a committee of referees—competent systematic botanists and zoologists—which could report on the quality of systematic work for which assistance towards publication was requested. Private publishers do not usually have an expert committee of this type to advise them, and as a result they may at times publish systematic work of a seriously defective kind.

There is a widespread current idea that the systematist and the collector are essentially the same kind of beast, with similar casts of mind and basic motivation. To apply recently fashionable criterion, that of the 'paradigm' beloved of linguistic philosophers, we may consider Aristotle himself, generally accepted as the father of systematics. There is not the slightest hint in the surviving records that he was a collector of anything but students. On the other hand, most of us are acquainted with at least one compulsive collector of something or other, be it of stamps or of alpines for the rockery, and how many such persons show any interest in the intellectual problems of classification? It is true that the pure systematist likes to have access to a reference collection, although theoretically it should be possible for him to operate entirely on second-hand information without ever studying the organisms directly. A true systematist, if he is not working in or

living in proximity to a great museum, will probably find it necessary
to build up some sort of reference collection for himself; conversely,
someone commencing as a pure collector may come to be dissatisfied
with the accepted classification of the specimens in his collection and
may thus be led into intellectual enquiry and systematic research.

Motivation in science, as in other spheres of human activity, is
generally of the ambivalent type symbolised by Plato's two horses;
the 'virtuous' spirit of disinterested intellectual enquiry is apt by itself
to provide an inadequate driving force, and in most successful scientists
it is harnessed with drives of ambition and/or acquisitiveness, perhaps
even with sadism in some experimental physiologists. The importance
of some of these less respectable motives in some recent 'big science'
aspects of biology has been emphasised in a recent book by Watson
[194b]. The acquisitive urge of the compulsive collector has played an
indispensable part in the development of systematic botany and zoology, just as the urge to impose his own intellectual concepts on others
is strong in many a 'pure' systematist.

We can therefore expect the collector to persist and to function in
the foreseeable future as an important adjunct to the systematist proper,
though it will probably become less common for the two to be the
same person. This does not mean that the systematist of the future
should, or could afford to, leave all the 'field work' to others. His
phylogenetic reasoning is continually likely to raise questions to answer
which he would need information of a kind which ordinary collectors
are unlikely to provide. Having learned from the collectors when and
where the relevant organisms are to be found, the systematist should
be enabled to make the required observation in nature. And in the
course of doing so he may well observe other new and significant
phenomena, and initiate further advances in botanical or zoological
knowledge.

As we have seen, botanical and zoological systematics are essentially
conservative, cumulative branches of science, in which a precondition
of real progress is the complete assimilation of the achievements of the
past. Furthermore, advances in systematic knowledge and understanding can generally be made at relatively small cost, in comparison with
the vast sums now being spent on the fashionable 'growing points'
of new biology. Both these circumstances should make systematics
peculiarly suited to be pursued in the old-established universities of
Europe. With world-wide museum and herbarium collections built up
over two centuries and more in several European cities, and with rich

botanical and zoological libraries readily available, Europe should have a built-in advantage over the U.S.A. in this field. At present, unfortunately, the universities of Europe, or at least their biological departments, manifest little awareness of the possibility and desirability of maintaining world leadership in systematic botany and zoology; engaged in a hopeless attempt to keep up with the Americans in the 'big science' aspects of biology, they are allowing the Americans now to outstrip them even in systematics, the one field in which there should be a reasonable possibility of maintaining European supremacy. It may still not be too late to restore the position, but time is running short.

In the long run, there are some grounds for optimism, and even in the short run the outlook for systematic botany and zoology is by no means as dark as some suppose. The great bodies of systematic knowledge of plants and animals, preserved in such works as Engler and Prantl's *Die naturlichen Pflanzenfamilien* and Kukenthal and Krumbach's *Handbuch der Zoologie*, are dispersed among numerous libraries, and are not likely to be destroyed by anything short of a world-wide holocaust (a possibility which, unfortunately, we cannot entirely exclude); the fact that these great systems of learning are largely forgotten and ignored by the present dominating generation of biologists makes it all the more likely that a future one will rediscover them with surprise and enthusiasm. One thing which New Biology has not discovered, and is unlikely to produce, is a cure for senescence, and psychoanalysis has not eliminated the 'oedipal' tendencies of younger generations. Meanwhile, it remains as necessary as ever for practical purposes that plants and animals should be identified, and there is little or no prospect that for this purpose computers will be able to replace botanists and zoologists.

If current intellectual fashions constitute only a short-term threat to systematic botany and zoology, there is a serious long-term one which we cannot afford to ignore. This is the effect of increasing human population pressure, by which a large part of the earth's fauna and flora is threatened with extinction. Fashions come and go, but extinction is an irreversible process. For all true natural historians, the causes of human population control and nature conservation should have overriding importance at the present time. The recent catastrophic increase of human populations, which menaces the survival of wild plants and animals, can in large measure be attributed to the advance of medical science, a major source also of the ultimate threat to the academic survival of botany and zoology. The struggle between the devotees of Nature and those whose god is Man is essentially a religious war, waged on all fronts.

However uncertain the future may look, the present time offers both unusual opportunities and stimulating challenges to the serious systematist. The ever-expanding army of professional biologists is producing an evergrowing flood of new information about animals and (unfortunately, to a much less degree) about plants. The serious systematist, whether he is of the Aristotelean or the phylogenetic persuasion, will always wish to have further comparative information about his organisms, as a test of the power of the system he is using and as a basis for improving it. The new kinds of information can serve both these purposes, even though as a rule they are exasperatingly scrappy and unsystematic. The challenge which faces the systematist is that of making valid generalisations from the results of the experimentalists, and of perceiving more significance in these results than the experimentalists themselves are aware of. The systematist should be in a position to perceive significant correlations which the ordinary experimental biologist could not be expected to notice. These correlations may be expected to throw light, not merely on phylogenetic relationships, but also on the functional and evolutionary significance of the experimentalists' data. If systematists can successfully meet this challenge, they should have little occasion to fear the future.

Meanwhile, if the threatened disintegration of our subjects is to be averted, I believe it will be necessary to give at least a proportion of our students a much more thorough systematic training than is to be had in the vast majority of our universities today. This training will need to include some familiarisation with the techniques and results (and, unfortunately, the technical vocabularies too) of modern laboratory disciplines in physiology, biochemistry, etc., and consideration in systematic terms of the data they provide. Thus a modern biologist trained on systematic lines will have a great deal more information to assimilate than had a similar student fifty years ago—but he will probably have no greater learning power or learning time than did his grandfather's generation. This confronts us with a difficulty which I fear can only be overcome by that bugbear of the liberal academics, greater specialisation. For example, if a botany student had the option of taking an Honours Degree in Dicotyledonae instead of Botany, he could give up the study of other groups of plants after his first year, and utilise the time thus gained for a much more thorough and many-sided study of his chosen group. One reason why proposals of this sort do not, as a rule, evoke much enthusiasm in present-day university departments is that there are likely to be few people on their staffs who would feel capable of planning and carrying out such courses— a 'vicious circle' exists which will not be easily broken.

The future of biological systematics will probably be determined in the botany and zoology departments of our universities rather than in our museums or our naturalists' societies—though both the museums and the societies have indispensable roles to play if systematics is to survive and flourish. All who care for systematics should be vitally concerned with current developments in the academic organisation of the so-called life sciences. At present, under the slogan of 'Cell Biology', the biochemists and geneticists are making a powerful bid for supremacy over mere botanists and zoologists; if they succeed, and botany and zoology lose their hard-won status as independent academic subjects, the consequences can hardly fail to be disastrous for systematics.

The Zoology and Botany departments of our universities are as much monuments to nineteenth-century idealism as is London's Natural History Museum or Wagner's 'Der Ring des Nibelungs'. In the practical, deistical eighteenth century, universities were thought of merely as training grounds for professional, administrative and social élites; such science as managed to find its way into the curriculum was justified on the grounds of its practical importance for medicine or the developing Industrial Revolution. The nineteenth century, reacting against the worldly ideology of its predecessor, developed idealistic modes of thought to an extent unprecedented in civilised Europe. The universities, like other social institutions, came to take a much more exalted view of their functions—symbolised, perhaps, in the soaring gothic lines of their new buildings. No longer were they content to turn out the requisite number of suitably trained and well-mannered men for the professions; a self-respecting university, it was thought, should be the guardian and transmitter of a priceless heritage of culture, the place in which the noblest activities of mankind were centred, and an instiller of high ideals in the young. Of course, the nineteenth century failed to live up to its ideals, thereby earning itself a reputation for hypocrisy, just as the eighteenth century had failed to live down its declared aims, and thus acquired its reputation for eccentricity.

In the thought of the nineteenth-century idealists, science appeared less as the basic theory of technology than as an enquiry into the Riddle of the Universe, and the scientists themselves tended to take a quasi-religious attitude to Nature (the name of the weekly journal reflected this outlook). In this climate of opinion, it was natural to remodel the structure of academic science to bring it more closely into correspondence with that of the natural world. It was this trend which led to the eventual acceptance of Botany and Zoology as academic subjects in their own right, independent of Medicine. The plant and animal kingdoms were objectively distinguishable realms in Nature, each containing sufficient diversity and complexity to provide unlimited scope for research, and an ever-growing mass of knowledge to be taught.

The period since the first World War has seen the catastrophic collapse of almost all manifestations of Victorian idealism. With the gothic towers falling all around, it is not surprising that Botany and Zoology are also in danger of falling from their previous high estates as fully independent academic disciplines or departments. Our century has not, however, simply reverted to the mundane attitudes of the eighteenth century. Vast sums of money are now being spent, and doubtless will continue to be spent, on researches whose practical utility is in the highest degree remote and speculative. Some aspects of pure science are coming to be looked on as a kind of intellectual Olympic Games, as field in which gifted individuals can perform dazzling and competitive feats of technique and ingenuity. A recent book by Watson [194b] reveals, among other things, the extent to which this is true in the 'big science' aspects of modern biology. Systematics is peculiarly ill-suited to fulfil such a function—though the numerical taxonomists (see Chapter 15) will no doubt try to introduce into it a more vigorously competitive spirit.

In the science faculties of our universities, the systematists ought to function as the main upholders of the old academic traditions of scholarship, dealing as they do with departments of knowledge within which the discoveries made by the present generation can hardly ever hope to match the importance of our inheritance from previous ones. The much publicised claim that the total of scientific knowledge is doubled in each decade is certainly untrue in respect of systematic botany and zoology. The ideal of scholarship itself is commonly ignored or slighted in current discussions about the functions of universities; it is a characteristic paradox of our day that at the very time when the History of Science is rapidly gaining recognition as an academic subject in its own right, the scientists are tending to attribute ever less value to their heritage from the past.

If systematics is eventually restored to its central position in academic botany and zoology, this will probably be one manifestation of a change in the general 'intellectual climate'. Other signs of such a change will be the official recognition of natural history as an aspect of science entitled to parity of esteem with natural philosophy, and the reinstatement of the view that a prime function of universities is the preservation and transmission of a vast heritage of culture and knowledge. Much more effort will be made to inculcate in the young a profound respect for their cultural heritage and correspondingly less emphasis placed on the cleverness of the present. Instead of being taught to despise or patronise previous generations, students would be made to feel that only by tremendous efforts could they hope to match the achievements of their predecessors. If and when such changes become manifest in the science faculties, we may expect comparable ones to have affected the arts subjects, whereby the split between C. P. Snow's 'two cultures' will virtually disappear.

Bibliography

1. AHRENDT, L. W. A. 1961. Berberis and Mahonia: a taxonomic revision. *J. Linn. Soc. Botany* **57**: 1–410.
2. ALEXANDER, C. P. 1931. Crane-flies of the Baltic Amber. *Bernstein-Forsch.* **2**.
3. ALSTON, R. E. & TURNER, B. L. 1963. *Biochemical Systematics*. Prentice-Hall.
4. AMES, L. M. 1961. A Monograph of the Chaetomiaceae. *U.S. Army Res. & Development, Series 2*.
5. ANFINSEN, C. 1963. *The Molecular Basis of Evolution*. John Wiley.
6. ASHBY, W. R. 1962. *Design for a Brain*. Chapman & Hall.
7. BACON, F. 1623. *De Dignitate et augmentis Scientiarum*. London.
8. BAKER, C. M. H. & HANSON, H. C. 1966. Molecular genetics of avian proteins. *Comp. Biochem. Physiol.* **17**: 997–1006.
9. BALDWIN, E. 1949. *Introduction to Comparative Biochemistry*, 3rd edition. Cambridge University Press.
10. BATE SMITH, E. C. & SWAIN, T., 1963. In Alston & Turner, loc. cit.
11. BEESON, C. F. C. 1941. *The Ecology and Control of the Forest Insects of India*. Dehra Dun.
12. BELL, T. R. D. & SCOTT, F. B. 1937. *Fauna of British India, Moths*, vol. 5. Taylor & Francis, London.
13. BISBY, G. R. 1945. *Taxonomy and Nomenclature of Fungi*. Imperial Mycological Institute, Kew.
14. BLACKWELDER, R. E. 1967. *Taxonomy*. John Wiley.
15. BLAKE, W. 1793. The Marriage of Heaven and Hell. In Keynes, G. (ed.), *Blake's Poetry and Prose*. Nonesuch Press, London, 1941, p. 185.
16. BLEST, A. D. 1957. The evolution of protective displays in the Saturnioidea and Sphingidae. (Lepidoptera). *Behaviour* **11**: 257–309.
17. BOYDEN, A. 1963. Precipitin testing and classification. *Syst. Zool.* **12**: 1–7.
18. BRACE, L. C. 1963. Structural reduction in evolution. *Amer. Naturalist* **63**: 274–279.
18a. BRAITHWAITE, R. B. 1953. *Scientific Explanation*. Cambridge University Press.
19. BRAUN-BLANQUET, J. 1964. *Pflanzensoziologie*. Springer Verlag.
20. BREED, R. S., MURRAY, E. G. D. & SMITH, N. R. 1967. *Bergey's Manual of Determinative Bacteriology*, 7th edition. Baltimore.

20*a*. BRITTEN R. J. & KOHNE D. E. 1968. Repeated Sequences in D.N.A. *Science* **161**: 529–540.

21. BRUCE, E. A. & LEWIS, J. 1960. *The Flora of Tropical East Africa: Loganiaceae*. Crown Agents for Overseas Governments, London.

22. BRYSON, V. & VOGEL, H. J. 1965. *Evolving Genes and Proteins*. Academic Press.

23. BUETTNER-JANUSCH, J. & HILL, R. L. 1965. Molecules and monkeys. *Science* **147**: 836–842.

24. BURTT, B. L. 1954–63. *Studies on the Gesneriaceae of the Old World*. Not. Roy. bot. Gardens, Edinburgh, 21–24.

25. CAIN, A. J. 1959. Taxonomic concepts. *Ibis* **101**: 302–318.

26. CAIN, A. J. 1959. The post-Linnaean development of taxonomy. *Proc. Linn. Soc. Lond.* **170**: 234–244.

27. CARR, S. G. M. & CARR, D. J. 1966. Cotyledonary stipules in Myrtaceae. *Nature Lond.* **210**: 185–186.

28. CHAPMAN, J. A. 1958. Studies on the physiology of the Ambrosia beetle Trypodendron in relation to its ecology. *Proc. X Intern. Cong. Entomology*, vol. 4, Montreal.

29. CHESTER, K. S. 1937. A critique of plant serology. *Q. Rev. Biol.* **12**: 19–46.

30. CHÛJO, M. & KIMOTO, S. 1961. Systematic catalog of Japanese Chrysomelidae. *Pacific Insects* **3**: 117–202.

31. CLAY, T. 1960. A new genus and species of Menoponidae (Mallophaga) from Apteryx. *Ann. Mag. nat. Hist. Lond.* (13)**3**: 571–576.

32. CLERCK, C. 1757. *Svenska Spindlar*. Stockholm.

32*a*. COIFFAIT, H. 1965. Sectophilonthus—rémarquable genre de Philonthini nouveau pour la région Paléarctique. *Zool. Zhurn.* **44**: 615–617.

33. COLLESS, D. H. 1966. A note on Wilson's consistency test for phylogenetic hypotheses. *Syst. Zool.* **15**: 358–359.

33*a*. CONWENTZ, H. W. 1886. *Die Flora des Bernsteins*, 2. Danzig.

34. CROWLE, A. J. 1961. *Immunodiffusion*. Academic Press.

35. CROWSON, R. A. 1953. On a possible new principle in taxonomy. *Nature, Lond.* **171**: 883.

36. CROWSON, R. A. 1955. *The Natural Classification of the Families of Coleoptera*. Nathaniel Lloyd, London. Reprint 1968, E. W. Classey.

37. CROWSON, R. A. 1965. Classification, statistics and phylogeny. *Syst. Zool.* **14**: 144–148.

38. CUMMINS, G. B. 1959. *Illustrated Genera of Rust Fungi*. Burgess, Minneapolis.

39. DAMIAN, R. T. 1964. Molecular mimicry: antigen sharing by parasite and host and its consequences. *Amer. Naturalist* **98**: 129–150.

40. DARLINGTON, C. D. & MATHER, K. 1949. *The Elements of Genetics*.

41. DARLINGTON, C. D. & WYLLIE, A. P. 1955. *Chromosome Atlas of the Flowering Plants*. Allen & Unwin.

41*a*. DARLINGTON, C. D. 1939. *The Evolution of Genetic Systems*. Cambridge.

42. DARLINGTON, P. J. 1950. Paussid beetles. *Trans. Amer. ent. Soc.* **76**: 47–142.

43. DARWIN, C. 1842. *Monograph of the Cirripedia.* Ray Society: London.

44. DARWIN, C. 1859. *The Origin of Species.* London.

45. DAUBENMIRE, R. 1966. Vegetation: identification of Typal communities. *Science* **151**: 291–298.

46. DAVIS, G. L. 1966. *Systematic Embryology of the Angiosperms.* John Wiley.

47. DAVIS, P. H. & HEYWOOD, V. H. 1963. *Principles of Angiosperm Taxonomy.* Oliver & Boyd.

47a. DAVIDSON, J. N. 1967. A sceptical chemist in a biological wonderland. *Proc. R. Soc. Edinburgh* (B) **70**: 169–191.

47b. DAYHOFF, M. O. & ECK, R.V. 1968. *Atlas of Protein Sequence and Structure.* National Biochemical Research Foundation. Silver Spring, Maryland U.S.A.

48. DICE, L. R. 1952. *Natural Communities.* University of Michigan Press.

48a. DEBOUTEVILLE, C. D. & MASSOUD, Z. 1967. Un groupe panchronique. Les Collemboles. *Ann. soc. ent. France* (N.S.) **3**: 625–630.

49. DOBZHANSKY, T. 1958. Species after Darwin. In S. A. Barnett (ed.), *A Century of Darwin.* Heinemann.

49a. DOBZHANSKY, T., HECHT, M. K. & STEERE, W. C. 1967. *Evolutionary Biology*, vol. I. Amsterdam.

50. DOLLO, L. 1922. Les Céphalopodes déroulés et l'irreversibilité de l'evolution. *Bijdr. dierk. k. zool. Genootschap Amsterdam* **22**: 215–226.

51. DONK, M. A. 1964. A conspectus of the families of Aphyllophorales. *Persoonia* **3**: 199–324.

51a. DOWRICK, G. J. 1961. Biology of reproduction in *Rubus. Nature Lond.* **191**: 680.

52. EAMES, A. J. 1961. *Morphology of the Angiosperms.* McGraw Hill.

53. ECKARDT, T. 1963. Zum Blutenbau der Angiospermen in Zusammenhang mit ihrer Systematik. *Ber. deutschen Bot. Ges.* **76**: 38–49.

54. EHRENDORFER, F. 1964. Cytologie, Taxonomie und Evolution. In Turrill, W. (ed.), *Vistas in Botany*, vol. IV. 99–186.

54a. EHRENDORFER F., KRENDL F., HABELER E. & SAUER W. 1968. Chromosome numbers and evolution in primitive Angiosperms. *Taxon*, **17**: 337–353.

55. EHRLICH, P. R. & RAVEN, P. H. 1964. Butterflies and plants: a study in coevolution. *Evolution* **13**: 586–608.

55a. EHRLICH, P. R. 1961. Systematics in 1970: some unpopular predictions. *Syst. Zool.* **10**: 157–158.

56. EMDEN, F. van, 1938. On the taxonomy of Rhynchophora larvae. *Trans. R. ent. Soc. London* **87**: 1–37.

56a. EMDEN, F. van 1957. The taxonomic significance of the characters of immature insects. *Ann. Rev. Entomology* **2**: 91–106.

57. EMERSON, A. E. 1961. Vestigial characters of termites and processes of regressive evolution. *Evolution* **15**: 115–131.

58. ENGLER, A. 1894. Proteaceae. *Die Natürlichen Pflanzenfamilien* **3**(1): 119–156. Leipzig.

59. ENGLER, A. 1964. *Syllabus der Pflanzenfamilien II*, ed. H. Melchior. Berlin.

60. ERDTMANN, G. 1952. *Pollen Morphology and Plant Taxonomy: Angiospermae.* Stockholm.

61. ERDTMANN, G. 1954. Pollen morphology and plant taxonomy. *Bot. Notiser* 65–81.

62. ERDTMANN, G. 1963. Palynology. *Advances in Botanical Research* **1**: 149–208.

63. FAEGRI, K. & VAN DER PIJL L. 1966. *Pollination Ecology.* Pergamon Press.

64. FERRIS, G. F. 1951. The sucking lice. *Mem. Pacific Coast ent. Soc.* 1.

64a. FINCHAM J. R. S. & DAY P. R. 1963. Fungal Genetics. Blackwall, Oxford.

65. FLORKIN, M. 1952. Caractères Biochemiques des Categories Supraspécifiques de la Systèmatique Animale. *Ann. Soc. zool. Belg.* **83**: 111–130.

66. FLORKIN, M. 1966. *A Molecular Approach to Phylogeny.* Elsevier.

66a. FLORKIN, M. & MASON H. S. 1960–1964. *Comparative Biochemistry*, vols. 1–6. Academic Press.

67. FORD, E. B. 1964. *Ecological Genetics.* London.

67a. FRIES, E. M. 1821–1832. *Systema Mycologicum.*

68. GALBRAITH, J. K. 1958. *The Affluent Society.* Hamish Hamilton.

69. GARSTANG, W. 1951. The ballad of the Veliger. In *Larval Forms.* Blackwell, Oxford.

70. GEOLOGICAL SOCIETY OF LONDON 1967. The fossil record: a Symposium.

71. GÉRY, J. 1962. Le Problème de la Sous-espèce et sa Définition Statistique. *Vie et Milieu* **13**: 521–541.

72. GIBBS, R. D. 1961. Comparative chemistry of plants as applied to problems of systematics. *Recent advances in Botany*, 1. Toronto.

73. GISLEN, T. 1924. Echinoderm studies. *Zool. bidr. Uppsala* **9**: 1–316.

74. GLAESSNER, M. F. & WADE, M. 1966. The late precambrian fossils from Ediacara, So. Australia. *Palaeontology* **9**: 599–628.

75. GODWIN, H. 1956. *The History of the British Flora.* Cambridge University Press.

76. GÖHLKE, K. 1913. *Serumdiagnostik in Pflanzenreiche.* Berlin.

77. GOODALL, D. W. 1963. The continuum and the individual association. *Vegetatio* **11**: 297–315.

78. GOUDGE, T. A. 1959. Causal explanations in natural history. *Brit. J. Phil. Sci.* **9**: 194–202.

78a. GOZIS, M. des, 1886. Récherche de l'espèce typique de quelques anciens genres. Montluçon, 35 pp.

79. GRAHAM, A. 1962. The role of fungal spores in palynology. *J. Palaeont.* **36**: 60–68.

80. GRANT, VERNE. 1951. The fertilisation of flowers. *Sci. Amer.* **184**(6): 152–56.

81. HALDANE, J. B. S. 1949. Suggestions as to quantitative measurements of rates of evolution. *Evolution* **3**: 51–56.

82. HALLIER, H. 1912. L'origine et le système phylétique des Angiospermes. *Arch. Neerl. Sc. exact. nat.* **3** B: 146–234.

83. HALTENORTH, T. 1963. Klassifikation der Saugetiere: Artiodactyla 1 (18): 1–167. *Handbuch der Zoologie* Band 8, Lfg. 32. Berlin.

84. HARBORNE, J. B. 1963. In Swain [185].

84a. HARBORNE, J. B. 1966. The evolution of flavonoid pigments in plants. In T. Swaine (ed.), [185a].

85. HARRISON, J. H. 1954. A synopsis of the Dactylorchids of the British Isles. In E. Rubel & W. Ludi, *Bericht über Geobotanische Forschungsinstitut Rubelin.* Zurich 1953.

86. HARTL, D. 1963. Das Placentoid der Pollensäcke, ein Merkmal der Tubifloren. *Ber. deutsch. bot. Ges.* **76**: 70–72.

87. HAYEK, F. A. 1964. The theory of complex phenomena. In Bunge, M., *The Critical Approach to Science and Philosophy*, ch. 22.

88. HEGNAUER, R. 1962. *Chemotaxonomie der Pflanzen.* Basel, Stuttgart.

89. HEGNAUER, R. 1965. Chemotaxonomy: past and present. *Lloydia* **28**: 267–278.

89a. HEGNAUER, R. 1965. A. P. de Candolle, fondateur de la chimiotaxomie moderne, et quelques aspects récents de cette branche de la science. *Bull. Soc. biol. France, Mémoires* 1965.

90. HEISER, C. B., SORIA, J. & BURTON, D. L. 1965. A numerical taxonomic study of Solanum species and hybrids. *Amer. Naturalist* **99**: 471–488.

91. HENNIG, W. 1950. *Grundzüge einer Theorie von phylogenetischen Systematik.* Berlin.

92. HENNIG, W. 1965. Einige allgemeine Gesichtspunkte fur die phylogenetische Deutung des Flügelgeaders der Dipteren. *Proc. XII Intern. Congr. Entomology, London.*

92a. HENNIG, W. 1968. *Phylogenetic Systematics.* University of Illinois Press.

93. HILL, R. L., BUETTNER-JANUSCH, J. & BUETTNER-JANUSCH, V. 1963. Evolution of hemoglobin in Primates. *Proc. nat. Acad. Sci. Washington* **50**: 885–893.

94. HINTON, H. E. 1947. On the reduction of functional spiracles in the aquatic larvae of the Holometabola. *Trans. R. ent. Soc. Lond.* **98**: 449–473.

94a. HIRST, S. 1923. On some arachnid remains from the Old Red Sandstone. *Ann. Mag. nat. Hist.* (9) **12**: 455–474.

95. HOLLAND, G. P. 1964. Evolution, classification and host relationships of the Siphonaptera. *Ann. Rev. Entom.* **9**: 122–164.

95a. HOPKINS, G. H. E. 1949. The host associations of the lice of mammals. *Proc. zool. Soc. Lond.* **191**: 387–604.

96. HOYER, E. T., MCCARTHY, B. J. & BOLTON, E. T. 1964. A molecular approach to the systematics of higher organisms. *Science* **144**: 959–967.

97. HOYLE, G. 1962. Neuromuscular physiology. In Lowenstein, O. (ed.), *Advances in Comparative Physiology and Biochemistry*, vol. 1.

98. HUGHES, N. F. 1961. Fossil evidence for Angiosperm ancestry. *Sci. Progr.* **49**: 84–106.

99. HUHN, R. 1927. Über die Verwertbarkeit der Serodiagnostik in der Botanik, erläutert an den Sympetalen. *Beitr. Biol. Pfl.* **15**: 228–262.

100. HULL, F. M. 1945. A revisional study of the fossil Syrphidae. *Bull. Mus. comp. Zool. Harvard* **95**: 249–355.

101. HUTCHINSON, J. 1926. *The Families of Flowering Plants*: vol. I. MacMillan.

102. HUXLEY, A. 1932. *Brave New World*. London.

103. HUXLEY, J. 1939. Clines: an auxiliary method in Taxonomy. *Bijdr. Dierk.* **27**: 491–520.

104. HUXLEY, J. (ed.) 1940. *The New Systematics*. London.

105. *Index Kewensis. Plantarum phanerogamorum*. Oxford 1893– .

105a. *Index Filicum*. Copenhagen 1906– .

106. INGER, R. F. 1958. Comments on the definition of genera. *Evolution* **12**: 370–384.

107. *International Code of Botanical Nomenclature*, adopted at the 8th International Botanical Congress, Paris 1944. Utrecht, 1956.

108. *International Code of Nomenclature of Bacteria and Viruses* 1958. Editorial Board of the International Committee on Bacteriological Nomenclature. Ames, Iowa.

109. *International Code of Zoological Nomenclature* 1961. International Trust for Zoological Nomenclature, London.

110. JARVIK, E. 1965. Specialisation in early vertebrates. *Ann. Soc. R. zool. Belg.* **94**: 11–95.

111. JEFFRIES, R. P. S. 1967. Some fossil chordates with Echinoderm affinities. *In Symp. zool. Soc. Lond.* no. 20, 'Biology of Echinoderms'. Academic Press.

111a. JENSEN, U. 1967. Serologische Beiträge zur Frage der Verwandtschaft zwischen Ranunculaceae und Papaveraccae. *Ber. Dtsch. bot. Ges.* **80**: 621–624.

111b. JENSEN, U. 1968. Serologische Beiträge zur Systematik der Ranunculaceae. *Bot. Jahrb.* **88**: 204–310.

112. JOUVENEL, B. de 1967. Transl. Nikita Lary. *The Art of Conjecture*. Weidenfeld & Nicolson.

113. KAUSIK, S. B. 1943. The distribution of Proteaceae: past and present. *J. Indian bot. Soc.* **22**: 105–123.

114. KERKUT, G. A. 1960. *The Implications of Evolution*. London.

114a. KIAUTA, B. 1968. Considerations on the evolution of the chromosome complement in Odonata. *Genetica* **38**(4): 430–468.

115. KHANNA, K. R. 1964. Differential evolutionary activity in Bryophytes. *Evolution* **18**: 652–670.

115*a*. KIRCHHEIMER, F. 1957. *Die Laubgewächse der Braunkohlenzeit*. Halle.

116. KLEBS, R. 1910. Uber Bernsteineinschlüsse in Allgemeine und die Käfer meiner Bernsteinsammlung. *Schr. phys.-ökon. Ges. Königsberg* **51**: 217–242.

117. KLEIN, R. M. & CRONQUIST, A. 1967. A consideration of the evolutionary and taxonomic significance of some biochemical, micromorphological and physiological characters of the Thallophyta. *Q. Rev. Biol.* **42**: 108–296.

118. KOESTLER, A. 1964. *The Act of Creation*. London.

119. KOJIMA, A. 1922. Serobiologische Untersuchungen über die Verwandt-schaft-verhältnisse zwischen Dikotyledonen und Gymospermen. *Mitt. mediz. Fak. k. Kyushu-Universitat Japan* **6**: 223–254.

120. KOKETSU, R. 1917. Serodiagnistische Untersuchungen über die Verwandt-schaftverhältnisse der Gymnospermen. *Mitt. mediz. Fak. k. Kyushu-Universitat Japan* **4**: 61–130.

121. KROHN, V. 1935. Eine kritische Nachprüfung der Sympetalen des Königs-berger serodiagnostische Stammbaum. *Bot. Archiv.* **87**: 328–372.

122. LADRIÈRE, J. 1957. *Les Limitations Internes des Formalismes*. Paris.

123. LAMB, I. L. 1959. Lichens. *Sci. Amer.* **201**(4): 144.

124. LAMBERT, J. M. & DALE, M. B. 1964. *The Use of Statistics in Phytosociology*. Advances in Ecological Research, II. Academic Press.

125. LEONE, C. (ed.), 1962. *Taxonomic Biochemistry and Serology*. Ronald Press.

125*a*. LEONE, C. A. 1949. Comparative serology of some brachyuran Crustacea and studies in Hemocyanin correspondence. *Biol. Bull.* **97**: 273–286.

126. LEONE, C. A. & WIENS, A. L. 1956. Comparative serology of Carnivora. *J. Mammal.* **37**: 11–23.

127. LINNAEUS, C. 1735. *Carolus Linnaei, Sveci, Methodus*.

128. LINNAEUS, C. 1740. *Systema Naturae*, ed. I.

129. LINNAEUS, C. 1753. *Species Plantarum*.

130. LINNAEUS, C. 1758. *Systema Naturae*, ed. X.

131. LINNAEUS. C. 1764. *Ordines Naturales*.

132. MABRY, T. 1962. The Betacyanins: a new class of red-violet pigments. In Leone [125].

133. MCKITTRICK, F. J. & MCKERRAS, M. J. 1965. Phyletic relations in the Blat-tidae. *Ann. ent. Soc. Amer.* **58**: 224–229.

134. MAHESHWARI, P. 1964. Embryology in Relation to Taxonomy. In Turrill, W. (ed.), *Vistas in Botany*, vol. IV.

135. MAIRS, D. F. & SINDERMANN, C. J. 1962. A serological comparison of five species of Atlantic clupeoid fishes. *Biol. Bull.* **123**: 330–343.

136. MANSKI, W., HALBERT, S. P. & AUERBACH, T. P. 1962. Immunological analysis of the phylogeny of lens proteins. In Leone [125].

137. MANTON, I. 1950. *Problems of Cytology and Evolution in the Pteridophyta*. Cambridge University Press.

138. MANTON, S. M. 1952. The evolution of arthopodan locomotory mechanisms, part 2. *J. Linn. Soc. Zool.* **42**: 94–166.
139. MARGOLIASH, E. & SMITH, E. L. 1963. Structural and functional aspects of cytochrome in relation to evolution. *Proc. nat. Acad. Sci. Washington.*
140. MAYR, E. 1960. Evolutionary novelties. In Sol Tax (ed.), *Evolution After Darwin*, vol. I.
140a. MAYR, E. 1963. *Animal Species and Evolution.* Cambridge. (Mass.).
140b. MAYR, E. 1942. *Systematics and the Origin of Species.* New York.
141. MAYR, E., LINSLEY, E. G. & USINGER, R. L. 1953. *Methods and Principles of Systematic Zoology.* McGraw-Hill.
141a. MERXMULLER, H. 1967. Chemotaxonomie? *Ber. Dtsch. Bot. Ges.* **80**: 608–620.
142. MEDAWAR, P. B. 1957. *The Uniqueness of the Individual.* London.
142a. MEYER-ABICH, A. 1964. The Historico-philosophical Background of the Modern Evolution Biology. *Acta Biotheoretica*, Suppl. 2, Leiden.
143. METCALFE, C. R. & CHALK, L. 1950. *The Anatomy of the Dicotyledons.* Oxford.
143a. MICHENER, C. D. 1963. Some future developments in taxonomy. *Syst. Zool.* **12**: 151–172.
144. MICKS, D. W. 1956. Paper chromatography in insect taxonomy. *Ann. ent. Soc. Amer.* **49**: 576–581.
144a. MORITZ, O. 1962. Some special features of serobotanical work. In Leone [125].
145. MOSES, I. & YERGANIAN, G. 1952. DNA content and cytotaxonomy of several Cricetinae. *Rec. genet. Soc. America* **21**: 51.
145a. NEAVE, S. A. 1939–   . *Nomenclator Zoologicus*, vols. I–V and Supplements. Zoological Society of London.
145b. NAGL W. 1969. Banded Polytene Chromosomes in the Legume *Phaseolus vulgaris*. *Nature* **221**: 70–71.
145c. NEI; M. 1969. Gene duplication and Nucleotide substitution in evolution. *Nature* **221**: 40–42.
146. NEWELL, N. D. 1962. Palaeontological gaps and geochronology. *J. Palaeont.* **36**: 592–610.
146a. NEWELL, N. D. 1947. Infraspecific categories in invertebrate palaeontology. *Evolution* **1**: 163–171.
146b. OAKESHOTT, M. 1962. *Rationalism in Politics.* Methuen, London.
147. OSBORNE, D. V. 1963. Some aspects of the theory of dichotomous keys. *New Phytologist* **62**: 144–160.
148. PALUŠKA, R. & KORINEK, J. 1960. Studium der antigenen Eiweiss Verwandtschaft zwischen Menschen und einigen Primaten mit Hilfe neuer immunobiologischer Methoden. *Zts. Immunitätsforsch.* **119**: 244–257.
149. PARSONS, P. S. & WILLIAMS, E. E. 1963. The relationships of the modern Amphibia: a re-examination. *Q. Rev. Biol.* **38**: 26–53.

149a. PERSOON, C. H. 1801. *Synopsis Methodica Fungorum.*

150. PATTERSON, J. T. 1943. The Drosophilidae of the south west. *Univ. Texas Publ.* **4313**: 7–216.

151. PATTERSON, J. T. & STONE, W. S. 1952. *Evolution in the Genus* Drosophila. MacMillan, New York.

152. PEYERIMHOFF, P. de 1911–26. Notes sur le Biologie de quelques Coléoptères Phytophages du nord-Africain. *Ann. ent. Soc. Fr.* **80**: 283–314, **84**: 19–61, **88**: 169–258, **95**: 319–390.

153. PLOUVIER, V. 1963. The distribution of aliphatic polyols and cyclitols. In Swain [185].

154. POKORNY, V. 1965. *Principles of Zoological Micropalaeontology*, vol. 2. Pergamon Press.

154a. POLANYI, M. 1968. Life's Irreducible Structure. *Science* **160**: 1308–1312.

155. *Premier Symposium sur la Spécificité parasitaire des Parasites des Vertébrés.* Université de Neuchâtel, Institut de Zoologie, 1957.

156. RENSCH, B. 1959. *Evolution Above the Species Level.* Methuen.

157. RILEY, H. P. & HOPKINS, J. D. 1962. Paper chromatographic studies in the Aloinae II. In Leone [125].

157a. RITCHIE, A. D. 1958. *Studies in the History and Methods of the Sciences.* Edinburgh, University Press.

158. ROBIN, Y. 1964. Biological distributions of guanidines and phosphagens in marine annelids and related phyla from California, with a note on Pluriphosphagens. *Comp. Biochem. Physiol.* **12**: 347–367.

159. ROHLF, F. J. 1963. Congruence of larval and adult classifications in *Aedes* (Diptera Culicidae). *Syst. Zool.* **12**: 97–117.

160. ROMER, A. S. 1941. *Man and the Vertebrates.* University of Chicago Press.

160a. ROMER, A. S. 1956. *The Vertebrate Body.* Saunders.

161. ROMER, A. S. 1945. *Vertebrate Paleontology.* University of Chicago Press.

161a. ROHDENDORF, B. B. 1962. *Fundamentals of Palaeontology (Osnovi Paleontologii): Arthropoda, Tracheata and Chelicerata.* Moscow, U.S.S.R. Academy of Sciences.

162. SACHS, L. 1952. Polyploid evolution and mammalian chromosomes. *Heredity* **6**: 357–364.

163. SCHEDL, K. 1947. Die Borkenkäfer des baltischen Bernsteins. *Zentralbl. Gesamtgeb. Entom.* **2**: 12–45.

164. SCHMALHAUSEN, I. I. 1949. Transl. I. Dordick. *Factors of Evolution.* Blakiston Co., Toronto.

164a. SCHUBERT, K. 1939. Mikroskopische Untersuchung pflanzlicher Einschlüsse des Bernsteins I. *Bernsteinforsch.* **4**: 23–44.

165 SCHUSTER, C. N. 1962. Serological correspondence among horseshoe crabs. *Zoologica* **47**: 1–7.

165a. SCHOONHOVEN, L. M. 1968. Chemosensory bases of host plant selection. *Ann. Rev. Entom.* **13**: 115–136.

166. SHORLAND, F. B. 1963. *The Distribution of Fatty Acids in Plant Lipids.* In Swain [185].

167. SIMONS, E. L. 1963. A critical reappraisal of Tertiary Primates. In Buettner-Janusch, J. (ed.), *Evolutionary and Genetic Biology of Primates* vol I.

168. SIMPSON, G. G. 1948. The beginning of the age of mammals in South America. *Bull. Amer. Mus. nat. Hist.* **91**: 1–232.

168a. SIMPSON, G. G. 1953. *The Major Features of Evolution.* Columbia Biological Series no. 17.

169. SIMPSON, G. G. 1951. *Horses.* Oxford University Press.

170. SIMPSON, G. G. 1961. *The Principles of Animal Taxonomy.* Oxford University Press.

170a. SIMPSON, G. G., ROE, A. & LEWONTIN, R. C. 1960. *Quantitative Zoology.* Harcourt, Brace.

171. SMITH, S. G. 1953. Chromosome numbers of Coleoptera. *Heredity* **7**: 31–48.

171a. SMITH, S. G. 1960. Chromosome numbers of Coleoptera II. *Canad. J. Genet. Cytol.* **2**: 67–88.

171b. SMITH, S. G. 1959. The cytogenetic basis of speciation in Coleoptera. *Proc. X Int. Congr. Genetics* **1**: 444–450.

172. SNEATH, P. H. A. 1962. *Comparative Biochemical Genetics in Bacterial Taxonomy.* In Leone [125].

173. SNEATH, P. H. A. & SOKAL, R. R. 1962. Numerical taxonomy. *Nature Lond.* **193**: 855–860.

174. SOKAL, R. R. & SNEATH, P. H. A. 1963. *Principles of Numerical Taxonomy.* Freeman.

175. SONDHI, K. H. 1962. The evolution of a pattern. *Evolution* **16**: 186–191.

176. SPENCER, H. 1864. *The Classification of the Sciences.* London.

177. SPIEGELMANN, S. 1964. Hybrid nucleic acids. *Sci. Amer.* **210** (5): 48–56.

178. SPORNE, K. 1948. Correlation and classification in the Dicotyledons. *Proc. Linn. Soc. Lond.* **160**: 40–58.

178a. STAMMER, H. J. 1957. Gedanken zu den parasitophyletischen Regeln. *Zool. Anz.* **159**: 255–267.

179. STEBBINS, G. L. 1950. *Variation and Evolution in Plants.* New York.

180. STEBBINS, G. L. 1966. Chromosomal variation and evolution. *Science* **152**: 1463–1469.

181. STEPHENSON, J. 1930. *The Oligochaeta.* Oxford University Press.

182. STEVENS, F. C., GLAZER, A. N. & SMITH, E. L. 1967. Amino-acid sequence of wheat cytochrome-*c*. *J. biol. Chem.* **242**: 2764–2779.

183. STURTEVANT, A. H. 1942. The classification of the genus *Drosophila. Univ. Texas Publ.* **4213**.

183a. SUBAK-SHARPE, J. H. 1967. Base doubtlet frequency patterns in the nucleic acid and evolution of viruses. *Brit. med. Bull.* **23**: 161–168.

183b. SUEOKA, N. 1961. *J. molec. Biol.* **3**: 31.

184. SUTCLIFFE, D. W. 1963. The chemical composition of the haemolymph in insects and some other Arthropods. *Comp. Biochem. Physiol.* **9**: 121–135.

185. SWAIN, T. (ed.) 1963. *Chemical Plant Taxonomy.* Academic Press.

185a. SWAIN, T. 1966. *Comparative Phytochemistry.* Academic Press, London & New York.

186. SYMMONS, S. 1952. Comparative anatomy of the Mallophagan head. *Trans. zool. Soc. Lond.* **27**: 349–436.

186a. TAKHTAJAN, A. L. 1963. *Osnovi Paleontologii* 15. Moscow.

186b. TEILHARD DE CHARDIN, P. 1925. L'Histoire Naturelle du Monde. *Scientia* **37**: 15–24.

187. TAYLOR, A. E. R. (ed.) 1965. *The Evolution of Parasites.* Blackwell.

187a. TEMPÈRE, G. 1967. Un Critère Meconnu des Systematiciens Phanerogamistes: l'Instincte des Insectes. *Le Botaniste* **50**: 477–482.

188. TEJERA, V. 1961. The nature of aesthetics. *Brit. J. Aesthetics* **1**: 209–216.

188a. THAXTER, R. 1896–1931. Contribution towards a monograph of the Laboulbeniaceae. *Mem. Am. Acad. Arts Sci.* **12**: 187–149; **13**: 217–469; **14**: 309–426; **15**: 427–580; **16**: 1–435.

189. THOAI, N. VAN & ROCHE, J. 1962. In Leone. [125].

190. THOMPSON, J. F. *et al.* 1959. Partition chromatography and its uses in the plants sciences. *Bot. Rev.* **25**: 1–263.

191. THOMSON, K. S. 1964. The ancestry of tetrapods. *Sci. Progr.* **52**: 451–459.

191a. THROCKMORTON, L. H. 1968. Biochemistry and taxonomy. *Ann. Rev. Entom.* **13**: 99–114.

192. TINBERGEN, N. 1959. Behaviour, systematics and natural selection. *Ibis* **101**: 318–330.

193. TIPPO, O. 1942. A modern classification of the plant kingdom. *Chron. Bot.* **7**: 203–206.

193a. TOULMIN, S. 1953. *The Philosophy of Science.* London.

193b. TOMLINSON, P. B. 1966. Anatomical data in the classification of Commelinaceae. *J. Linn. Soc. Botany* **59**: 371–396.

193c. TURRILL, W. B. *et al.* 1942. Differences in the systematics of plants and animals and their dependence on differences in structure, function and behaviour in the two groups. *Proc. Linn. Soc. Lond.* **153**: 272–287.

194. WADDINGTON, C. H. 1953. The Baldwin effect, genetic assimilation and homoeostasis. *Evolution* **7**: 386–387.

194a. WATTS, R. L. & D. C. 1968. Gene duplication and the evolution of enzymes. *Nature, Lond.* **217**: 1125–1130.

194b. WATSON, J. 1968. *The Double Helix.* Weidenfeld & Nicolson.

195. WAGENITZ, G. 1963. Taxonomie und Evolutionforschung im Bereich höheren Kategorien. *Ber. dtsch. bot. Ges.* **76**: 91–97.

195a. WATSON, W. C. R. 1958. *Handbook of the Rubi of Great Britain and Ireland.* Cambridge University Press.

195*b*. WALSH, W. H. 1967. *An introduction to Philosophy of History.* 3rd edition. Hutchinson, London.

196. WEST, A. S., HORWOOD, R. H., BOURNS, T. K. R. & HUDSON, A. 1959. Systematics of *Neodiprion* sawflies. I. A preliminary report on serological and chromatographic studies. *Rep. ent. Soc. Ontario* **89**: 59–65.

197. WETTSTEIN, R. 1933–5. *Handbuch der systematischen Botanik.* Wein, Leipzig.

198. WHEELER, W. M. 1914. The ants of the Baltic Amber. *Schrift. physik.-ökon. Ges. Könisberg* **55**: 1–142.

199. WHITE, G. 1784. *The Natural History of Selborne.*

199*a*. WHITE, M. J. D. 1949. Cytological evidence on the phylogeny and classification of the Diptera. *Evolution* **3**: 252–260.

200. WHITE, M. J. D. 1954. *Animal Cytology and Evolution.* Cambridge University Press.

201. WHITE, M. J. D. 1957. Cytogenetics and systematic entomology. *Ann. Rev. Entom.* **2**: 71–90.

202. WHITE, M. J. D. 1957. Some general problems of chromosomal evolution and speciation in animals. *Survey biol. Progr.* **3**: 109–147.

203. WILLIAMS, C. A. 1960. Immunoelectrophoresis. *Sci. Amer.* **202**(3): 130–140.

204. WILLIAMS, C. A. & WEMYSS, C. T. 1961. Experimental and evolutionary significance of similarities among serum protein antigens of man and the lower Primates. *Ann. N.Y. Acad. Sci.* **94**: 77–92.

204*a*. WILLIAMS, W. T. 1967. Numbers, taxonomy and judgment. *Bot. Rev.* **33**: 379–386.

204*b*. WILLIAMS, W. 1964. Genetical Principles and Plant Breeding. Blackwell, Oxford.

205. WILLIS, J. C. 1922. *Age and Area.* Cambridge.

206. WILSON, A. C. & KAPLAN, N. D. 1962. Enzyme structure and its relation to Taxonomy. In Leone [125].

207. WILSON, E. O. 1965. A consistency test for phylogenies based on contemporary species. *Syst. Zool.* **14**: 221–236.

207*a*. WILSON, E. O., CARPENTER, F. M. & BROWN, W. L. 1967. The first mesozoic ants, with the description of a new subfamily. *Psyche* **74**: 1–19.

208. WINDELBAND, W. 1894. *Geschichte und Naturwissenschaft.*

209. WRIGHT, S. 1964. Pleitropy in the evolution of structural reduction and of dominance. *Amer. Naturalist* **98**: 65–69.

210. ZEUNER, F. E. 1952. *Dating the Past.* Methuen, London.

211. *Zoological Record.* Zoological Society of London, 1864– .

212. ZUCKERKANDL, E. 1965. The evolution of hemoglobin. *Sci. Amer.* **212**: 110–118.

213. ZUCKERKANDL, E., JONES, R. T. & PAULING, L. 1960. A comparison of animal hemoglobins by tryptic peptide pattern analysis. *Proc. U.S. nat. Acad. Sci.* **46**: 1349–1360.

214. VAUGHAN, J. G. 1968. Serology and other protein separation methods in studies of angiosperm taxonomy. *Sci. Prog. Oxford* **56**: 205–222.

215. Symposium on Newer Trends in Taxonomy, Sheschachar B. R. (ed.), *Bull. nat. Inst. Sci. India* **34**: 1967.

216. Symposium on Microbial Classification, 1962. *Symp. Soc. gen. Microbiol.* **12**.

217. HARBORNE, J. B. 1967. *Comparative Biochemistry of the Flavonoids.* Academic Press.

Subject Index

Index of Taxa

*Index of Taxa*337

Cycadaceae 59, 107, 214
Cycadales 256
Cycadophyta 259
Cycas 214
  antiserum 177
Cyclosporeae 259
Cyclostomata 99, 190
  absence of limbs 203
Cydnidae 127
Cylindrotominae 151
Cynocrambaceae 80

Dactylorchis 52
Dactylorchis fuchsi 41, 42
Dascillidae 245
Dascilloidea 234, 240, 245, 246
Datiscaceae, chromosomes 159
Daucus, antiserum 177
Decapoda 123
Degeneria 207
Dermestoidea 235, 245
Derodontidae 244, 245
Deuterostomia 258
Diatomaceae 219, 223
Dicotyledonae 51, 73, 74, 75, 79,
  80, 81, 102, 108, 109, 110, 111,
  129, 193, 208, 229, 256, 257, 259,
  272
  serology 176, 177, 184
  spores 219
Dictyoptera 105
Didieraceae 80
Digenea 124
Digitalis 128
Dinosauria 134, 247
Dioscorea 130
Diplopeltis 218
Diplopoda, neoteny 187
Dipnoi 96, 97, 247, 254, 255
Diprotodontia 214
Dipsacaceae 158
  serology 175
Diptera 35, 102, 112, 122, 127, 147,
  150–2, 190, 238, 254, 272
  chromosomes 150, 151
  giant chromosomes 146, 148
  larvae 190
Dixidae 151, 152

Dizygomyza verbasci 127
Docophthirus 119, 120
Dolichandrone 126
Doronicum, antiserum 176
Dracaena 130
Dragon-flies 155 (*see* Odonata)
Dromia vulgaris, serology
Dromiaceae, serology 178
Drosophila 30, 43, 52, 73, 112, 145, 147,
  148, 201, 203
  legs 203
  setae 197
  unit characters 200
Drosophila pseudoobscura 43
Drosophila subobscura 162
Drosophilidae 162, 163
Dryopoldea 235, 240
Ducks 34

Echinodermata 83, 84, 99
  base doublet figures 167
  development 188, 189
  larvae 192
Echiuroid 84
Elasmobranchii 99, 103, 124, 190, 254
Elateridae 112, 191
  chromosomes 159
Elateriformia 233, 246
Elateroidea 235, 240
Elatinaceae 80
Embothriae 139, 140
Emu 227
Endopterygota 84, 190, 191, 258
  chromosomes 151
Enteropneusta, larvae 189
Entomostraca 190
Eohippus 250, 251
Eopaussus 138
Ephedra 109
Epiceratodus 96
Epiophlebia 155
Equidae 67, 250
Equisetaceae 208
Equisetales 256
Equisetum 248
Equus 58
Ericaceae 69, 194
  antiserum 176

Paris 70, 103
Parka 100
Parus major 92
Pasania, antiserum 177
Passeriformes 195
Passiflora, serology 175
Paussid beetles 136
Paussidae 63, 64, 136, 137, 139, 140, 141
Paussina 136, 137, 138
Paussini 136, 137, 141
Pecora 214, 215
Pedaliaceae 82, 128
Pedicinus 119, 120
Pedicularis, as host plant 128
Pediculidae 118, 119
Pediculus 118, 119, 120
Pediculus humanus 118
Peltogastridae 123
Pentaplatarthrina 136, 137, 138
Pentaplatarthrus 136
Peramelidae 214
Pereskia, leaves of 211
Periophthalmus 227
Perissodactyla 250, 251
  cytochrome-*c* 169
Persooniae 139
Persoonioideae 139, 140, 141
Petaluridae 155
Petauristes 279
Petrobius 84
Petroselinum sativum 174
Phaeophyceae 256
Phaeophyta 258
  diploids 157
Phalonia degeyrana 127
Phaseolus vulgaris 148
Phillyrea 126
Philonthus decorus 36
Philopteridae 120
Phthiraptera 121
Phthirpediculus 119, 120
Phthirus 119, 120
Phthirus pubis 118
Phygelius 126
  as host plant 128
Phylloscopus 78
Phylloscopus collybita 30, 31, 77
Phylloscopus trochilus 31, 77

Phyllotreta 125
Physea 137
Phytolaccaceae 80
Pieris 125, 126
Pigs 60
  serology 182
Pinaceae, chromosomes 158
Pinnipedia, serology 179
Pinus 83
  serology 177
Pinus succinifera 62
Pisces 96
Pityogenes 64
Pityophthorus 64
Plantae 256
Plantaginaceae 82, 113, 127, 194
Plantago 82, 113, 127, 128, 129
Plants, vascular 100
Plasmodium 33
Platyhelminthes 50, 117, 192, 218
Platypodinae 64
Platyrhopalina 136, 137, 138
Platyrrhina 61, 119
  cytochrome-*c* 169
Plectonycha 129
Plumbagineae 218
  antiserum 176
Podocarpus, chromosomes 158
  serology 177
Podophyllum, serology 175
Polemoniaceae 82, 194
Polychaeta 84
  parapodia 202
Polygonaceae, serology 175
Polyommatus, haploid numbers 154
Polyommatus bellargus and *corydon* 154
Polypetalae 194
  serology 176
Polyphaga 73, 74, 75, 233, 243, 244
  coxae 242
Polyplacinae 119
Polyprotodontia 214
Polyzoa 30
  larvae 192
Pontia daplidice 125
Porifera 83
Portulacaceae 80
Portunidae, serology 178

Rosales 256
Roses 81
Roupala 140
Rubiaceae 82, 128, 194
  antiserum 176
  serology 175
Russula 77

Sacculinidae 123
Salicaceae 107, 218
  stipules 193
Salix, serology 174
Salvia officinalis, serology 174
Santalaceae 127
Sappaphis bicolor, biguttacus, dubius, gallica, luctosis and *plantaginea* 127
Saxifragaceae 244
Scarabaeiformia 246
Scarabaeoidea 234, 240
Scatopsidae 151
Sciadopitys, serology 177
Sciara 147
Sciaridae 151
Scolytidae 64, 213
Scolytinae 64
Scolytus destructor 117
Scorpion 65
Scrophularia 126, 127
  as host plant 128, 129
Scrophulariaceae 81, 82, 102, 113, 126, 127, 128, 214
  embryonic development 194
Sea-gulls 90
Sectophilonthus 36
Sehirus 127
Selaginaceae 82, 194
Selaginellaceae 208
Sequoia sempervirens, polyploid 158
Sharks 86, 103 (*see* Elasmobranchii)
  serology 184
Simuliidae, chromosomes 151
Siphononaptera 122, 203
Siphunculata 118, 119, 120, 121
Sirenia 70
Smilax 130
Snakes 96, 102 (*see* Ophidia)
  absence of limbs 203

Solanaceae 82, 103, 128, 129, 194
Solanum 202
Spartina townsendi 157
Spermatophyta 39, 256, 259
Sphecomyrma 138
Spiders 71, 83, 84
Sponges 256 (*see* Porifera)
Squamata 96, 101, 255
Staphyliniformia 233, 246
Staphylinoidea 234, 240
Stegnospermaceae 80
Stephanocircidae 122
Stephanocircinae 122
Stereonychus 126, 129
Stethopachys 129
Sticklebacks 90
Strepsiptera (*see* Stylopoidea)
Strychnos 128
Stylidium, serology 176
Stylopoidea 233
Styracaceae, antiserum 176
Synapsida 254, 258
Syringa 125, 129
Syrphidae 64

Tachyplocus 248
Tamaricaceae 109
Tanyderidae 151, 152
Tapirs 247, 250
Tapiridae 250
Taxodiaceae, chromosomes 158
Taxus, chromosomes 158
Teleostei 123, 149
Termites 91 (*see* Isoptera)
Tetracentron, vessels lacking in 208
Tetrapoda 96, 97, 102, 105, 111, 209, 247, 249, 254, 256, 258
Thallophyta 259
Thaumaleidae 151, 152
Thaumapsyllinae 122
Theropithecus, serology 183
Theropithecus gelada, serology 182
Thesium humifusum 127
Thorictidae 245
Thysanoptera 66
Thysanura 65
Tinamiformes 121
Tinamous 121

Index of Authors